Methods in Molecular Biology

Series Editor
John M. Walker
School of Life and Medical Sciences
University of Hertfordshire
Hatfield, Hertfordshire, AL10 9AB, UK

For further volumes:
http://www.springer.com/series/7651

Bacterial Polysaccharides

Methods and Protocols

Edited by

Inka Brockhausen

Department of Biomedical and Molecular Sciences, Queen's University, Kingston, ON, Canada

Humana Press

Editor
Inka Brockhausen
Department of Biomedical and Molecular Sciences
Queen's University
Kingston, ON, Canada

ISSN 1064-3745 ISSN 1940-6029 (electronic)
Methods in Molecular Biology
ISBN 978-1-4939-9153-2 ISBN 978-1-4939-9154-9 (eBook)
https://doi.org/10.1007/978-1-4939-9154-9

Library of Congress Control Number: 2019933564

This Humana Press imprint is published by the registered company Springer Science+Business Media, LLC, part of Springer Nature.
The registered company address is: 233 Spring Street, New York, NY 10013, U.S.A.

Preface

Bacteria are surrounded by polysaccharides and glycoconjugates that function in the maintenance of cell wall structure and function. The complex glycans are involved in protection from host immunity and phagocytosis and promote adherence, colonization, biofilm formation, and virulence. The structures of these polymers have great diversity and complexity, and their functions are dictated by their sizes and charges and hydrophilicity of their sugars. Extracellular polysaccharides include lipopolysaccharides, large capsules, and secreted unbound slime that function in the maintenance of the cell shape, regulation of the permeability of membranes, and protection from dehydration, environmental stresses, and toxic materials. Many of the glycans are unique to bacteria and are antigenic. Mono- and oligosaccharides that are mainly found in vertebrate glycoconjugates are often mimicked in bacteria or are present as cryptic structures with various effects on pathogenicity. Polysaccharides or their fragments have the potential to improve the antigenicity and potency of a vaccine and to regulate the immune response. Although polysaccharides themselves are often not highly immunogenic, they can stimulate immune responses by conjugation to immunogens.

Gram-positive bacteria have one cell membrane that is surrounded by a thick layer of peptidoglycans (murein) which can have teichoic acids or other negatively charged polymers bound to it. Teichoic acids play important roles in virulence and antibiotic resistance. These cross-linked polymers contribute to the rigidity and mechanical strength of the cell and form a protective barrier against osmotic pressure. The GlcNAc-MurNAc linkages of peptidoglycans can be cleaved with lysozyme that weakens cells and enhances bacterial disruption methods. Successful antibiotics such as penicillin and vancomycin block the assembly of the peptidoglycan layer and thus reduce bacterial survival.

The periplasmic space between the two membranes of Gram-negative bacteria also contains peptidoglycans, but the layer is relatively thin and contains other glycans such as beta-glucans, biosynthetic precursors of polysaccharides, as well as enzymes and other proteins. On the surface, the main polysaccharides are derived from lipopolysaccharides (LPS) that carry external O antigenic polysaccharides. Hundreds of different serotype-specific structures of O antigens are known. O antigens have up to 50 repeating units of 2 to 8 sugars in length and provide a barrier to phagocytosis, certain antibiotics, and toxins. Certain bacteria lack the repeated oligosaccharide structure of O antigens and instead have short lipo-oligosaccharides.

Large, external capsular polysaccharides cover the surface of some mucoid bacterial strains, both Gram-positive (e.g., *Streptococcus* and *Staphylococcus* species) and Gram-negative (e.g., *Salmonella*, *Neisseria*, *E.coli*, *Pseudomonas*, and *Klebsiella*). Capsules contain major antigenic determinants and are significant virulence factors since they block complement phagocytosis. Some capsules (e.g., polysialic acids and hyaluronans) share structural features with human glycoconjugates and glycolipids. Other external polysaccharides include N-acetylheparosans and secondary cell wall polymers.

The cell envelope of mycobacteria consists of a complex array of polymers associated with the external peptidoglycan layer, including arabinogalactans and lipoarabinomannans. These bacteria are covered with a lipid-rich mycolic acid layer and capsule-like structures that are also found on their membrane-bound glycosylphosphatidylinositol anchors.

Modifications of the sugar residues of polysaccharides can affect their antigenicity and ability to be recognized by glycohydrolases. For example, o-acetylation helps to define the identity of a strain and is important in regulating the antigenicity of polysaccharides.

The structures of many polysaccharides have been determined, and a vast number of bacterial genes are known to be involved in the biosynthesis of polysaccharides. However, the specific action of individual enzymes and specific steps and mechanisms in the assembly pathways remain obscure. Polysaccharides are assembled by a restricted number of general pathways, often utilizing common enzyme substrates. Most bacterial glycosyltransferases have very little sequence identity with human enzymes that form similar linkages. Therefore, they do not appear to be directly derived from humans. However, the catalysis and mechanisms by which bacteria form glycan linkages are often similar to those in humans.

The first step in the biosynthesis of LPS-bound O antigenic and other polysaccharides is the transfer of a sugar phosphate from nucleotide sugar to P-Und to form sugar-PP-Und. This step resembles that in mammalian N-glycan synthesis where dolichol phosphate is the lipid intermediate and both reactions are blocked by tunicamycin. The chemical synthesis of GlcNAc-PP-Und and GalNAc-PP-Und analogs has opened the door to in vitro characterization of glycosyltransferases that form many of the oligosaccharide repeating units.

In LPS synthesis, GlcNAc-PP-Und/GalNAc-PP-Und is extended by glycosyltransferases to form an O antigen polymer in the cytoplasmic compartment that is transported by an ABC transport system to the periplasm. In a second major synthesis pathway, repeating units are assembled by sequentially acting glycosyltransferases at the cytoplasmic face of the inner membrane and are flipped across the membrane followed by polymerization in the periplasmic space. The polysaccharide is then transferred from PP-Und to a specific sugar residue of an acceptor oligosaccharide. The PP-Und is dephosphorylated and recycled as lipid intermediate. The antibiotic bacitracin blocks this process. A number of membrane-bound transport proteins and protein complexes are involved in localizing the polysaccharides to the external surface of the bacterial cell wall.

The synthesis of peptidoglycans, secondary cell wall polymers, and certain capsules also utilizes P-Und as a lipid intermediate. In peptidoglycan synthesis, repeating GlcNAc-MurNAc-pentapeptide units are synthesized in the cytoplasm on a PP-Und lipid carrier and transported across the inner membrane. The repeating units are then polymerized and transferred from PP-Und to the growing peptidoglycan chain by linking peptides together.

Bacteria produce a wealth of glycohydrolases that are secreted and can degrade both bacterial and host glycoconjugates. These hydrolases also have significant medical and industrial importance. The book will help to assay and characterize both glycosyltransferases and glycohydrolases.

The protocols in this book describe the isolation and analyses of polysaccharides, their biosynthesis, and chemical modifications. The biosynthetic enzymes are encoded in gene clusters and operons. Cloning of the genes involved in polysaccharide biosynthesis is a difficult task but is the basis for studying the enzymes involved and for carrying out functional analysis of genes. The first two chapters describe new methodologies for cloning of these genes to allow their functional analysis and manipulation.

Chapters 3, 4, 5, and 6 deal with methods of polysaccharide extraction from Gram-negative and Gram-positive bacteria as well as mycobacteria for the analysis of their quality and structures and for immunological studies. Lipid-linked glycans are common in mycobacteria, and protocols to extract glycosylphosphatidylinositols and other phospholipids are described in Chapter 6.

Vaccine development and knowledge of interaction of polysaccharides with the immune system are becoming increasingly important. To design new vaccine candidates and to prepare conjugates for biological studies, methods for chemical conjugation to proteins are described in detail in Chapters 7 and 8. Hyaluronans can stimulate host cytokine production, and de-N-acetylation may reduce this effect. A method for the chemical modification of hyaluronic acids for the purpose of modifying cytokine production is described in Chapter 9. Another common modification of microbial polysaccharides is O-acetylation that can reduce immune recognition and polymer degradation. The protocol in Chapter 10 allows the assaying of enzymes that add and remove O-acetyl groups on sugars which can lead to the development of O-acetylation inhibitors.

Chapters 11, 12, 13, 14, 15, 16, 17, 18, and 19 address methodologies to characterize the individual glycosyltransferases that assemble polysaccharides. In Chapter 11, the in silico analyses of glycosyltransferases are described. A number of programs are available as bioinformatic tools that can analyze the domain structure essential for the glycosyl transfer reaction, protein folding and function, and substrate binding. These tools allow the design of optimal constructs for the expression of enzyme proteins.

The synthesis of acceptor substrates has provided essential tools to characterize enzymes and understand substrate specificities, biosynthetic mechanisms, and pathways. Since natural acceptors based on P-Und or oligosaccharides are difficult to isolate, the chemical or chemo-enzymatic synthesis of acceptor substrates as analogs of natural substrates is a prerequisite for assaying glycosyltransferases (Chapters 12 and 13). Depending on the substrate specificity and activity of the glycosyltransferase, various assaying methods have been suggested in Chapters 12, 13, 14, 15, 16, 17, 18, and 19. While glycosyltransferase assays that transfer radioactive sugars provide high sensitivity, a number of nonradioactive assays are described that utilize color reagents (Chapter 18) or fluorescently labeled acceptor substrates (Chapters 12, 17, and 19). Using a combination of glycosyltransferases acting in the same pathway of O antigen synthesis, the repeating units of oligosaccharides of pathogenic bacteria can be assembled in vitro (Chapter 15). Thus, chemo-enzymatic synthesis of O antigens could lead to a successful vaccine. A number of enzymes have two glycosyltransferase domains that can be assayed separately (Chapters 12 and 19). The organization and topology of enzymes and membranes are instrumental for efficient assembly of complex polysaccharides. Methods to investigate the topology and functional domains of a membrane-bound protein and specific amino acids exposed to the solvent are suggested in Chapter 16.

Sugar-modifying enzymes include 4-epimerases and glycosidases. In Chapter 20, assays are described that allow the identification of the nucleotide sugar specificity of 4-epimerases. The assays for fucosidase (Chapter 21) show an example of how to assess hydrolytic enzymes.

Based on our current knowledge of bacterial polysaccharides, a number of novel antibacterial methodologies are proposed. New aspects of antibiotic development by glycosylating antimicrobial peptides are described in Chapter 22. Screening methods for finding inhibitors of polysaccharide synthesis could lead to therapeutic drugs (Chapter 23).

With this book, we wish to contribute our knowledge of bacterial polysaccharide structures, biosynthesis, and functions by providing the detailed protocols for important aspects in this field. This knowledge will help to develop novel antibacterial strategies against pathogens and antibiotic-resistant bacteria and has potential biotechnological applications. Unraveling the mechanisms of polysaccharide assembly could provide polysaccharides with specific structures for studies in biology, immunology, and medicine and as vaccine candidates and to identify potential antibacterial targets for drug development. Because of the medical importance of bacterial polysaccharides, it is essential that we continue to extend our knowledge in this field. My sincere thanks to all authors for their hard work and for sharing their specific expertise with other scientists, researchers, and clinicians to stimulate successful progress in studying essential aspects of bacteria.

Kingston, ON, Canada *Inka Brockhausen*

Contents

Contributors

TASNIM ABUKAR • *Department of Chemistry and Biology, Ryerson University, Toronto, ON, Canada*

CARLOS A. ALVAREZ • *Case Western Reserve University School of Medicine, Cleveland, OH, USA*

TASSOS ANASTASSIADES • *Division of Rheumatology, Department of Medicine, Arthritis Research Centre, Queen's University, Kingston, ON, Canada*

SHIVA KUMAR ANGALA • *Mycobacteria Research Laboratories, Department of Microbiology, Immunology and Pathology, Colorado State University, Fort Collins, CO, USA*

SIZIWE BEBE • *Division of Rheumatology, Department of Medicine, Arthritis Research Centre, Queen's University, Kingston, ON, Canada*

INKA BROCKHAUSEN • *Department of Biomedical and Molecular Sciences, Queen's University, Kingston, ON, Canada*

ASHLEY S. BROTT • *Department of Molecular and Cellular Biology, University of Guelph, Guelph, ON, Canada*

ERIC D. BROWN • *Department of Biochemistry and Biomedical Sciences, McMaster University, Hamilton, ON, Canada; Michael G. DeGroote Institute of Infectious Disease Research, McMaster University, Hamilton, ON, Canada*

NAKITA BUENBRAZO • *Department of Chemistry and Biology, Ryerson University, Toronto, ON, Canada*

ALICE CHATEAU • *Howard Taylor Ricketts Laboratory, Argonne National Laboratory, Lemont, IL, USA; Department of Microbiology, University of Chicago, Chicago, IL, USA*

ANTHONY J. CLARKE • *Department of Molecular and Cellular Biology, University of Guelph, Guelph, ON, Canada*

BRIAN A. COBB • *Case Western Reserve University School of Medicine, Cleveland, OH, USA*

CAROLE CREUZENET • *Department of Microbiology and Immunology, University of Western Ontario, London, ON, Canada*

DIANA CZUCHRY • *Department of Biomedical and Molecular Sciences, Queen's University, Kingston, ON, Canada*

LEONID DANILOV • *N.D. Zelinsky Institute of Organic Chemistry, Russian Academy of Sciences, Moscow, Russia*

LIAM DOYLE • *Department of Molecular and Cellular Biology, University of Guelph, Guelph, ON, Canada*

OMAR M. EL-HALFAWY • *Department of Biochemistry and Biomedical Sciences, McMaster University, Hamilton, ON, Canada; Michael G. DeGroote Institute of Infectious Disease Research, McMaster University, Hamilton, ON, Canada; Microbiology and Immunology Department, Faculty of Pharmacy, Alexandria University, Alexandria, Egypt*

LANA EL OSTA • *Department of Molecular and Cellular Biology, University of Guelph, Guelph, ON, Canada*

AMY FORD • *Centre for Infection and Immunity, Queen's University Belfast, Belfast, UK*

TAYLOR J. B. FORRESTER • *Department of Molecular and Cellular Biology, University of Guelph, Guelph, ON, Canada*

YANN GUÉRARDEL • *Univ. Lille, CNRS, UMR 8576—UGSF—Unité de Glycobiologie Structurale et Fonctionnelle, Lille, France*

MARY JACKSON • *Mycobacteria Research Laboratories, Department of Microbiology, Immunology and Pathology, Colorado State University, Fort Collins, CO, USA*
BETTINA JANESCH • *Department of Chemistry and Biology, Ryerson University, Toronto, ON, Canada*
KENNETH F. JARRELL • *Department of Biomedical and Molecular Sciences, Queen's University, Kingston, ON, Canada*
LAURA KELL • *Department of Chemistry and Biology, Ryerson University, Toronto, ON, Canada*
MATTHEW S. KIMBER • *Department of Molecular and Cellular Biology, University of Guelph, Guelph, ON, Canada*
ALEXANDER KOCEV • *Department of Biomedical and Molecular Sciences, Queen's University, Kingston, ON, Canada*
PAVOL KOVÁČ • *Laboratory of Bioorganic Chemistry, National Institutes of Health, Bethesda, MD, USA*
JOSEPH S. LAM • *Department of Molecular and Cellular Biology, University of Guelph, Guelph, ON, Canada*
MICHAEL A. LIU • *School of Life and Environmental Sciences, The University of Sydney, Sydney, NSW, Australia*
XIAOWEI LU • *National Institutes of Health, NIDDK, LBC, Bethesda, MD, USA*
ZHONGRUI MA • *Department of Chemistry, Georgia State University, Atlanta, GA, USA*
ZOË ANNE MEGSON • *Sandoz GmbH, Langkampfen, Austria*
PAUL MESSNER • *NanoGlycobiology Unit, Department of NanoBiotechnology, Universität für Bodenkultur Wien, Vienna, Austria*
DOMINIQUE MISSIAKAS • *Howard Taylor Ricketts Laboratory, Argonne National Laboratory, Lemont, IL, USA; Department of Microbiology, University of Chicago, Chicago, IL, USA*
YASU S. MORITA • *Department of Microbiology, University of Massachusetts, Amherst, MA, USA*
RUPA NAGAR • *CSIR-Institute of Microbial Technology (CSIR-IMTECH), Chandigarh, India*
OLGA G. OVCHINNIKOVA • *Department of Molecular and Cellular Biology, University of Guelph, Guelph, ON, Canada*
HÉLÈNE B. PFISTER • *National Institutes of Health, NIDDK, LBC, Bethesda, MD, USA*
JULIA PUFFAL • *Department of Microbiology, University of Massachusetts, Amherst, MA, USA*
WENSHENG QIN • *Department of Biology, Lakehead University, Thunder Bay, ON, Canada*
KATHRYN C. RAHLWES • *Department of Microbiology, University of Massachusetts, Amherst, MA, USA*
ALKA RAO • *CSIR-Institute of Microbial Technology (CSIR-IMTECH), Chandigarh, India*
PETER R. REEVES • *School of Life and Environmental Sciences, The University of Sydney, Sydney, NSW, Australia*
XIANG RUAN • *Department of Microbiology and Immunology, University of Western Ontario, London, ON, Canada*
IRINA SADOVSKAYA • *Équipe BPA, Univ. Littoral Côte d'Opale, convention ANSES, EA 7394, ICV Charles Violette, Univ. Lille, Univ. Artois, INRA, ISA F-62321, Boulogne-sur-mer, France*
CHRISTINA SCHÄFFER • *NanoGlycobiology Unit, Department of NanoBiotechnology, Universität für Bodenkultur Wien, Vienna, Austria*
OLAF SCHNEEWIND • *Howard Taylor Ricketts Laboratory, Argonne National Laboratory, Lemont, IL, USA; Department of Microbiology, University of Chicago, Chicago, IL, USA*

M. Shafiqur Rahman • *Department of Biology, Lakehead University, Thunder Bay, ON, Canada; Department of Microbiology, University of Chittagong, Chittagong, Bangladesh*

Sulav Sharma • *Department of Biomedical and Molecular Sciences, Queen's University, Kingston, ON, Canada*

Sameh E. Soliman • *National Institutes of Health, NIDDK, LBC, Bethesda, MD, USA*

David Sychantha • *Department of Molecular and Cellular Biology, University of Guelph, Guelph, ON, Canada*

Faviola Tavares-Carreón • *Department of Microbiology and Immunology, University of Western Ontario, London, ON, Canada*

Vladimir Torgov • *N.D. Zelinsky Institute of Organic Chemistry, Russian Academy of Sciences, Moscow, Russia*

Natalia Utkina • *N.D. Zelinsky Institute of Organic Chemistry, Russian Academy of Sciences, Moscow, Russia*

Miguel A. Valvano • *Centre for Infection and Immunity, Queen's University Belfast, Belfast, UK; Department of Microbiology and Immunology, University of Western Ontario, London, ON, Canada*

Vladimir Veselovsky • *N.D. Zelinsky Institute of Organic Chemistry, Russian Academy of Sciences, Moscow, Russia*

Warren Wakarchuk • *Department of Chemistry and Biology, Ryerson University, Toronto, ON, Canada*

Peng George Wang • *Department of Chemistry, Georgia State University, Atlanta, GA, USA*

Chris Whitfield • *Department of Molecular and Cellular Biology, University of Guelph, Guelph, ON, Canada*

Danielle M. Williams • *Department of Molecular and Cellular Biology, University of Guelph, Guelph, ON, Canada*

Peng Xu • *Laboratory of Bioorganic Chemistry, National Institutes of Health, Bethesda, MD, USA*

Chapter 1

Customizable Cloning of Whole Polysaccharide Gene Clusters by Yeast Homologous Recombination

Michael A. Liu and Peter R. Reeves

Abstract

Cloning of whole polysaccharide biosynthesis gene clusters for expression in a common *Escherichia coli* tester strain has the major advantage of enabling direct functional comparisons between gene clusters that are normally found in different strains, where their expression is potentially under differential regulatory control. However, due to the large size of many of these gene clusters, classical cloning methods are highly inefficient, time-consuming, and/or labor-intensive. Here we describe a recently developed system, called the operon assembly protocol (OAP), in which yeast homologous recombination pathways are used to assemble overlapping PCR fragments onto a specially engineered yeast *E. coli* shuttle vector, resulting in full-length customizable gene cluster clones on single-copy plasmids. Multiple versions of the same gene cluster can also be assembled in parallel with genes deleted, replaced, or rearranged, allowing the function and/or specificity of individual genes to be examined. Since the vector can be easily modified to include other bacterial replicons, it can also be broadly applied to the functional analysis of a wide range of bacterial gene clusters and operons.

Key words Cloning, Surface polysaccharides, Gene clusters, Homologous recombination, Yeast, *E. coli*, Gene deletions, Gene replacements, Customized operon assembly

1 Introduction

Direct functional comparisons between highly diverse surface polysaccharide gene clusters expressed in different bacterial strains, such as those for the biosynthesis of capsules [1–3] and O antigens [4–8], are often complicated by the difficulty of adequately controlling for the differential regulation of their expression by externally encoded regulatory factors. One way to overcome this problem is to clone each gene cluster onto a common vector and express them all in a common tester strain. Unfortunately, successful cloning of entire polysaccharide gene clusters, which can be greater than 20 kb in length, can be difficult using classical directional cloning approaches, while gene cluster clones isolated from cosmid libraries typically contain large segments of unwanted

Inka Brockhausen (ed.), *Bacterial Polysaccharides: Methods and Protocols*, Methods in Molecular Biology, vol. 1954, https://doi.org/10.1007/978-1-4939-9154-9_1, © Springer Science+Business Media, LLC, part of Springer Nature 2019

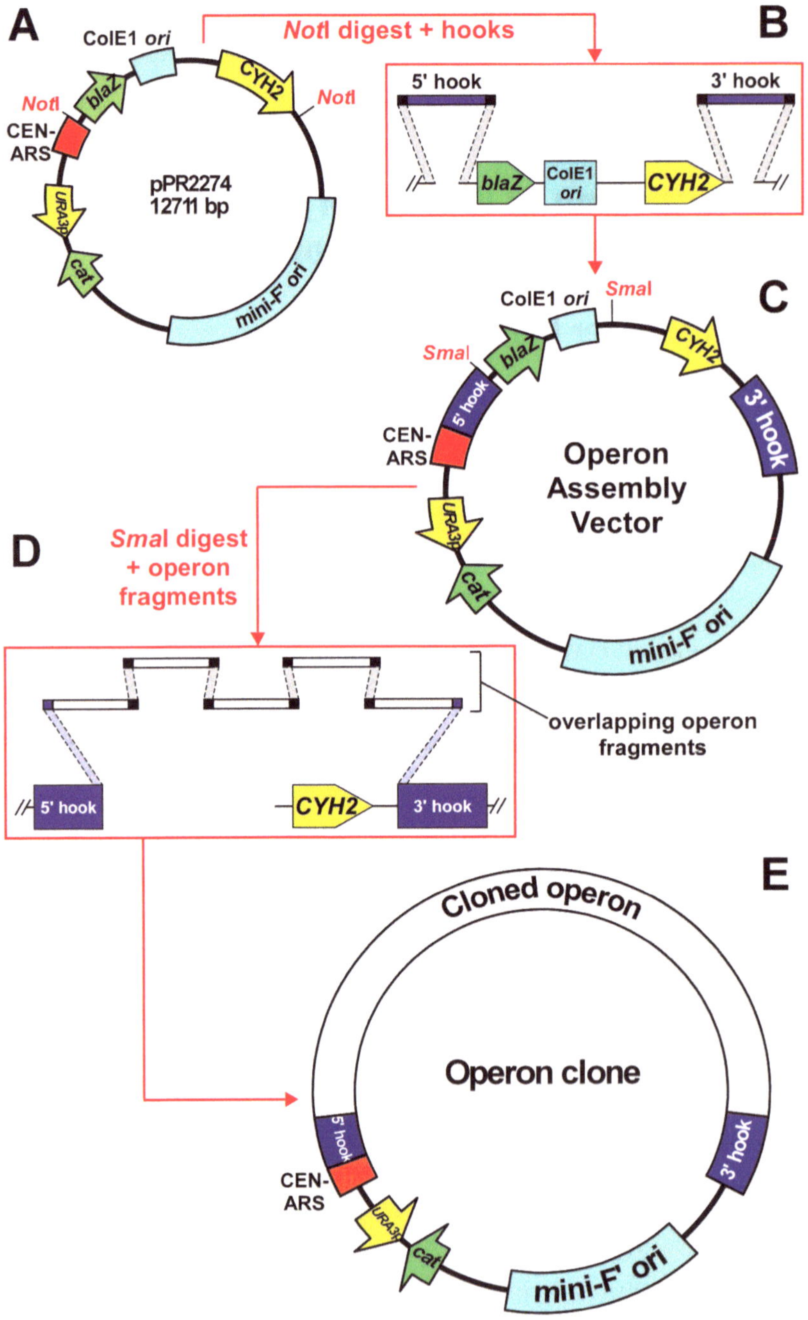

Fig. 1 Workflow for the operon assembly protocol. (**a**) Yeast–*E. coli* shuttle vector pPR2274 with single-copy (mini-F *ori*) and multicopy (ColE1 *ori*) replicons for *E. coli*, and a CEN-ARS replicon for yeast. The vector has selective markers for *E. coli* (*blaZ* ampicillin resistance, *cat* chloramphenicol resistance) and yeast (*URA3p* uracil synthesis, *CYH2*, cycloheximide sensitivity). (**b**) *Not*I-digested pPR2274 is reassembled in yeast by homologous recombination with PCR fragments containing 5′ and 3′ hook sequences (purple bars) flanked by 40 bp sequences (black boxes) that overlap with either side of the *Not*I sites (indicated by the shaded grey

flanking DNA [9–12]. To overcome these problems, we have adapted and modified a previously described system for cloning entire polysaccharide gene clusters, in which the homologous recombination DNA repair pathways of *Saccharomyces cerevisiae* are used for the targeted assembly of overlapping DNA fragments onto a specially engineered yeast–*Escherichia coli* shuttle vector [13, 14].

This new methodology, known as the operon assembly protocol (OAP) [15], is a two-stage process (Fig. 1). The first stage involves constructing an operon assembly vector (OAV) by incorporating targeted "hook" sequences into the vector (Fig. 1a–c). These hooks are designed to "capture" the target gene cluster(s), and thus typically contain sequences that flank these gene clusters in their original chromosomal location. When properly designed, a single OAV can be used to clone a wide array of different polysaccharide gene clusters. While the default vector is designed specifically for expression in *E. coli*, it is also possible to incorporate an additional replicon and/or selective marker into the vector at this stage to facilitate replication and expression in other gram-negative or even gram-positive bacteria.

The second stage of the OAP involves assembly of full-length polysaccharide gene clusters onto the OAV between the hook sequences using PCR fragments containing gene cluster sequences that overlap with one another and/or the hooks (Fig. 1d, e), with multiple different clones able to be assembled in parallel. By engineering artificial homology sequences onto the ends of these PCR fragments, it is also possible to delete and/or replace individual genes, or even assemble gene clusters with entirely novel gene arrangements (Fig. 2). The resultant gene cluster clones, which replicate as single-copy plasmids in *E. coli* to mimic chromosomal copy numbers, can all be expressed in the same strain, thus facilitating direct functional comparisons.

We have successfully used the OAP to clone and express a series of *E. coli* O antigen gene clusters [15], and are continuing to utilize this system for ongoing research within our group. By enabling the parallel assembly of several different genes knockouts and/or

Fig. 1 (continued) areas). These hooks contain sequences that flank the target operons in their chromosomal location. (**c**) The resultant operon assembly vector (OAV) contains the incorporated 5′ and 3′ hooks, which flank the *blaZ* and *CYH2* genes and the ColE1 *ori*. (**d**) The *Sma*I-digested OAV is reassembled in yeast by homologous recombination with a series of overlapping PCR fragments containing the target operon sequences (white bars). The two outermost fragments overlap with the OAV hooks (indicated by the purple shaded areas), while also overlapping with the adjacent PCR fragments (indicated by the grey shaded areas). (**e**) The resultant operon clone contains the entire operon sequence flanked by the 5′ and 3′ hooks on a single-copy plasmid (via the mini-F *ori*)

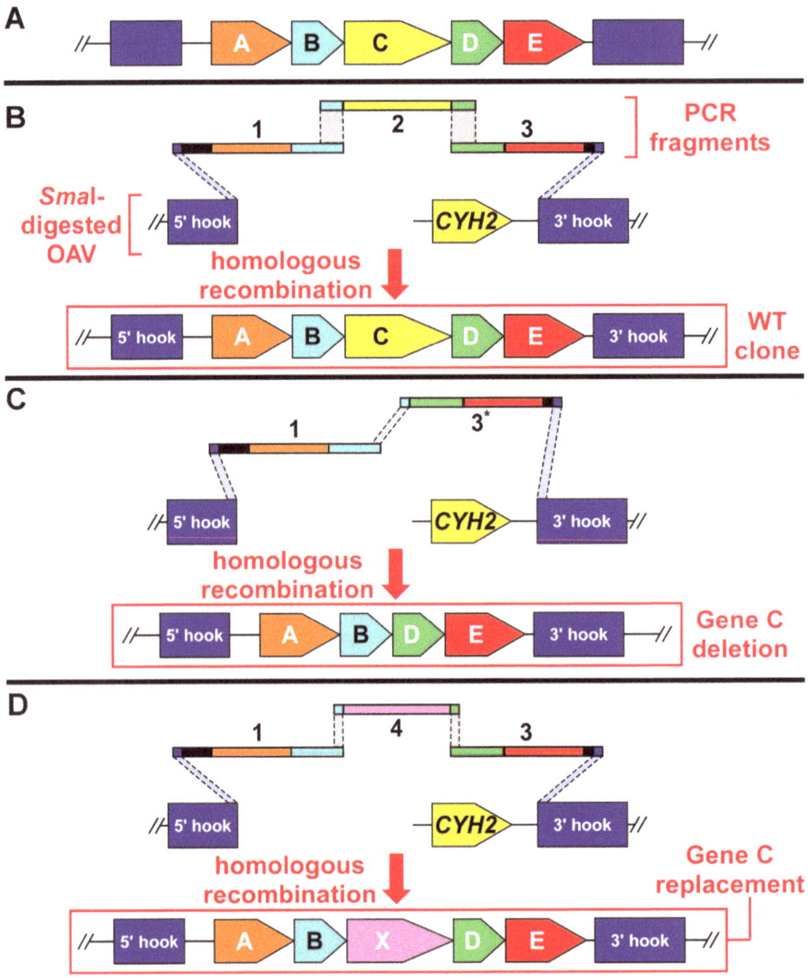

Fig. 2 Types of gene cluster clones assembled by the OAP. (**a**) The original chromosomal organization of a hypothetical 5-gene cluster (genes A to E) flanked by sequences (purple boxes) used as hooks in the OAV. (**b**) To assemble a fully WT clone, three overlapping PCR fragments (colored bars 1 to 3) are amplified. The colors of the fragments indicate which gene sequences they contain, with the overlapping regions shaded in grey. Fragments 1 and 3 also overlap with the 5′ and 3′ hooks, respectively (overlapping regions shaded in purple). These PCR fragments and the *Sma*I-digested OAV are assembled in yeast, resulting in a gene cluster clone with a fully WT arrangement (boxed in red). (**c**) A clone with gene C deleted can be assembled in the same manner as the WT clone, but using only fragment 1 and fragment 3*, which is an alternate form of fragment 3 with an additional 40 bp of sequence that overlaps with fragment 1 (indicated by the grey shading) introduced via the forward primer. (**d**) A clone with gene C replaced by a new gene (gene X) can also be assembled similarly to the WT clone, but with fragment 2 replaced by a new fragment 4 containing the gene X sequence and 40 bp of sequence overlap with fragments 1 and 3 (indicated by the grey shading) introduced via the forward and reverse primers, respectively. This figure is adapted from [15]

replacements for the same gene cluster, this system can rapidly provide critical insights into the function and specificity of the glycosyltransferases, translocases, and polymerases involved in surface polysaccharide biosynthesis. Furthermore, by using this system to assemble new gene arrangements and compositions, it is possible to synthesize entirely novel polysaccharide structures in vivo. Due to its flexibility and ease of use, the OAP can also be applied more generally to the cloning, analysis, and manipulation of a broad range of bacterial gene clusters and operons.

2 Materials

Prepare all solutions using sterile ultrapure water at room temperature. All autoclaving is performed at 121 °C and 100 kPa above atmospheric pressure for 15 min. Ensure that all biological GMOs and hazardous substances are disposed of via appropriate waste streams.

2.1 Strains

1. CRY1-2: *S. cerevisiae* recombinational cloning strain; *MATα*, *ura3Δ*, *cyh2*^{R} (*see* **Note 1**). Available on request from Peter Reeves.
2. DH10B: *E. coli* strain designed for transformation with large plasmids; Δ(*ara leu*)*7697*, *araD139*, *fhuA*, Δ*lacX74*, *galK16*, *galE15*, *e14-*, φ80d*lacZ*Δ*M15*, *recA1*, *relA1*, *endA1*, *nupG*, *rpsL*, *rph*, *spoT1*, Δ(*mrr-hsdRMS-mcrBC*). Available from New England Biolabs (NEB) and several other commercial suppliers.

2.2 Plasmids

1. pPR2274: yeast–*E. coli* shuttle vector for recombinational cloning of large DNA fragments; *E. coli* mini-F *ori* and ColE1 *ori* (*see* **Note 2**), yeast CEN-ARS replicon, *URA3p* and *CYH2* selective markers for yeast, *blaZ* and *cat* selective markers for *E. coli* (*see* **Note 3**). Available on request from Peter Reeves.

2.3 Media

1. YPD broth/agar: 2% (w/v) bacteriological peptone, 1% (w/v) yeast extract, 2% (w/v) D-glucose, 2% (w/v) agar (for solid media only) (*see* **Note 4**).
2. SD Ura^{-} broth/agar: 0.67% (w/v) yeast nitrogen base (Sigma-Aldrich), 0.192% (w/v) dropout supplement without uracil (Sigma-Aldrich), 2% (w/v) D-glucose, 2% (w/v) agar (for solid media only) (*see* **Note 5**).
3. Nutrient broth/agar (NB/NA): 1% (w/v) bacteriological peptone, 0.5% (w/v) yeast extract, 0.5% (w/v) NaCl, 1.5% (w/v) agar (for solid media only) (*see* **Note 6**).

2.4 Enzymes and Reagents

1. Restriction endonucleases *Not*I-HF and *Sma*I from NEB.
2. DNA polymerase I, large (Klenow) fragment from NEB.
3. Q5 High-Fidelity DNA polymerase from NEB.
4. CutSmart buffer (10×) from NEB.
5. dNTP mix (10 mM) from NEB.

2.5 Lithium Acetate Yeast Transformations

1. 10× LiAc: 1 M lithium acetate (pH 7.5). Dissolve 10.2 g of lithium acetate dihydrate in 70 mL of water and adjust pH to 7.5 with dilute acetic acid. Make up to 100 mL with water and autoclave.
2. 50% PEG: Dissolve 50 g of polyethylene glycol (avg. mol. wt. = 3350) in 50 mL of water. Make up to a final volume of 100 mL with water and autoclave (*see* **Note 7**).
3. 10× TE buffer: 0.1 M Tris, 10 mM EDTA, pH 7.5. Dissolve 1.21 g of Tris and 0.29 g of EDTA (*see* **Note 8**) in 90 mL of water and adjust pH to 7.5 with 1 M HCl. Make up to 100 mL with water and autoclave.
4. 1× TE buffer: Add 10 mL of 10× TE buffer to 90 mL of water.
5. TE–LiAc solution: Thoroughly mix 0.5 mL of 10× TE buffer, 0.5 mL of 10× LiAc, and 4 mL of water. Prepare fresh immediately prior to use.
6. PEG–LiAc solution: Thoroughly mix 8 mL of 50% PEG, 1 mL of 10× TE and 1 mL of 10× LiAc. Prepare fresh immediately prior to use.
7. Carrier DNA: single-stranded DNA from Salmon testes (Sigma-Aldrich) (*see* **Note 9**).

2.6 Yeast Plasmid Isolations

1. Tris buffer: 100 mM Tris (pH 9.3). Dissolve 2.42 g of Tris in 190 mL of water and adjust pH to 9.3 with 1 M HCl. Make up to 200 mL with water and autoclave (*see* **Note 10**).
2. MP buffer: 1 M sorbitol, 1 M NaCl, 10 mM acetic acid, pH 5.5. Dissolve 36.4 g of sorbitol, 11.68 g of NaCl and 11.4 mL of 1% (v/v) glacial acetic acid in 130 mL of water and adjust pH to 5.5 with 1 M NaOH. Make up to 200 mL with water and autoclave.
3. Protoplasting solution: 10 mg mL^{-1} yeast lysing enzymes (Glucanex, Sigma-Aldrich) in MP buffer. Dissolve 30 mg of Glucanex per 3 mL of MP buffer.
4. Purelink Quick Plasmid Miniprep Kit (Thermo-Fisher Scientific).

3 Methods

Carry out all procedures at room temperature unless otherwise indicated. All wash steps using water were carried out with sterile ultrapure water.

3.1 Construction of a Targeted Operon Assembly Vector

1. Streak out yeast strain CRY1-2 onto YPD agar and incubate at 30 °C for 2–3 days (*see* **Note 11**).

2. Perform a *Not*I-HF digestion in 1× CutSmart buffer (37 °C for 2 h; *see* **Note 12**) on 5 μg of the yeast–*E. coli* shuttle vector pPR2274, then heat-inactivate the enzyme (65 °C for 20 min).
3. Add 5 units of DNA polymerase I, large (Klenow) fragment and 33 μM dNTPs to the mixture, and then incubate at 25 °C for 15 min to blunt the sticky ends of the vector (*see* **Note 12**). Stop the reaction by adding 10 mM EDTA and heat-inactivate the enzyme at 75 °C for 20 min.
4. PCR-amplify 1 μg of the 5′ and 3′ hook fragments for incorporation into the pPR2274 *Not*I sites using Q5 High-Fidelity DNA polymerase (*see* **Note 13**), ensuring that each primer contains at least 40 bp of sequence overlapping with the appropriate site on the vector (*see* **Notes 14** and **15**).
5. Use several CRY1-2 colonies (*see* **step 1**) to inoculate 50 mL of YPD broth and incubate overnight at 30 °C with shaking at 250 rpm.
6. Transfer sufficient overnight culture (10–30 mL) to 300 mL of fresh YPD broth (prewarmed to 30 °C) to obtain a starting OD_{600} of 0.2–0.3.
7. Incubate culture at 30 °C with shaking at 230 rpm until an OD_{600}~0.4–0.6 is obtained (*see* **Note 16**).
8. Harvest cells by centrifugation (1000 × *g*, 5 min) and resuspend the cells in 25–50 mL of water.
9. Harvest cells by centrifugation (1000 × *g*, 5 min) and resuspend the cells in 1.5 mL of freshly prepared TE/LiAc.
10. Set up the following DNA transformation mixes into separate 1.5 mL microfuge tubes. (a) no DNA, (b) undigested pPR2274 (100 ng); (c) *Not*I-digested pPR2274 (100 ng; *see* **steps 2** and **3**); (d) *Not*I-digested pPR2274 (100 ng) + hook PCR fragments (1 μg each; *see* **step 4**) (*see* **Note 17**).
11. Add 0.1 mg of carrier DNA into each tube (*see* **Note 9**), followed by 100 μL of competent CRY1-2 cells (*see* **step 9**) and mix by pipetting.
12. Add 600 μL of freshly prepared PEG/LiAc solution to each tube, mix vigorously by vortexing, and incubate at 30 °C for 30 min with shaking at 200 rpm (*see* **Note 18**).
13. Add 70 μL of dimethyl sulfoxide to each tube, mix by gentle inversion, and then incubate in a 42 °C water bath for 15 min.
14. Chill cells on ice for 1–2 min, harvest cells by brief centrifugation (15,000 × *g*, 10 s), and resuspend in 100 μL of TE buffer.
15. Spread all transformation mixes onto SD Ura^- agar plates and incubate at 30 °C for 2–3 days.
16. Inspect plates after incubation, ensuring that colonies can be observed on the transformation mix (d) plate (*see* **Note 19**).

17. Dispense 0.5 mL of SD Ura^{-} broth onto the transformation mix (d) plate, and use a spreader to remove and resuspend the colonies into the broth.
18. Tilt the plate at an angle and collect the cell suspension with a pipette, and then transfer the cells into 8 mL of fresh SD Ura^{-} broth.
19. Incubate the culture overnight at 30 °C with shaking at 230 rpm.
20. Harvest cells by centrifugation (1800 × *g*, 5 min), wash cells twice with 5 mL of water and resuspend in 3 mL of Tris buffer containing 16 mM β-mercaptoethanol.
21. Incubate cells at 30 °C for 30 min, then harvest the cells by centrifugation (1100 × *g*, 5 min) and wash cells twice with 5 mL of water.
22. Resuspend cells in 3 mL of protoplasting solution and incubate overnight at 37 °C (*see* **Note 20**).
23. Harvest cells by centrifugation (300 × *g*, 5 min), wash once with 3 mL of MP buffer, and then resuspend the cells in 250 μL of Resuspension Buffer from the Purelink Quick Plasmid Miniprep kit.
24. Extract plasmid DNA from the yeast protoplasts using the Purelink Quick Plasmid Miniprep kit as per the manufacturer's instructions (*see* **Note 21**).
25. Transform the isolated plasmid DNA into electrocompetent DH10B cells by electroporation (12.5 kV/cm, 25 μFD, 200 Ω) as previously described [16], and then select for transformants on NA containing 12.5 μg mL^{-1} chloramphenicol (Cml) (*see* **Note 22**).
26. Perform colony PCR on several transformants using primers flanking the two hook insertion sites to identify DH10B colonies containing pPR2274 with both hook sequences incorporated into the two *Not*I sites (*see* **Note 23**).
27. Isolate plasmid DNA from a positively identified colony using the Purelink Quick Plasmid Miniprep kit, and then sequence the plasmid across the hooks to confirm successful OAV assembly (*see* **Notes 24** and **25**).

3.2 Gene Cluster Cloning into the Operon Assembly Vector

1. Streak out yeast strain CRY1-2 onto YPD agar and incubate at 30 °C for 2–3 days (*see* **Note 11**).
2. Perform a *Sma*I digestion in 1× CutSmart buffer (25 °C for 2 h; *see* **Note 12**) on 5 μg of the OAV, and then heat-inactivate the enzyme (65 °C for 20 min) (*see* **Note 26**).
3. PCR-amplify each of the overlapping DNA fragments containing the desired gene cluster sequences using Q5 High-Fidelity

DNA polymerase (*see* **Note 13**), ensuring that the fragments contain at least 40 bp of sequence overlap (*see* **Note 27**).

4. Perform **steps 5–9** from Subheading 3.1.
5. Set up the following DNA transformation mixes into separate 1.5 mL microfuge tubes: (a) no DNA; (b) 2 tubes of undigested OAV (100 ng); (c) *Sma*I-digested OAV (100 ng; *see* **step 2**); (d) *Sma*I-digested OAV (100 ng) + gene cluster PCR fragments (1 μg each; *see* **step 3**) (*see* **Note 28**). At least 11 different transformation mixes (d) can be set up in parallel with different combinations of gene cluster PCR fragments (*see* **Note 29**).
6. Perform **steps 11–14** from Subheading 3.1.
7. Spread transformation mix (a) and one tube of transformation mix (b) onto SD Ura^- plates. Spread the second tube of transformation mix (b), as well as transformation mixes (c) and (d), onto SD Ura^- plates containing 2.5 μg mL^{-1} cycloheximide (Chx) to select for recombinant clones.
8. Incubate all plates at 30 °C for 2–3 days, and then inspect the plates to confirm the presence of colonies for each transformation mix (d) plate (*see* **Note 30**).
9. Dispense 0.5 mL of SD Ura^- broth onto each transformation mix (d) plate, and use a spreader to remove and resuspend the colonies.
10. Perform **steps 18–24** from Subheading 3.1 for the cells from each different transformation mix (d) (*see* **Note 31**).
11. Transform plasmid DNA into electrocompetent DH10B cells by electroporation (12.5 kV/cm, 25 μFD, 200 Ω) as previously described [16], and select for transformants on NA containing 3 μg mL^{-1} Cml (*see* **Note 22**).
12. Perform colony PCR on several transformants using primers for identification of DH10B colonies harboring recombinant gene cluster clones (*see* **Note 32**).
13. Isolate plasmid DNA from positively identified DH10B transformants using the Purelink Quick Plasmid Miniprep kit, and then transform the plasmids into the desired *E. coli* tester strain by electroporation [16] for functional testing (*see* **Note 33**).

4 Notes

1. CRY1-2 is unable to synthesize uracil (*ura3*Δ) and is also Chx-resistant ($cyh2^R$).
2. The ColE1 *ori* [17] is a multicopy *E. coli* replicon, which allows large quantities of the vector to be readily isolated throughout

the OAV assembly process (*see* Subheading 3.1). This replicon is eliminated during gene cluster clone assembly (*see* Subheading 3.2), after which replication in *E. coli* is dependent on the single-copy mini-F *ori* [18]. Hence, recombinant gene cluster clones mimic chromosomal copy numbers in *E. coli*.

3. The *URA3p* gene confers the ability for CRY1-2 to synthesize uracil and is used to select for transformants. The *CYH2* gene confers Chx sensitivity to CRY1-2, with the elimination of this gene during the OAP (*see* Subheading 3.2) enabling selection for recombinant clones. The vector also has two *E. coli* selective markers: *blaZ*, which confers ampicillin resistance and is eliminated during the OAP; and *cat*, which confers Cml resistance.
4. Add the peptone, yeast extract and agar (for solid media only) to water and autoclave. Allow the media to cool to 55 °C prior to adding 5 mL of filter-sterilized 40% (w/v) D-glucose per 100 mL of media.
5. Add the yeast nitrogen base, dropout supplement and agar (for solid media only) to water and autoclave. Allow the media to cool to 55 °C prior to adding 5 mL of filter-sterilized 40% (w/v) D-glucose per 100 mL of media. When selecting for recombinant gene cluster clones, add 25 μL of filter-sterilized 10 mg mL^{-1} cycloheximide (prepared in water) per 100 mL of media (final concentration of 2.5 μg mL^{-1}) at this stage.
6. Add the peptone, yeast extract, NaCl and agar (for solid media only) to water and autoclave. Allow the media to cool to 55 °C prior to adding 12 to 50 μL of 25 mg mL^{-1} Cml (prepared in 100% (v/v) ethanol) per 100 mL of media (final concentration of 3 to 12.5 μg mL^{-1}, depending on the *E. coli* strain; *see* **Note 33**).
7. The solution can be heated to 50 °C while stirring to help the PEG to dissolve. After autoclaving, store in a tightly sealed bottle to prevent changes in the PEG concentration due to evaporation, which can result in reduced transformation efficiencies. It is recommended that a fresh solution be prepared every 6 to 12 months.
8. An alternative option is to prepare a 0.4 M EDTA (pH 8.0) solution, and then add 2.5 mL of this solution per 100 mL of 10× TE buffer. Note that EDTA is only soluble in water at pH ~8.0.
9. Just prior to use, denature the carrier DNA by heating to 100 °C for 20 min, then immediately transfer to ice to cool.
10. When 16 mM β-mercaptoethanol is required in the Tris buffer, add 50 μL of 1 M β-mercaptoethanol per 3 mL of Tris buffer immediately prior to use.

11. CRY1-2 colonies on the YPD plates are usable for up to 3 weeks (stored at 4 °C), after which the plate should be discarded and a fresh streak prepared.
12. All enzymatic reactions are performed under the conditions described by the manufacturer with the supplied reaction buffers.
13. All PCRs using Q5 High-Fidelity DNA polymerase are performed as per the manufacturer's instructions. PCR products should be purified using a commercial PCR cleanup kit, and then analyzed by agarose gel electrophoresis to confirm amplification of the desired fragment and to estimate the purity and yield.
14. The hooks should contain sequences that flank the target gene cluster(s) in the chromosome. For instance, when cloning *E. coli* O antigen gene clusters, our 5′ and 3′ hooks included the *galF* and *gnd* genes, respectively [15], which flank most *E. coli* O antigen gene clusters. It is also critical that the 5′ hook sequence is designed to capture the target gene cluster's promoter sequence, as pPR2274 itself has no preexisting promoter at the cloning site.
15. To introduce homology between the hook PCR fragments and pPR2274, the 5′ ends of the forward and reverse primers for both hooks should include 40 bp of sequence overlap with either side of the target pPR2274 *Not*I sites (Fig. 1b).
16. It should take around 3 h for the starting OD_{600} to double. If the growth rate is significantly slower, the protocol should be restarted using fresh CRY1-2 colonies.
17. Transformation mixes (a) and (b) are negative and positive controls, respectively, for yeast transformation. Transformation mix (c) is a negative control for repair of *Not*I-digested pPR2274, which should not occur without addition of the hook PCR fragments. Large colony counts on this plate will likely indicate a significant number of false-positive colonies on the transformation mix (d) plate.
18. Position the microfuge tubes horizontally during shaking for efficient mixing.
19. Ideally, there should be no colonies on the plates for transformation mixes (a) and (c), and several hundred colonies on the plates for transformation mixes (b) and (d). Ultimately however, the most critical observation is the presence of significantly more colonies on the transformation mix (d) plate relative to the transformation mix (c) plate.
20. To quickly confirm protoplast formation, 50 μL of the protoplasting cell suspension can be transferred into separate 0.5 mL

aliquots of Tris buffer and MP buffer. Protoplasts will burst in the Tris buffer but not the MP buffer, resulting in greater turbidity in the MP buffer mixture. If no difference in turbidity is apparent, we recommend adding a further 10 mg mL^{-1} of Glucanex and incubating at 37 °C for a further 24 h.

21. Yeast protoplasts lyse in a similar fashion to gram-negative bacterial cells, so follow the manufacturer's protocol beginning from the resuspension of the cells in Resuspension buffer.
22. We also recommend performing **steps 17–24** of Subheading 3.1 on the colonies from the transformation mix (b) plate, as pPR2274 DNA isolated from yeast serves as an ideal positive control for DH10B electroporation. A no-DNA negative control should also be included for the electroporation setup.
23. A protocol for performing colony PCRs is available from [19]. Primers should be designed for sites approximately 100–200 nt upstream or downstream of the hooks, so that colonies containing the vector without hooks can be readily identified on the basis of the smaller PCR fragments that they produce.
24. The primers from the colony PCR screen can also be used for sequencing, although additional internal primers may also be necessary if the hooks are larger than ~1.5 kb.
25. While the OAV described here is specifically designed for expression in *E. coli*, it is also possible to construct an OAV with the ability to replicate in other gram-negative or even gram-positive bacteria. We would recommend undertaking the following general steps prior to incorporation of the targeted hook sequences:
    - Identify a restriction site in the pPR2274 backbone, between the mini-F *ori* and either *cat* or *CYH2* (*see* Fig. 1a), that is either unique or found at only a few other locations on the vector.
    - Digest pPR2274 with the relevant restriction enzyme, blunt the sticky ends with DNA polymerase I, large (Klenow) fragment (if required), and then heat-inactivate the enzyme (*see* **steps 2** and **3** of Subheading 3.1).
    - PCR-amplify a replicon and selective marker that functions in the organism of interest (either as a single fragment or as two overlapping fragments). Include 40 bp sequences at the ends of the fragment(s) that overlap with each end of the target restriction site on the vector (in a similar fashion to the hooks; *see* **Note 15**).
    - Transform the digested pPR2274 and the replicon-marker fragment(s) into CRY1-2 alongside the relevant controls (*see* **step 10** of Subheading 3.1). Any additional restriction sites on pPR2274 can be repaired by also including in the

transformation 80 bp oligonucleotides with 40 bp of sequence overlap to each end of the restriction sites (*see* [14] for further information).

- Follow **steps 11–27** of Subheading 3.1 to assemble, isolate and screen the modified vector.

26. Although not essential, we recommend gel-purifying the large fragment containing the vector backbone and hooks following *Sma*I digestion. This will eliminate the 2 kb *blaZ*-ColE1 *ori* fragment and any remaining undigested or partially digested OAV prior to yeast assembly. Any commercial gel purification kit will suffice.
27. Assembling fully WT gene clusters is a straightforward process, as the PCR fragments naturally overlap with one another. The primers for a given PCR fragment are simply designed to start close to the ends of the adjacent fragments (*see* Fig. 2b). Assembly of customized gene clusters with deleted, replaced, or rearranged genes requires homology between PCR fragments with no natural overlap. This is achieved by artificially incorporating 40 bp of sequence at the 5′ end of the forward and/or reverse primer for a PCR fragment, which overlaps with the desired adjacent fragment(s) (*see* Fig. 2c, d).
28. Transformation mix (a) is a negative control for yeast transformation. Transformation mix (b) on SD Ura^- agar is a positive control for yeast transformation, while the same mix on SD Ura^- + Chx agar is used to confirm that cells transformed with the parental pPR2274 are Chx-sensitive. Transformation mix (c) is a negative control for repair of the *Sma*I-digested pPR2274, which should not occur without the operon PCR fragments. A large number of colonies on this plate will likely indicate a significant number of false-positive colonies for transformation mixes (d).
29. The yeast transformation protocol described here is suitable for up to 15 simultaneous transformations (including controls); however the entire process can be scaled-up if required.
30. Ideally, there should be no colonies on the plates for transformation mixes (a) and (c), and hundreds of colonies for transformation mix (b) on the SD Ura^- plate, but none on the SD Ura^- + Chx plate. Several hundred colonies typically appear on the transformation mix (d) plates, however positive clones can also be found on plates with far fewer colonies.
31. Yeast cells with putative recombinant clones can be grown in SD Ura^- broth without Chx after **steps 7** and **8** of Subheading 3.2, as the absence of uracil provides sufficient selection for ongoing plasmid maintenance.
32. A protocol for performing colony PCRs is available from [19]. Due to the large size of many gene clusters, it may be

unfeasible to screen with flanking primers that amplify the entire cloned region. Hence, we recommend designing primers for genes within the target gene clusters.

33. The Cml concentration required for *E. coli* tester strain transformation is strain dependent. We recommend determining the minimal inhibitory concentration (MIC) for your chosen strain by testing growth on NA containing sequential twofold reductions in Cml concentrations starting from 25 μg mL^{-1}. Transformants should then be selected on NA with a Cml concentration that is double the MIC for that strain.

References

1. Whitfield C (2006) Biosynthesis and assembly of capsular polysaccharides in *Escherichia coli*. Annu Rev Biochem 75:39–68
2. Yother J (2011) Capsules of *Streptococcus pneumoniae* and other bacteria: paradigms for polysaccharide biosynthesis and regulation. Annu Rev Microbiol 65:563–581
3. Guerry P, Poly F, Riddle M et al (2012) *Campylobacter* polysaccharide capsules: virulence and vaccines. Front Cell Infect Microbiol 2:7
4. DebRoy C, Fratamico PM, Yan X et al (2016) Comparison of O-antigen gene clusters of all O-serogroups of *Escherichia coli* and proposal for adopting a new nomenclature for O-typing. PLoS One 11:e0147434
5. Liu B, Knirel YA, Feng L et al (2008) Structure and genetics of *Shigella* O antigens. FEMS Microbiol Rev 32:627–653
6. Reeves PR, Cunneen MM, Liu B et al (2013) Genetics and evolution of the *Salmonella* galactose-initiated set of O antigens. PLoS One 8:e69306
7. Liu B, Knirel YA, Feng L et al (2014) Structural diversity in *Salmonella* O antigens and its genetic basis. FEMS Microbiol Rev 38:56–89
8. Kenyon JJ, Cunneen MM, Reeves PR (2017) Genetics and evolution of *Yersinia pseudotuberculosis* O-specific polysaccharides: a novel pattern of O-antigen diversity. FEMS Microbiol Rev 41:200–217
9. Bastin DA, Romana LK, Reeves PR (1991) Molecular cloning and expression in *Escherichia coli* K-12 of the *rfb* gene cluster determining the O antigen of an *E. coli* O111 strain. Mol Microbiol 5:2223–2231
10. Kessler AC, Brown PK, Romana LK et al (1991) Molecular cloning and genetic characterization of the *rfb* region from *Yersinia pseudotuberculosis* serogroup IIA, which determines the formation of the 3,6-dideoxyhexose abequose. J Gen Microbiol 137:2689–2695
11. Brown PK, Romana LK, Reeves PR (1991) Cloning of the *rfb* gene cluster of a group C2 *Salmonella* strain: comparison with the *rfb* regions of groups B and D. Mol Microbiol 5:1873–1881
12. Brahmbhatt HN, Wyk P, Quigley NB et al (1988) Complete physical map of the *rfb* gene cluster encoding biosynthetic enzymes for the O antigen of *Salmonella typhimurium* LT2. J Bacteriol 170:98–102
13. Raymond CK, Sims EH, Kas A et al (2002) Genetic variation at the O-antigen biosynthetic locus in *Pseudomonas aeruginosa*. J Bacteriol 184:3614–3622
14. Raymond CK, Sims EH, Olson MV (2002) Linker-mediated recombinational subcloning of large DNA fragments using yeast. Genome Res 12:190–197
15. Liu MA, Kenyon JJ, Lee J et al (2017) Rapid customised operon assembly by yeast recombinational cloning. Appl Microbiol Biotechnol 101:4569–4580
16. Hong Y, Reeves PR (2014) Diversity of O-antigen repeat unit structures can account for the substantial sequence variation of Wzx translocases. J Bacteriol 196:1713–1722
17. del Solar G, Giraldo R, Ruiz-Echevarría MJ et al (1998) Replication and control of circular bacterial plasmids. Microbiol Mol Biol Rev 62:434–464
18. Kline BC (1985) A review of mini-F plasmid maintenance. Plasmid 14:1–16
19. Woodman ME, Savage CR, Arnold WK et al (2016) Direct PCR of intact bacteria (colony PCR). Curr Protoc Microbiol 42: A.3D.1–A.3D.7

Chapter 2

RecET Direct Cloning of Polysaccharide Gene Cluster from Gram-Negative Bacteria

Zhongrui Ma and Peng George Wang

Abstract

RecET direct cloning enables obtaining a large DNA region from genome. Here we describe its applications in cloning of polysaccharide gene cluster from gram-negative bacteria. Rapid and exact cloning of polysaccharide gene cluster can be achieved by this method.

Key words RecET direct cloning, Polysaccharide gene cluster, Gram-negative bacteria, Genome isolation, Restriction enzyme digestion, Linear cloning vector, Electroporation

1 Introduction

RecET direct cloning, developed by Prof. Zhang Y et al., enables exact cloning of a large DNA region from genomic DNA into an expression vector without mutations [1]. This technique is based on the discovery that Rac prophage protein RecE and its partner RecT mediate efficient homologous recombination between two linear DNA molecules in a replication-independent mechanism [2]. To date, several gene clusters for natural product biosynthesis including luminmycin, syringolin and salinomycin have been cloned using this method [2–5].

Bacterial polysaccharide (O antigen or capsular polysaccharide) biosynthesis pathway is a process involving multiple enzymes for sugar nucleotide biosynthesis, glycosyltransferase and polysaccharide processing [6]. Genes encoding these enzymes are typically clustered together on genome, forming a large DNA region ranging from ~10 kb to 30 kb.

Here, we design in detail to clone *Escherichia coli* (*E. coli*) O1 O antigen polysaccharide gene cluster into p15A-Cm vector using RecET direct cloning (Fig. 1). This method can be widely used in rapid (2–3 days) and high-fidelity (not rely on PCR amplification of gene cluster) cloning bacterial polysaccharide gene cluster with low cost.

Inka Brockhausen (ed.), *Bacterial Polysaccharides: Methods and Protocols*, Methods in Molecular Biology, vol. 1954, https://doi.org/10.1007/978-1-4939-9154-9_2,

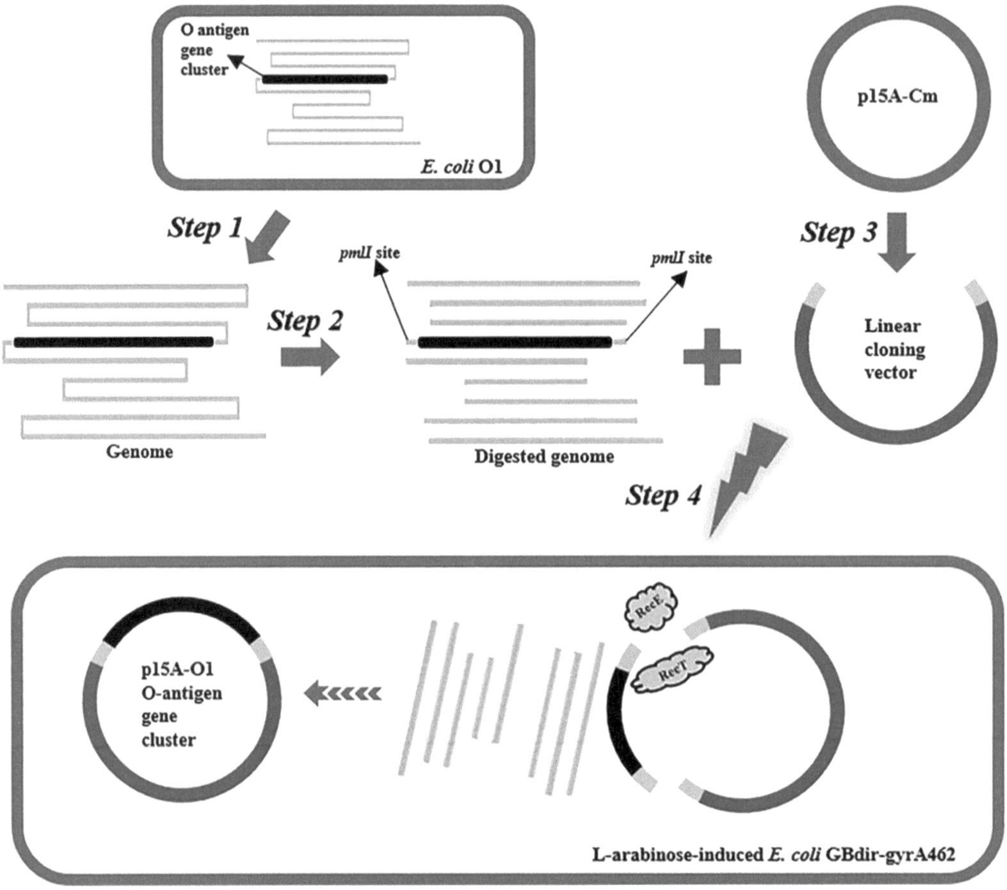

Fig. 1 Strategy diagram for RecET direct cloning of *E. coli* O1 O antigen gene cluster. Briefly, *E. coli* O1 genomic DNA is isolated (Step 1) and digested with restriction enzyme *pmlI* (Step 2). On the other hand, linear cloning vector is amplified with vector p15A-Cm as the template and primers containing homology arms flanking *E. coli* O1 O antigen gene cluster (Step 3). Then, both digested genome and linear cloning vector are coelectroporated into L-arabinose-induced *E. coli* GBdir-gyrA462 expressing RecE and RecT (Step 4). In the cells, RecE and RecT mediate the homologous recombination between O antigen gene cluster and linear cloning vector, to generate *E. coli* O1 O antigen gene cluster containing p15A-Cm (p15A-O1 O antigen gene cluster)

2 Materials

Prepare all solutions using deionized water (ddH_2O) and analytical grade reagents. Prepare and store all reagents and solutions at room temperature (unless indicated otherwise). Diligently follow all waste disposal regulations when disposing waste materials. We do not add sodium azide to reagents.

2.1 Strains and Plasmids

1. *E. coli* O1: Serotype O1 of the bacterial species *E. coli*. Store in 20% glycerol at −80 °C.

2. *E. coli* GBdir-gyrA462: DH10B, *fhu*A::IS2, ΔybcC, ΔrecET, P_{BAD}-*ETγA*, *gyrA*$_{Arg462Cys}$. *ETγA* operon (full-length *recE*, *recT*, *redγ*, and *recA*) is L-arabinose inducible. Store in 20% glycerol at −80 °C.
3. p15A-Cm: p15 origin and conveys chloramphenicol resistance. Store in DH5α strain in 20% glycerol at −80 °C. p15A-Cm is isolated from DH5α with commercial plasmid extraction kit for use.

2.2 Genome Isolation and Digestion

1. Luria–Bertani (LB) liquid media: 10 g/L tryptone, 10 g/L NaCl, 5 g/L yeast extract. Autoclave at 121 °C for 15 min before use.
2. 10 mM Tris–HCl (pH 8.0): Weigh 1.21 g Tris base and add ddH_2O to a volume of 900 mL. Mix and adjust pH with HCl. Make up to 1 L with ddH_2O.
3. 20 mg/mL proteinase K: Weigh 20 mg proteinase K and add ddH_2O to a volume of 1 mL. Store at −20 °C. It is better to prepare this fresh each time.
4. 10% (wt/vol) SDS: Weigh 1 g SDS and add ddH_2O to a volume of 10 mL. Wear a mask when weighing SDS.
5. Phenol: chloroform (1,1): Mix 5 mL liquified phenol and 5 mL chloroform. Prepare this at fume hood. It is better to prepare this fresh each time.
6. 3 M Na-acetate (NaAc) (pH 7.5): Weigh 2.46 g NaAc and add ddH_2O to a volume of 10 mL.
7. Restriction enzyme PmlI, KpnI, SalI: Store at −20 °C.
8. RNase A: Dissolve at a concentration of 1 mg/mL in 10 mM Tris–HCl (pH 8.0). Store at −20 °C.

2.3 PCR

1. Primers p15A-O1-F, p15A-O1-R, p15A-S-F, p15A-S-R: Dissolve at a concentration of 25 μM in ddH_2O.
2. Phusion High-Fidelity DNA Polymerase: Store at −20 °C.
3. Gel purification kit.

2.4 Electroporation

1. 10% (wt/vol) L-arabinose: Weigh 5 g L-arabinose and add ddH_2O to a volume of 50 mL. Pass the solution through 0.2-μm syringe filter to sterilize it inside a sterile laminar hood. Stored at −4 °C.
2. 1-mm electroporation cuvette.
3. LB agar: 10 g/L tryptone, 10 g/L NaCl, 5 g/L yeast extract, 5 g/L agar. Autoclave at 121 °C for 15 min before use.
4. 15 mg/mL chloramphenicol: Weigh 0.15 g chloramphenicol and add ethanol to a volume of 10 mL. Pass the solution

through 0.2-μm syringe filter to sterilize it inside a sterile laminar hood. Store at −20 °C.

5. Glycerol: Autoclave at 121 °C for 15 min. Package 200 μL sterile glycerol in sterile 1.5 mL centrifuge tube.

3 Methods

Carry out all procedures at room temperature, unless otherwise specified.

3.1 Genomic DNA Isolation

1. Culture 5 μL of *E. coli* O1 in 5 mL LB liquid media at 37 °C overnight (*see* **Note 1**).
2. Centrifuge 1.4 mL of the cultures in 1.5 mL centrifuge tube at 11,000 × *g* for 1 min. Take away the supernatant and wash the cells with 1.4 mL ddH_2O. Centrifuge again at 11,000 × *g* for 1 min. Take away the supernatant and resuspend the cells in 450 μL of 10 mM Tris–HCl (pH 8.0). Add 30 μL of 20 mg/mL proteinase K and mix well by inverting the tube for several times. Add 40 μL of 10% (wt/vol) SDS and mix well by inverting the tube for several times. Incubate the tube at 40 °C with occasional inversion for ~2 h until the solution becomes clear (*see* **Note 2**).
3. Add 500 μL of phenol: chloroform (1:1) and invert the tube for more than 30 times until the mixture becomes complete white. Centrifuge at 11,000 × *g* for 15 min.
4. Transfer the aqueous phase into a new tube using a blunt-end pipette tip (*see* **Note 3**). Add 35 μL of 3 M NaAc (pH 7.5) and mix well by inverting the tubes for several times. Add 1.2 mL ethanol and mix well by vortex (*see* **Note 4**).
5. Carefully transfer the cotton-like precipitate with pipette tip to a new tube. Rinse the precipitate with 1 mL of 75% ethanol. Centrifuge at 11,000 × *g* for 2 min. Take away the supernatant carefully and leave the tube cap open until no liquid left. Dissolve the pellet in 300 μL ddH_2O (*see* **Note 5**).

3.2 Restriction Enzyme Digestion

1. Mix 20–30 μg of *E. coli* O1 genomic DNA (*see* **Note 6**), 1 × buffer, 2 μL of restriction enzyme PmlI (*see* **Note 7**) and ddH_2O to totally 100 μL in a centrifuge tube and incubate the tube at 37 °C for 1 h.
2. Add 1 μg of RNase A to the reaction and incubate the tube for 30 min (*see* **Note 8**).
3. Add 20 μL of 3 M NaAc (pH 7.5) and mix well by inverting the tube for several times. Add 750 μL ethanol and mix well by vortex (*see* **Note 9**).

4. Centrifuge at 11,000 × *g* for 2 min. Take away the supernatant and rinse the pellet with 1 mL of 75% ethanol. Centrifuge at 10,000 × *g* for 2 min. Take away the supernatant carefully and leave the tube cap open until no liquid left. Dissolve the pellet in 15 μL ddH_2O (*see* **Note 10**).

3.3 Amplification of Linear Cloning Vector

1. Design primers (p15A-O1-F, p15A-O1-R) containing 80-nt homology arms flanking the digested *E. coli* O1 O antigen gene cluster sequence (immediately adjacent to the restriction site PmlI) and 20-nt standard PCR primers to amplify p15A-Cm vector at the 3′ end (*see* **Note 11**).
2. Set up PCR reaction (200 μL in total) using DNA polymerase (*see* **Note 12**) with vector p15A-Cm as template and primers p15A-O1-F, p15A-O1-R.
3. Purify the 200 μL PCR products using gel purification kit and elute into 30 μL ddH_2O (*see* **Note 13**).

3.4 Preparation of Electrocompetent Cells

1. Culture 5 μL of *E. coli* GBdir-gyrA462 in 5 mL LB liquid media at 37 °C overnight.
2. Transfer 125 μL of the cultures into 5 mL LB liquid media and incubate at 37 °C for 2.5 h.
3. Add 125 μL of 10% (wt/vol) L-arabinose to the cultures and incubate at 37 °C for 15 min (*see* **Note 14**).
4. Centrifuge 1.4 mL of the cultures in 1.5 mL centrifuge tube at 6000 × *g* for 30 s. Wash the cells twice with 1 mL ddH_2O and resuspend the cells in 40 μL ddH_2O (*see* **Note 15**).

3.5 Electroporation of Digested Genomic DNA and Linear Cloning Vector

1. Add 10 μL (~3–5 μg) of digested genomic DNA and 10 μL (~0.5 μg) of linear cloning vector (*see* **Note 16**) to 40 μL of electrocompetent cells, and transfer the mixture into 1-mm electroporation cuvette.
2. The condition for electroporation is 1350 V, 10 μF, 600 Ω using Eppendorf Electroporator 2510.
3. After electroporation, add 1 mL LB liquid media (store at 4 °C) into the cuvette immediately (*see* **Note 17**). Transfer the mixture into a 1.5 mL centrifuge tube and incubate at 37 °C for 1–2 h.
4. Spin down the cells by centrifugation at 1000 × *g* for 2 min. Discard the supernatant and leave ~100 μL for spreading on LB agar plate containing 15 μg/mL chloramphenicol. Upside down the LB agar plate at 37 °C overnight until the single colonies are visible (*see* **Note 18**).
5. Pick up a single colony from LB agar plate into 5 mL LB liquid media containing 15 μg/mL chloramphenicol, and incubate at 37 °C overnight.

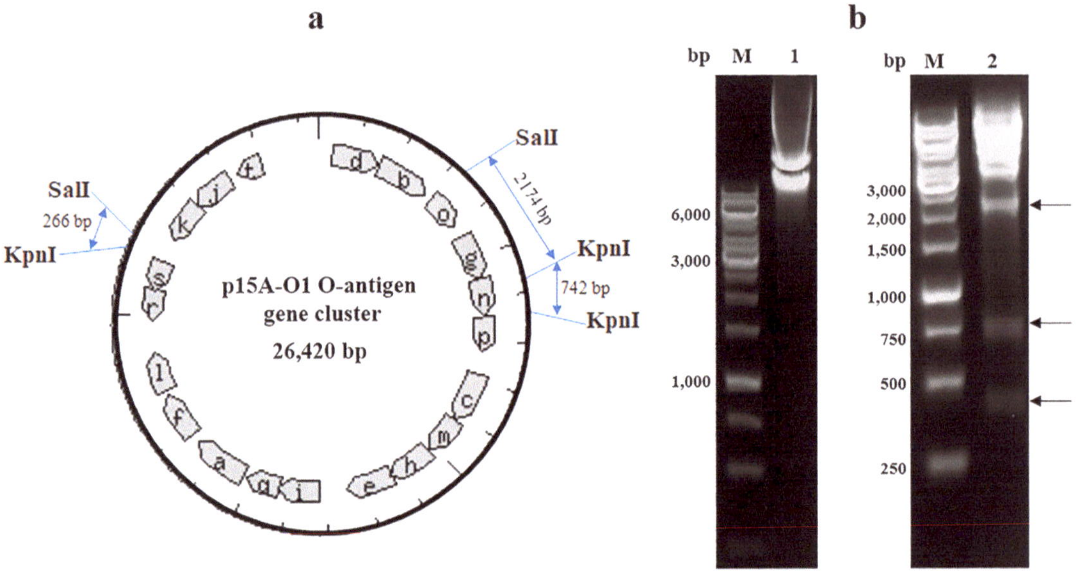

Fig. 2 Restriction analysis of recombinant plasmid. (**a**) The plasmid profile of p15A-O1 O antigen gene cluster. KpnI and SalI cut the recombinant plasmid into three pieces less than 3000 bp. (**b**) Agarose gel electrophoresis of recombinant plasmid (1) and KpnI and SalI-digested recombinant plasmid (2)

6. Add 600 μL of the cultures into 200 μL sterile glycerol and mix well with pipette. Store at −80 °C.

3.6 Characterization of Recombinant Plasmid

1. Culture 5 μL of single colony in 5 mL LB liquid media at 37 °C overnight.
2. Isolate the plasmid from the cultures using plasmid extraction kit.
3. Use test primers (p15A-S-F, p15A-S-R) to sequence the candidate plasmid (*see* **Note 19**).
4. Use restriction enzyme KpnI and SalI to digest the candidate plasmid according to manufacturer's instruction. Run agarose gel electrophoresis, and compare the size of digested DNA pieces with theoretical size (*see* **Note 20**) (Fig. 2).

4 Notes

1. The OD_{600} of the overnight cultures is ~2.1, 5 mL of the cultures is suitable for three-time genome isolation.
2. SDS is added to break the cell membrane of bacteria, proteinase K is added to digest the released proteins (only suitable for gram-negative bacteria). Under incubation with the two reagents, the milky solution will become clear. Sometimes, it

needs a longer or shorter incubation time due to variations in bacteria cell concentration.

3. There are three layers from top to bottom: aqueous phase, white layer, and organic phase. The aqueous phase is very sticky because of the genome. It has potentials to take the white layer out. Recommend using a blunt-end pipette tip to carefully take the aqueous phase and avoid take the white layer.
4. Under this step, cotton-like precipitate (genomic DNA) will suspend in the solution.
5. The time for air-drying the pellet is important, normally for 10–15 min. Overdried pellet may have difficulties to dissolve. If partly not dissolved in ddH_2O, incubate the solution at 65 °C for 1 h or at 4 °C overnight to rehydrate the genomic DNA.
6. The genomic DNA is quantified by NanoDrop UV spectrophotometer. Normally ~20–30 μg of genomic DNA equals to ~50 μL.
7. The principle for restriction enzyme(s) selection is that the restriction enzyme sites are upstream and downstream, but not within the target gene sequence (Fig. 1). Input *E. coli* O1 O antigen gene cluster sequence together with its upstream (~2 kb) and downstream (~2 kb) gene sequence to online restriction enzyme finding tool NEBcutter V2.0 (http://nc2.neb.com/NEBcutter2/), PmlI is selected to entirely cut *E. coli* O1 O antigen gene cluster from the genomic DNA. Restriction enzymes like AlwI and DpnII, always blocked by Dam methylation, should not be used as most *E. coli* strains were Dam$^+$ strains.
8. RNA may also be extracted by genomic DNA isolation. RNA may interfere with subsequent recombination between digested genomic DNA and linear cloning vector.
9. Under this step, digested genomic DNA will precipitate, but not cotton-like, so it is hard to take them by pipette tip. Instead, the digested genomic DNA will pellet with centrifugation.
10. After this step, the concentration of digested genomic DNA is ~50–100 ng/μL.
11. p15A-O1-F: TCCTGTACGGTCATGCGTTTCTGGAAATCTGCCAGCCCGAGGCTG,GAACAGGTGGCGGTGTAACCGTAAGTCGGCAGCACTTACGCCCCGCCCTGCCACTC; p15A-O1-R: TGCAGGCGCGTACCGACGGTGAGTTCACAGGCCCCCAGAATTTTGCCCATTTCCAGATCGTTGAGTTCATCCATCACCACGGTAACGAATCAGACAATTG (Single underlined sequences are homology arms; double underlined sequences are PCR primers to amplify p15A-Cm).

80-nt homology arms that lie immediately adjacent to the restriction sites are needed for optimal efficiency.

12. High-fidelity DNA polymerase is recommended for PCR.
13. Purification of the PCR products is necessary to remove non-specific amplification products, primer dimers, or large quantities of unused PCR primers. The final concentration of linear cloning vector is ~20–50 ng/μL.
14. RecE and RecT are expressed under the induction of L-arabinose.
15. The speed for centrifugation in this step should not exceed 6000 × *g*. Too fast speed centrifugation may lead to cell lysis and not effective for electroporation. The prepared electro-competent cells should be used for electroporation in 30 min for better transformation efficiency.
16. If the concentration of digested genomic DNA and linear cloning vector is lower than that recommended, they could be concentrated by vacufuge.
17. Electroporation can generate in instant heat, which might lead to cell death. Adding cooled LB liquid media into the cuvette immediately after the electroporation is helpful to cool down the cuvette.
18. Spreading the LB agar plate until the agar surface is dry. Liquid on the agar surface comprises the formation of single colonies.
19. p15A-S-F: CTGCGATGAGTGGCAGGGCGGG; p15A-S-R: GTCAATTGTCTGATTCGTTACC. Confirm the homology arms by sequencing.
20. Analyze the theoretical sequence of p15A-O1 O antigen gene cluster, and select restriction enzymes KpnI and SalI to digest the recombinant plasmid into pieces with theoretical size 266 bp, 742 bp and 2174 bp. Other suitable restriction enzymes can also be chosen to get different size of pieces. After restriction enzyme digestion, three pieces of DNA with similar size is found by agarose gel electrophoresis (Fig. 2).

Acknowledgments

This work was supported by NIH grant R01AI083754. Thank Prof. Zhang Y and Dr. Wang H for kindly providing p15A-Cm and *E. coli* GBdir-gyrA462.

References

1. Wang HL, Li Z, Jia RN et al (2016) RecET direct cloning and red alpha beta recombineering of biosynthetic gene clusters, large operons or single genes for heterologous expression. Nat Protoc 11(7):1175–1190
2. Fu J, Bian XY, Hu SB et al (2012) Full-length RecE enhances linear-linear homologous recombination and facilitates direct cloning for bioprospecting. Nat Biotechnol 30(5):440–446
3. Bian XY, Huang F, Wang HL et al (2014) Heterologous production of glidobactins/luminmycins in *Escherichia coli* Nissle containing the glidobactin biosynthetic gene cluster from Burkholderia DSM7029. Chembiochem 15 (15):2221–2224
4. Bian XY, Huang F, Stewart FA et al (2012) Direct cloning, genetic engineering, and heterologous expression of the syringolin biosynthetic gene cluster in *E. coli* through red/ET recombineering. Chembiochem 13:1946–1952
5. Yin J, Hoffmann M, Bian XY et al (2015) Direct cloning and heterologous expression of the salinomycin biosynthetic gene cluster from *Streptomyces albus* DSM41398 in *Streptomyces coelicolor* A3(2). Sci Rep 5
6. Kalynych S, Morona R, Cygler M (2014) Progress in understanding the assembly process of bacterial O-antigen. FEMS Microbiol Rev 38:1048–1065

Chapter 3

Purification of Capsular Polysaccharide Complex from Gram-Negative Bacteria

Carlos A. Alvarez and Brian A. Cobb

Abstract

Capsular polysaccharides are a dominant class of antigens from bacteria, both pathogenic and symbiotic or commensal. With the rise of awareness for the influence of the microbiota over immune system development and immune homeostasis, analysis of the antigens is more important than ever. Here we describe a method for the isolation of capsular polysaccharide from gram-negative bacteria, with the purification of polysaccharide from the commensal bacterium *Bacteroides fragilis* serving as an example. The method efficiently removes all detectable endotoxins and other lipid components, proteins, and nucleic acids, providing a source of capsular polysaccharide for immunologic study.

Key words Capsule, Polysaccharide, Bacteria, Purification, Carbohydrate, Glycan, Extraction, Endotoxin

1 Introduction

A wide range of microorganisms, each occupying a specific niche and role, populates the human body and is collectively known as the microbiome. Over the last couple of decades, it has become clear that the composition of the microbiome plays a critical role in the maintenance of homeostasis. The mechanisms underlying this homeostatic function are quite diverse, ranging from nutrient metabolism and niche competition to immunologic development [1–4]. From an immune perspective, the microbiota produces many immunogens and antigens that have been shown to drive immune system development and bias the host against deleterious inflammatory responses common to autoimmune and allergic diseases [5–9]. These include proteins, lipoproteins, nucleic acids, endotoxins (lipopolysaccharide; LPS), and capsular polysaccharides.

Capsular polysaccharides represent the outermost surface of encapsulated bacteria, such as *Bacteroides fragilis*, making these molecules the first point of contact with the immune system. Of

Inka Brockhausen (ed.), *Bacterial Polysaccharides: Methods and Protocols*, Methods in Molecular Biology, vol. 1954, https://doi.org/10.1007/978-1-4939-9154-9_3,

the eight different capsular polysaccharides found on *B. fragilis* [10], polysaccharide A (PSA) and B (PSB) are zwitterionic, possessing both positive and negative charges in their repeating units [11]. This property has been shown to enable their ability to activate CD4$^+$ T cells through major histocompatibility complex class II-mediated presentation [12, 13], which leads to immune inhibition and resistance to inflammatory disease [5, 6]. However, it is critical to separate the capsular polysaccharide from underlying LPS, protein and nucleic acid for immunologic evaluation and study.

Here, we describe the purification of capsular polysaccharide from gram-negative bacteria, which is based on previously reported methods [11, 14] and our experience with *B. fragilis* and PSA [12, 15, 16]. The procedure begins with a hot phenol extraction, which breaks down the bacteria and leads to the precipitation of nucleic acids and proteins, leaving the lipid and carbohydrate fractions in the aqueous phase. Residual phenol is removed from the aqueous phase by ethyl ether extraction followed by evaporation of residual ethyl ether using a rotary evaporator. The extracted lipids/LPS and polysaccharides are then treated with DNase, RNase, and proteinase K to digest any remaining traces of nucleic acids and proteins respectively. The resulting LPS and capsular polysaccharide mixture is separated by size exclusion chromatography in deoxycholate, which disaggregates the LPS so that it runs as a relatively small molecule compared to the very large (>300 kDa) polysaccharides. LPS-free polysaccharide is pooled and deoxycholate removed through dialysis to yield between 100 and 400 mg of polysaccharide from 20 L of log-phase liquid culture.

2 Materials

In this method, the bacterial growth step is skipped, since culture conditions can vary wildly, depending on the bacterial species/strain. The extraction is suitable for any capsular polysaccharide from a gram-negative bacterium. Thus, the starting material for this purification is pelleted bacteria from 20 L of log-phase growth culture. The large scale is due to the relative expense and time required for the preparation, making repeated preparations needlessly inefficient. Moreover, all solutions are made with ultrapure (18 MΩ/cm) water, and hazardous waste disposal is done according to the appropriate local regulations.

2.1 Solutions, Reagents, and Supplies

1. 1.0 M Tris, pH 7.5: 1.0 M solutions of both Tris–HCl (157.60 g in 1 L water; do not adjust pH) and Tris-base (121.14 g in 1 L water, do not adjust pH). Titrate 500 mL Tris base with Tris–HCl in a 1 L beaker with stirring until the pH reaches 7.5. Filter-sterilize and store at room temperature.

2. 1.0 M $MgCl_2$: Dissolve 95.21 g $MgCl_2$ in 1 L of water. Filter-sterilize and store at room temperature.
3. 1.0 M $CaCl_2$: Dissolve 110.98 g $CaCl_2$ in 1 L of water. Filter-sterilize and store at room temperature.
4. 1.0 M EDTA, pH 8.0: Dissolve 146.12 g anhydrous ethylenediaminetetraacetic acid (EDTA) in 250 mL water. Adjust pH to 8.0 with NaOH under constant stirring. The EDTA will not completely go into solution until near neutral pH is reached. Dilute to 500 mL, filter-sterilize and store at room temperature.
5. 5 L sterile water stored at room temperature.
6. 2 M NaCl: Dissolve 116.8 g NaCl in 1 L of water. Filter-sterilize and store at room temperature.
7. Approximately 4 L of 95% ethanol: Store at room temperature, unless indicated otherwise.
8. Phosphate buffered saline (PBS): Dissolve 8 g NaCl (137 mM), 0.2 g KCl (2.7 mM), 1.44 g Na_2HPO_4 (10 mM), and 0.24 g KH_2PO_4 (1.8 mM) in 800 mL water. Adjust pH to 7.4 with HCl. Bring to 1 L with water, filter-sterilize, and store at room temperature.
9. 2 L Ethyl ether. Store at room temperature.
10. Glacial acetic acid. Store at room temperature.
11. 500 g phenol (*see* **Note 1**). Store at room temperature.
12. 1.0 g Pronase (Millipore Sigma). Store at −20 °C.
13. 500 mg Ribonuclease A (Worthington Biochemical Corp.). Store at −20 °C.
14. 100 mg Deoxyribonuclease Type I (Worthington Biochemical Corp.). Store at −20 °C.
15. Column running buffer: Dissolve 60 g deoxycholic acid, ultrapure grade (30 g/L), 7.5 g glycine (50 mM), and 7.44 g EDTA (10 mM) into 1.5 L water (*see* **Note 2**). Adjust pH to 9.8 with NaOH. Bring to 2 L with water. Do not filter and store at room temperature.
16. Pro-Q Emerald 300 lipopolysaccharide Gel Stain Kit (ThermoFisher). Store components as instructed by the manufacturer.
17. Sephacryl S-400 column (100 cm × 5 cm) fitted to a peristaltic pump and fraction collector.
18. Dialysis membrane tubing (3 in. diameter; 12–14 kDa MWCO) (*see* **Note 3**).
19. Glass pipettes (25 mL).

2.2 Special Equipment

1. Heavy-duty stir plate.
2. Glass beads (~5 mm diameter).
3. Six 250 mL phenol/chemical resistant centrifuge bottles with gaskets.
4. Stand mixer.
5. Separatory funnel (2 L) with large ring stand.
6. Rotary evaporator with 1 L sample flask.
7. Lyophilizer.
8. Peristaltic pump and fraction collector.

3 Methods

3.1 Harvest and Phenol Extraction

1. Setup extraction equipment in a fume hood, including a water bath prewarmed to 68 °C and a stand mixer (Fig. 1).
2. Warm 1 L of water and the 500 g bottle of solid phenol to 68 °C in the water bath until all of the phenol is melted.
3. Once up to temperature, add 167 mL of prewarmed water to the phenol to make it 75%.
4. Resuspend the frozen bacterial pellet in 667 mL of 68 °C water, making sure that no chunks of bacterial pellet remain.
5. Transfer the resuspended bacteria into a 4 L glass beaker, and place into the 68 °C water bath.

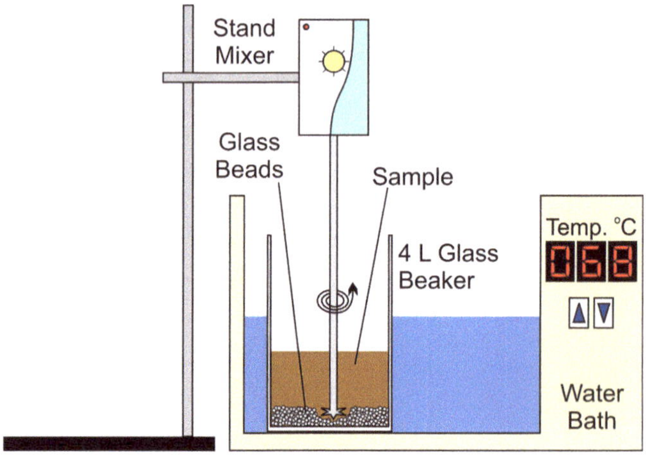

Fig. 1 Phenol extraction assembly for the initial extraction of LPS and capsular polysaccharide from intact gram-negative bacteria. Resuspended bacteria are combined with 75% phenol in a 4 L beaker at 68 °C in a water bath. Glass beads are placed in the beaker and agitated with a stand mixer. The extraction should be performed within a fume hood

6. Add about 250 mL equivalent volume of glass beads to the sample.
7. Place the mixer blades into the sample so that the blades are just above the bottom of the beaker (Fig. 1), and turn on at a speed fast enough that the glass beads are well agitated.
8. Add all of the 75% prewarmed phenol to the sample mixture and allow robust mixing for 30 min (*see* **Note 4**).
9. Remove the stand mixer and add a large stir bar to the sample, cover with several layers of Parafilm and aluminum foil, and stir overnight on a heavy duty stir plate at 4 °C. Be sure to use secondary containment.
10. Using the chemically resistant 250 mL centrifuge bottles, centrifuge the sample at 10,000 × *g* for 20 min at 4 °C.
11. Very carefully (*see* **Note 5**) decant the top aqueous layer using a glass pipet, placing the extract into a clean 1 L glass bottle.
12. Repeat the centrifugation until the entire sample has been centrifuged, combining the top aqueous phase in the 1 L glass bottle, and the waste in the waste vessel.

3.2 Ether Extraction

1. Prepare a 2 L separatory funnel by applying stopcock grease to the fittings.
2. Add approximately 400 mL of sample and an equal volume of ethyl ether to remove dissolved phenol.
3. Close the funnel and shake vigorously for 5 min (CAUTION, *see* **Note 6**).
4. Allow the mix to separate into two layers (top, ethyl ether; bottom, aqueous) for about 20 min by placing the funnel on a ring stand.
5. Slowly allow the bottom aqueous phase, which contains the lipids and carbohydrates, out of the funnel into a clean 1 L glass bottle (*see* **Note 7**). Discard the top layer into an appropriate waste vessel for ethyl ether disposal.
6. Repeat until the entire phenol extract has been extracted with ethyl ether, combining the bottom phase into one or more 1 L glass bottles.
7. Repeat the ethyl ether extraction on the entire sample once more.
8. Setup a rotary evaporator in a fume hood, circulating cool water through the condenser and prewarming the water bath to 60 °C.
9. Place approximately 500 mL of the ether-extracted sample into the sample flask, and evaporate the residual ethyl ether and concentrate the sample for approximately 1 h, being careful to guard against boiling bumps in the first 10 min.

10. Remove the sample, saving it in a fresh 1 L glass bottle.
11. Repeat the evaporation until all of the sample is complete, combining the final product together.
12. Place the sample in prehydrated dialysis tubing (*see* **Note 3**), and dialyze against water supplemented with 0.05% sodium azide at 4 °C (*see* **Note 8**).

3.3 Digestions and Precipitation

1. Remove the sample from dialysis using aseptic technique into sterile 1 L bottle(s).
2. Measure the approximate volume of the sample and calculate 6.5% of this value.
3. In a 50 mL conical tube, add calculated volume of 1.0 M Tris, pH 7.5.
4. Add 1.0 M $MgCl_2$ and 1.0 M $CaCl_2$ to make the Tris solution contain 20 mM each.
5. Add RNase A and DNase Type I to the Tris solution such that the final concentration in the sample will be 0.33 mg/mL and 0.07 mg/mL respectively (*see* **Note 9**).
6. Filter-sterilize the enzyme mixture, then add to the sample.
7. Cover/close the bottle and incubate overnight at 37 °C.
8. Readjust the pH of the sample to 7.5 with 1.0 M NaOH if necessary (*see* **Note 10**).
9. Weigh enough Proteinase K to achieve a final concentration of 0.33 mg/mL, dissolve it in 5 mL 1.0 M Tris, pH 7.5, filter-sterilize the enzyme solution, and add it to the sample.
10. Incubate overnight at 37 °C.
11. Place several liters of 95% ethanol in the −20 °C freezer overnight.
12. Repeat the Proteinase K digest with fresh enzyme prepared as before.
13. Incubate 2 h 37 °C.
14. Add enough 1.0 M EDTA, pH 8.0 to the sample to make the final concentration 50 mM EDTA.
15. Add a sterile stir bar and stir for 30 min at 4 °C.
16. Add 11% of the sample volume of 2 M NaCl (200 mM final concentration).
17. Precipitate the LPS and capsular polysaccharide by dividing the sample into 200 mL aliquots in as many 1 L glass bottles as necessary. Add 800 mL ice cold (−20 °C) ethanol, close and mix well (*see* **Note 11**).
18. Incubate overnight at −20 °C.

19. Collect precipitate by centrifugation of the samples at 10,000 × *g* at 4 °C for 1 h (*see* **Note 12**). It is okay to use the same centrifuge bottles over and over, stacking the sample in the pellet.
20. Decant the ethanol/water and dry the pellets by inversion for 2 h.
21. Resuspend all of the pellets in a total of no more than 30 mL Column Running Buffer (*see* **Note 13**).

3.4 Chromatography

1. Wash the S-400 column with 600 mL of 0.1 M NaOH at 1.5 mL/min.
2. Equilibrate the column with 2 L Column Running Buffer at 1.5 mL/min.
3. Load the equilibrated column with no more than 30 mL of sample. If the sample is in greater volume, separate the sample into multiple column runs.
4. Run the column with 2 L Column Running Buffer, collecting 20 mL fractions between 500 mL and 1500 mL elution volume.
5. Regenerate the column by repeating the 600 mL 0.1 M NaOH wash and re-equilibration in **steps 1** and **2** before running another sample to prevent LPS contamination.

3.5 Fraction Analysis and Sample Pooling

1. Using standard SDS-PAGE techniques and sample loading buffer, run 15 μL of each fraction on a 10% continuous polyacrylamide gel (*see* **Note 14**) until the blue loading dye has just exited the bottom of the gel.
2. Stain the gel using the Pro-Q Emerald 300 lipopolysaccharide Gel Stain Kit according to the manufacturer's protocol. This kit actually stains carbohydrates and will detect both LPS and capsular polysaccharide with very high sensitivity (Fig. 2).
3. Pool clean capsular polysaccharide fractions that lack detectable LPS.
4. Place the entire sample into dialysis against continuous running water (*see* **Note 15**) for at least 1 week to remove the deoxycholate.
5. Collect and freeze-dry the final sample using a lyophilizer.
6. Collect and measure the dry weight on an analytical balance.

3.6 Quality Check

1. Make a 1 mg/mL solution of capsular polysaccharide in water.
2. Perform a wavelength absorbance scan between 200 and 350 nm in a 1 cm path length cuvette. The absorbance at 260 and 280 nm should be less than 0.1.

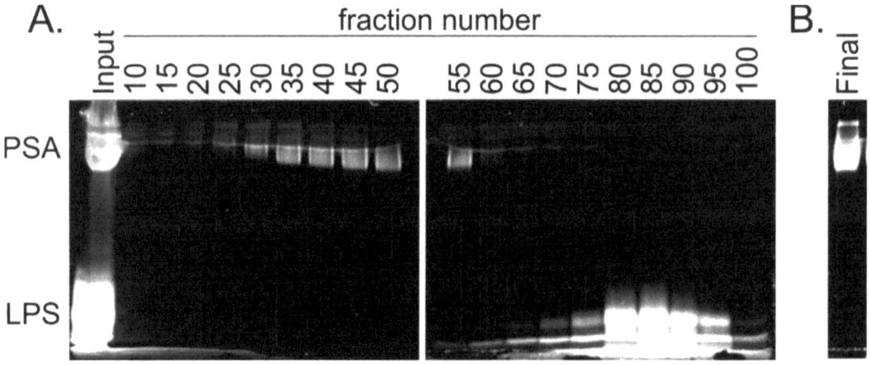

Fig. 2 Pro-Q Emerald stained SDS-PAGE gels from the purification of PSA from *B. fragilis*. (**a**) In the first lane, 2 μL of the sample loaded onto the column was run as the input control. All other lanes contain 20 μL of the fraction indicated, with each collected fraction being 20 mL. "Clean" PSA was pooled from fractions 30–50, giving a final PSA sample volume of 420 mL containing approximately 300 mg of PSA. (**b**) The final PSA sample after pooling and dialysis, showing the lack of detectable LPS

3. Run 20 μg of sample on SDS-PAGE as before, staining with Pro-Q. No detectable LPS should be visible.
4. For PSA from *Bacteroides fragilis*, H^1 NMR is performed for the known spectrum and a lack of non-PSA resonant peaks.

3.7 Storage

1. Dissolve the entire sample in water.
2. Make 1 and 5 mg aliquots in small glass vials and freeze-dry.
3. Seal the tubes and store the dried polysaccharide at −80 °C.

4 Notes

1. Phenol should be purchased as a single bottle of 500 g of solid phenol. The entire 500 g will be used in liquid phase, thus eliminating the need to measure the solid and increasing safety. Upon use, the glass bottle in which phenol is supplied can become the waste container for hazardous waste disposal.
2. Use caution when making this buffer. Begin by dissolving the glycine and EDTA. Then add the deoxycholate. Powered deoxycholate is very light and will disperse into the air with the slightest agitation. As a result, use slow and careful movements when weighing and dispensing the deoxycholate to minimize this effect. Use goggles and a respirator, as the dispersed deoxycholate is very irritating to mucous membranes. Use of a fume hood is incompatible with this procedure due to the air flow. Finally, deoxycholate will not dissolve in water until the pH is at least 8.5. Thus, NaOH must be added with constant mixing and pH monitoring as the deoxycholate is added. It can take

30 min or more for all deoxycholate to enter solution. Adjust the pH to 9.8 with NaOH, then bring the final volume to 2 L with water.

3. In order to prepare the dialysis tubing, cut the desired length, allowing for at least 50% sample swelling, boil in 5 mM EDTA, pH 8.0 for 10 min, rinse twice with room temperature water and store at 4 °C in 20% ethanol until use. Do not allow the membrane to dry out once rehydrated.
4. Addition of 75% phenol to the resuspended bacteria will cause an immediate lysis of the bacteria. The resulting solution will have a viscous chocolate milk appearance.
5. If things are going well, after the spin you should see a large viscous and sticky tan-colored pellet and a relatively small amount of a tan/greenish-colored liquid on the top, which is the desired fraction. The pellet is very loose, so caution is required in removing the top layer. In addition, it is better to get some of the pellet in the sample to get all the top aqueous phase, but if this happens, a repeat spin in clean bottles will be necessary to remove the bottom layer. Finally, the bottom layer is mostly phenol and must be discarded appropriately.
6. It is critically important to vent the separatory funnel often. Gas is released upon the first several shakes, and will build up in the funnel. This could lead to glass failure if not vented. Use a face shield in addition to goggles and other standard chemical and laboratory safety personal protective gear.
7. It is best to get all of the bottom phase in the first extraction, even if some of the top phase is collected. Upon the second ethyl ether extraction, the separation will be clearer and easier to prevent top layer contamination.
8. The sample is large at this point, thus splitting the sample into two parts is often easiest. Dialysis should have a sample–water ratio of at least 1:10, changing the water six times after at least 4 h has passed with stirring each time.
9. The amount of RNase A and DNase Type I is based on the sample volume, not the 1.0 M Tris volume calculated in **step 2**, Subheading 3.3.
10. Use a sterile pipet tip and drop sample onto pH paper to estimate the pH. Do not contaminate the sample by using a pH probe.
11. The ethanol must be very cold for optimal precipitation. The sample will become milky white immediately upon addition of the ethanol.
12. This will take more than one spin to collect the large volume of sample. It is okay to use the same centrifuge bottles over and over, stacking the sample in the pellet. Also, much of the

precipitate will stick to the walls of the original 1 L glass bottles. This is normal and desirable. Once all liquid has been removed for centrifugation, dry the stuck precipitate by inverting the bottles on paper towels for at least 2 h.

13. The Column Running Buffer contains a high concentration of detergent, and is therefore soapy. Avoid making bubbles. Also, all pellets and glass bottle precipitates should be combined into a single 30 mL sample. The final sample will have a brown tone, but should be free of visible particulates.
14. The SDS-PAGE gel does not need a stacking gel, and the samples do not need to be boiled before loading. The LPS will run just above the dye front, while the capsular polysaccharide will run high, just inside the wells at the top (Fig. 2).
15. A large plastic bucket from a hardware store, cleaned to laboratory standards, can be used for this purpose, making sure that there are no sharp edges or imperfections before use. Place the bucket in a sink, fill with dH_2O, add the sample in a dialysis bag, and allow the water to continuously flow, slowly. Do this at 4 °C.

Acknowledgments

This work was supported by NIH grants GM115234 and GM082916 to BAC, and AI089474 to CAA.

References

1. Falk PG, Hooper LV, Midtvedt T et al (1998) Creating and maintaining the gastrointestinal ecosystem: what we know and need to know from gnotobiology. Microbiol Mol Biol Rev 62:1157–1170
2. Hooper LV, Midtvedt T, Gordon JI (2002) How host-microbial interactions shape the nutrient environment of the mammalian intestine. Annu Rev Nutr 22:283–307
3. Nishio J, Honda K (2012) Immunoregulation by the gut microbiota. Cell Mol Life Sci 69:3635–3650
4. Peterson CT, Sharma V, Elmen L et al (2015) Immune homeostasis, dysbiosis and therapeutic modulation of the gut microbiota. Clin Exp Immunol 179:363–377
5. Johnson JL, Jones MB, Cobb BA (2015) Bacterial capsular polysaccharide prevents the onset of asthma through T-cell activation. Glycobiology 25:368–375
6. Johnson JL, Jones MB, Cobb BA (2018) Polysaccharide-experienced effector T cells induce IL-10 in FoxP3+ regulatory T cells to prevent pulmonary inflammation. Glycobiology 28:50–58
7. Mazmanian SK, Round JL, Kasper DL (2008) A microbial symbiosis factor prevents intestinal inflammatory disease. Nature 453 (7195):620–625
8. Ochoa-Reparaz J, Mielcarz DW, Wang Y et al (2010) A polysaccharide from the human commensal Bacteroides fragilis protects against CNS demyelinating disease. Mucosal Immunol 3:487–495
9. Tzianabos AO, Onderdonk AB, Rosner B et al (1993) Structural features of polysaccharides that induce intra-abdominal abscesses. Science 262(5132):416–419
10. Krinos CM, Coyne MJ, Weinacht KG et al (2001) Extensive surface diversity of a commensal microorganism by multiple DNA inversions. Nature 414(6863):555–558
11. Tzianabos AO, Pantosti A, Baumann H et al (1992) The capsular polysaccharide of

Bacteroides fragilis comprises two ionically linked polysaccharides. J Biol Chem 267:18230–18235

12. Cobb BA, Wang Q, Tzianabos AO et al (2004) Polysaccharide processing and presentation by the MHCII pathway. Cell 117:677–687
13. Johnson JL, Jones MB, Cobb BA (2015) Polysaccharide A from the capsule of Bacteroides fragilis induces clonal CD4+ T cell expansion. J Biol Chem 290:5007–5014
14. Tzianabos AO, Pantosti A, Baumann H et al (1991) Structural characterization of two surface polysaccharides of Bacteroides fragilis. Trans Assoc Am Phys 104:285–295
15. Kreisman LS, Friedman JH, Neaga A et al (2007) Structure and function relations with a T-cell-activating polysaccharide antigen using circular dichroism. Glycobiology 17:46–55
16. Cobb BA, Kasper DL (2008) Characteristics of carbohydrate antigen binding to the presentation protein HLA-DR. Glycobiology 18:707–718

Chapter 4

Simple Protocol to Purify Cell Wall Polysaccharide from Gram-Positive Bacteria and Assess Its Structural Integrity

Irina Sadovskaya and Yann Guérardel

Abstract

Cell wall polysaccharides (CWPS), which are usually covalently bound to the peptidoglycan and are closely associated with the cell wall, are considered as ubiquitous components of the cell envelope of gram-positive bacteria and play an important role as mediators of bacterial interactions with the environment. Here, we describe a simple method for purifying CWPS by extraction of bacterial cells with consecutive acid treatments. Purified CWPS are obtained by gel-filtration chromatography following treatment with HF. We also provide the methodology to easily assess the integrity of CWPS using high-resolution magic-angle spinning (HR-MAS) NMR.

Key words Cell wall polysaccharide, Purification, HR-MAS NMR

1 Introduction

Cell wall polysaccharides (CWPS) are considered as ubiquitous components of the cell envelope of gram-positive bacteria. Unlike capsular polysaccharides (CPS), forming a thick outermost shell of the cell wall, and exopolysaccharides (EPS), loosely bound to bacterial cells and released in the growth medium, CW PS are usually covalently bound to the peptidoglycan and are closely associated with the cell wall. Because of their localization at the bacterial surface, they are mediators of bacterial interactions with the environment such as adhesion to abiotic surfaces, specific interactions with other microorganisms, eukaryotic host cells, and infecting bacteriophages [1].

Knowledge of the detailed chemical structure of CWPS of gram-positive bacteria is essential for understanding of molecular mechanisms of their various functions. Structurally different CWPS form the basis of some serological typing systems, such as Lancefield classification of *streptococci* [1]. Precise structures of CWPS acting as coaggregating receptors for the formation of the dental plaque have been described for several oral bacteria, both gram

Inka Brockhausen (ed.), *Bacterial Polysaccharides: Methods and Protocols*, Methods in Molecular Biology, vol. 1954,
https://doi.org/10.1007/978-1-4939-9154-9_4,

positive and gram negative [1, 2]. It was shown that structural differences of CWPS of *Lactobacillus helveticus*, used in cheese manufacturing, are affecting its autolytic properties [3]. Another example is a recently described mechanism of action of glycerophosphodiester phosphodiesterase (GDPD) of bacteriophage 17 infecting *Lactobacillus delbrueckii* subsp. *bulgaricus* Ldb17, which hydrolyzes the Gro3P decoration of the major CWPS (Gro3P-substituted linear D-galactan) of Ldb17 [4].

It is important to point out that often several CWPSs in variable quantities are present in one bacterial strain, some more surface-exposed, others more tightly bound to the peptidoglycan (PG). As an example, the cell surface of *L. delbrueckii* subsp. *bulgaricus* Ldb17 was shown to contain at least two different cell surface polysaccharides, in addition to a CWTA [5]. In *Lactococcus lactis* MG1363, a CWPS referred to as PSP, "polysaccharide pellicle" forms a compact pellicle on the outer surface of the cell [6]. A rhamnan, another CWPS of *L. lactis*, is trapped inside the peptidoglycan and is covalently bound to it. It appears to be conserved component of the lactococcal cell wall and plays an essential role in cell division and morphogenesis in *L. lactis* cells [7].

The majority of other *L. lactis* strain studied to date have a CW architecture similar to MG1363 and have two distinct CWPSs: a PSP which is often composed of oligosaccharide repeating units linked with a phosphodiester bond, and a neutral rhamnan [8–10]. The two polysaccharides are often covalently linked. Crude trichloroacetic acid (TCA) extracts (*see* **Note 1**) contain both polysaccharides in different proportions, which often comigrate as a broad peak. HF treatment, cleaving the phosphodiester bonds in PSP leads to its depolymerization and liberation of free rhamnan and oligosaccharide (OS) fragments of PSP (Fig. 1).

The extraction method described here is adapted essentially from Prakobpol et al. [11], and allows preparing samples of purified CWPS strongly bound to PG after a pre-extraction of other surface polysaccharides. Briefly, the extraction is carried out by hot HCl solutions (0.01 N and 0.1 N, consecutively). Proteins and nucleic acid debris are precipitated by addition of TCA (*see* **Note 2**). Crude CWPS is obtained after dialysis and lyophilization. Treatment of the crude CW PS with 48% HF and following gel-filtration chromatography allow for efficient depolymerization and removal of contaminating nucleic acids, as well as teichoic and lipoteichoic acids (Fig. 2). However, for structural analysis of CWPS carrying phosphate groups, a different strategy should be adopted. This simple method afforded pure preparations, suitable for structural analysis, in case of various gram-positive bacilli and cocci [7, 12]. A panel of conventional methods including monosaccharide composition analysis, methylation analysis, mass spectrometry, and NMR may then be used to elucidate the structure of purified polysaccharides.

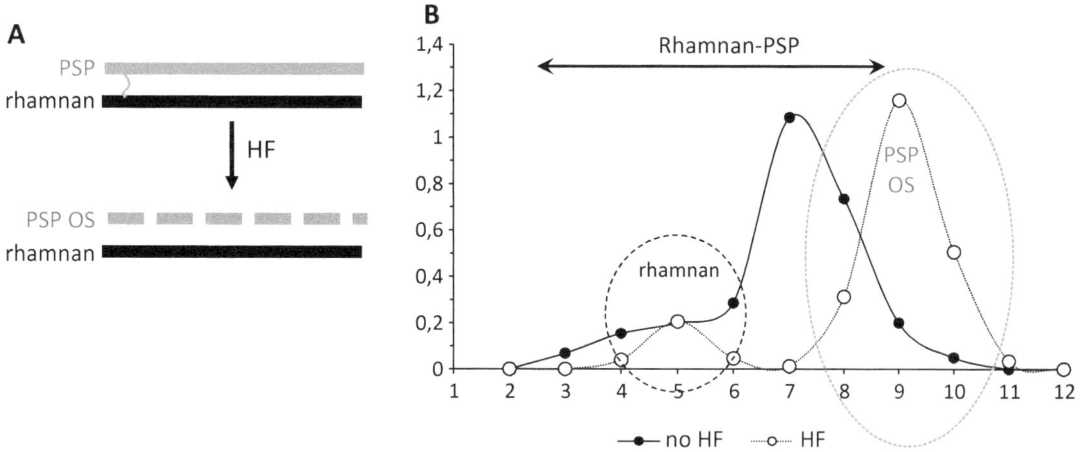

Fig. 1 The majority of *Lactococcus lactis* strain studied to date have two cell wall polysaccharides: a surface polysaccharide pellicle, PSP which is often composed of oligosaccharide repeating units linked with a phosphodiester bonds, and a neutral rhamnan CW PS [6, 8]. The two polysaccharides are often covalently linked. TCA extracts contain both polysaccharides in different proportions, which often migrate as a broad peak. HF treatment, cleaving the phosphodiester bonds in PSP leads to its depolymerization and liberation of free rhamnan and OS fragments of PSP (**a**). The figure shows a representative elution profile of the TCA extract of *L. lactis* SK11 on a Sephadex G-50 column (1 × 40 cm) before and after HF treatment (**b**)

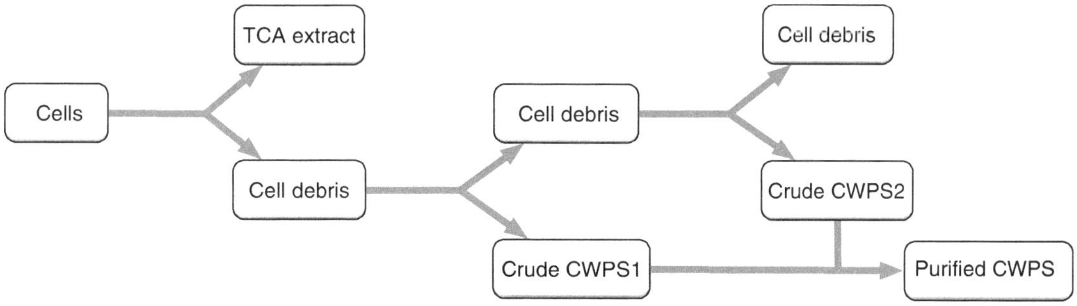

Fig. 2 Purification pipeline of CWPS from gram-positivem bacteria. Purification steps described in Subheading 3.2, **steps 1–4** may be applied either to CWPS1 and CWPS2 independently or to the pooled CWPS extracts from 0.01 (Subheading 3.1, **steps 4** and **5**) and 0.1 M (Subheading 3.1, **steps 6–10**) HCl extraction steps. TCA extract may also contain a certain amount of loosely bound CWPS and may be submitted to purification steps described in Subheading 3.2, **steps 1–4**

Not only that extraction procedure should be simple and consistent, but it should also keep the extracted polysaccharide as intact as possible. However, considering that the vast majority of structural analysis methodologies require polysaccharides to undergo extensive purification processes, it is very difficult to assess to what extent the final structure reflects the native cell-surface expressed polysaccharide. This can be achieved using high-resolution magic-angle spinning (HR-MAS) NMR that enables the acquisition of NMR spectra from small amounts of viscous or solid samples, with a resolution equivalent to liquid NMR analysis of soluble molecules. Indeed, by spinning at the magic angle, line broadening effects due to dipolar interactions and susceptibility differences within the sample are removed resulting in high-resolution quality spectra. In the field of glycobiology, this technique enables an easy and sensitive read out of surface components on intact cells. Among other applications, it was previously effectively used to analyze cell wall polysaccharides from yeasts and bacteria without proceeding to any potentially denaturing purification step [13–15]. We will provide here the methodology to easily assess the integrity of polysaccharides using HR-MAS NMR and show that the easy extraction method of CWPS from gram-positive bacteria does not modify the structure of purified polysaccharides by comparing the structural features of the polysaccharides before and after extraction procedure.

2 Materials

2.1 Extraction of the CWPS

1. Centrifuge and centrifuge tubes.
2. Dialysis tubing (Visking), dialysis reservoirs, closures.
3. Lyophilizer and adapted glassware.
4. Trichloroacetic acid (TCA) solution, 50% (w/w): in a glass beaker, weigh 100 g of TCA, and add 100 mL of water. Stir in a magnetic stirrer till the TCA is completely dissolved, and transfer into a glass bottle (*see* **Note 3**).
5. 0.01 M and 0.1 M HCl.

2.2 Chromatography

1. Empty columns for gel permeation chromatography: 1 × 40 cm; 2.6 cm × 1 m for larger quantities.
2. Chromatography media Sephadex G-50 fine (GE Healthcare).
3. 0.01% acetic acid (AcOH) in milli-Q water.

2.3 HR-MAS

1. 4 mM ZrO_2 MAS rotors (CortecNet, Paris, France).
2. Disposable inserts for 4 mM rotor with sealing screws and plugs (CortecNet, Paris, France).

3 Methods

3.1 Extraction of CWPS

Extraction and purification steps of CWPS are summarized in Fig. 2.

1. Cells (7–10 g) are washed twice with water (*see* **Note 4**).

2. Suspend cells in 5% TCA (~8 mL per 1 g of cells) in a 250 mL or 500 mL glass bottle. Stir on a magnetic stirrer for 48 h at 4–6 °C.
3. Transfer into appropriate 35 mL centrifuge tubes. Centrifuge (10,000 × *g*, 10 min). Clear supernatant (TCA extract) usually contains surface PS more loosely bound to the CW (if present) and CWPS in different proportions. Dialyze (*see* **Note 5**) and lyophilize to obtain the crude TCA extract.
4. Suspend the cells in 0.01 M HCl (~100 mL) and transfer them into a glass bottle. Stir for 20 min in a boiling water bath.
5. Cool down and transfer into centrifuge tubes. Centrifuge (10,000 × *g*, 10 min). Clear supernatant contains the crude 0.01 M HCl extract.
6. Suspend the cells in 0.1 M HCl (~100 mL) and proceed as in **step 4**. Clear supernatant contains crude 0.1 N extract. Discard the pellets.
7. Measure the volume of the clear supernatant and transfer it into a glass beaker. Add 1/10 volume of the 50% TCA, in order to precipitate eDNA and proteins (*see* **Note 2**).
8. Centrifuge (10,000 × *g*, 10 min). Discard the pellet and collect the supernatant.
9. Prepare dialysis tubing by cutting an appropriate length, soaking and closing one edge with a closure or by making a knot. Fill approx. two thirds of the volume of the tubing with the supernatant and close the tubing at the top.
10. Dialyze against water for at least 48 h (*see* **Note 5**).
11. Freeze-dry. The obtained material corresponds to a crude preparation of CWPS (*see* **Note 6**).

3.2 Purification of CWPS

1. Under the fume-hood, add 48% HF (50 μL) to the crude HCl extract (10 mg in an Eppendorf tube, *see* **Note 3**). Make sure that the extract is dissolved completely. Keep 24 h at 4 °C.
2. Evaporate HF under a stream of nitrogen at room temperature.
3. Take the extract in distilled H_2O (1 mL) and fractionate on a Sephadex G-50 column, eluted with 0.1% acetic acid. Fractions are collected and assayed for neutral [16] and, if necessary, amino sugars (Fig. 1).
4. Fractions corresponding to CWPS are pooled and lyophilized.

3.3 HR-MAS NMR Analysis of Intact Cells

1. Cells are washed twice with water and pelleted down by centrifugation in a 1 mL microcentrifuge tube (*see* **Note 7**).
2. Bacterial pellets are washed once with two volumes of D_2O then incubated for 2 h in two volumes of D_2O with 0.5% (v/v) acetone as internal standard.

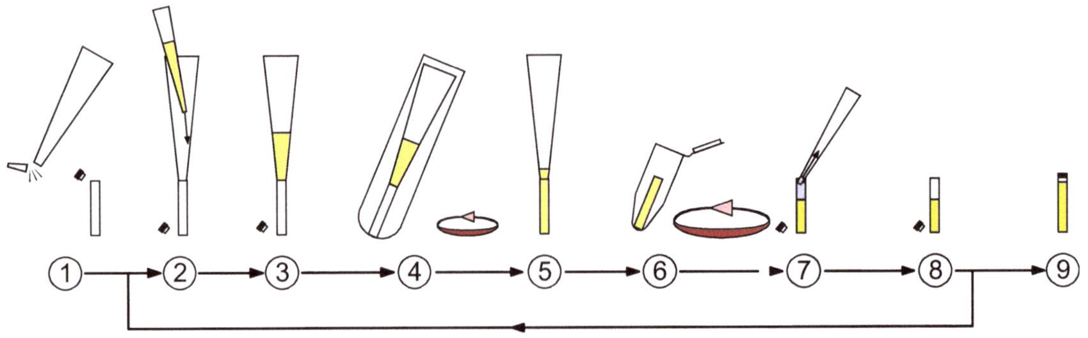

Fig. 3 HR-MAS rotors or inserts for HR-MAS rotor are loaded with bacteria as previously described by Hanoulle and collaborators [19]. Briefly, (1) 1 mL pipette tips are shorten and (2) inserted into a 4 mm ZrO_2 MAS rotors; (3) about 100 μL of bacterial paste is loaded into the tip with automatic pipette; (4) the rotor and tip are inserted into a 4 mL plastic tube and centrifuged at low speed (600 × *g*, 5 min); (5) the tip is removed and (6) the rotor is spun at high speed (14,000 × *g*, 10 min) into a 1 mL microcentrifuge tube; (7) the supernatant D_2O is removed by suction and the process is repeated from (2) until the rotor (8) or the insert is filled up (9)

3. HR-MAS rotor is loaded as described in Fig. 3. Bacterial paste is first loaded into 4 mM rotor mm ZrO_2 MAS rotors by low speed centrifugation (5 min at 600 × *g*) through a 1 mL pipette tip in a plastic 4 mL tube. The pipette tip is removed and the rotor is transferred into a 1 mL plastic microcentrifuge tube. The bacteria are then compacted in the rotor by higher speed centrifugation (5 min at 10,000 × *g*). Supernatant D_2O is removed by suction using a 100–200 μL automatic pipette and the process is repeated until the rotor is filled up with compacted bacterial paste. Depending on the consistency of the bacterial suspension, it may take two to four rounds to fill up the rotor.
4. Rotor is inserted into HR-MAS NMR probe, tilted at 54.7° and spun at 8 kHz. NMR experiments are run using dedicated HR-MAS-NMR pulses. The most useful pulses for analysis of intact cells are 1D ^{1}H, 2D ^{1}H-^{1}H COSY, and ^{1}H-^{13}C HSQC experiments, with or without water suppression.
5. HR-MAS NMR analysis of intact bacteria generates spectrum very similar to liquid NMR analysis of the purified CWPS isolated from the same bacterial strain. As exemplified in Fig. 4, the liquid NMR ^{1}H-^{13}C HSQC spectrum of CWPS isolated from *Lactococcus lactis* MG1363 by HCl extraction shows NMR signals (A–F) associated to the hexasaccharide repeat unit [-6)βGlcNAc1-3αRha1-3(αGlc1-6)βGlcNAc1-2βGal*f*1-6αGlc-P-]$_n$, as previously identified [6]. In addition to the signals associated to the intact polysaccharide, two minor signals A1'a and A1'b were identified as terminal reducing αGlc and βGlc residues resulting from the partial cleavage of αGlc-P bond. The HR-MAS NMR ^{1}H-^{13}C HSQC spectrum of the

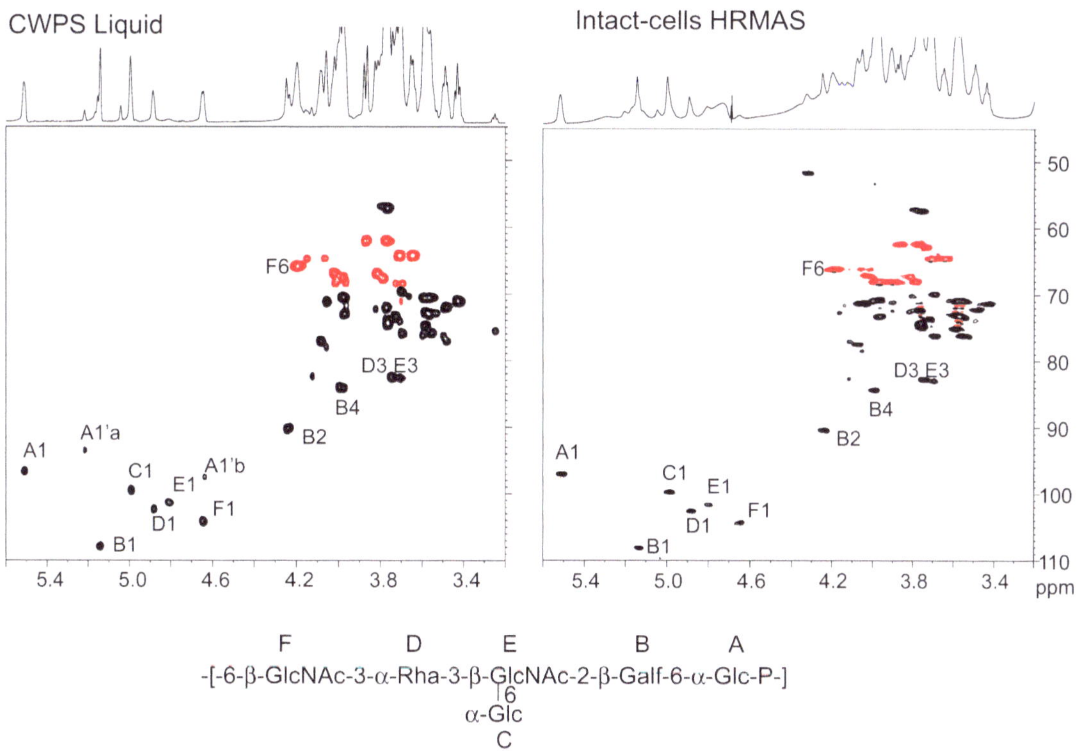

Fig. 4 Comparison of the ^{1}H-^{13}C HSQC spectra from (left panel) the liquid NMR analysis of purified CWPS purified from *Lactococcus lactis* cell wall and (right panel) the HR-MAS NMR analysis of total *Lactococcus lactis* [8]

intact *Lactococcus lactis* MG1363 bacteria shows identical signals than the purified CWPS, despite the absence of A1'a and A1'b signals.

4 Notes

1. It is recommended to treat bacterial CWs with hot SDS prior to extraction of CWTAs and CWPS [17] in order to eliminate LTA and other lipophilic compounds. However, in some strains, hot SDS was shown to extract also relatively hydrophobic CWPS, such as surface-associated polysaccharides sPS1 from *Lactobacillus delbrueckii* subsp. *bulgaricus* Ldb17, composed of repeating units containing Glc, Gal, Rha, and Rha2OAc [5]. Teichoic acids of *Listeria monocytogenes* were also partially extracted with SDS [18]. We therefore recommend to omit the SDS treatment step to prevent losses of CWPS.

2. In our experience, precipitation with TCA is an efficient method of removal of protein and nucleic acids. It can be used instead of enzymatic digestions.
3. TCA and HF are very aggressive agents causing burns. Always wear gloves and work in the fume hood. Please keep in mind to never use glass when manipulating HF.
4. The suspension of cells in water should not be very thick. We typically suspend the cells in water in large (350 mL) centrifuge tubes with a glass rod. 6000 × *g* for 20 min is usually sufficient to spin down the cells. However, if the pellet is "slimy," add more water to the suspension and/or increase centrifugation time.
5. Water should be changed several times during the first hours of dialysis, then twice a day. The efficiency of dialysis can be checked with the simple pH paper placed on the bottom of the dialysis reservoir, detecting the accumulation of acid. The volume of dialysate increases significantly during dialysis. Neutral pH of the solution indicates that dialysis is complete.
6. Sometimes the crude extract is not completely soluble in water. Insoluble material usually represents protein and lipid impurities. Take the material in water and remove the insoluble material by centrifugation (14,000 × *g*, 5 min).
7. 100 μL of packed bacteria is required to fill up a 4 mM rotor. This volume may be reduced down to 25 μL when disposable inserts are used with 4 mM rotors. Filling up the rotors or the insert requires about 2–3 times the final volume of bacterial pellets in order to ease the handling of biological material.

Acknowledgments

We acknowledge Douwe VanSideren (School of Microbiology, University College Cork, Ireland) and Marie-Pierre Chapot Chartier (Micalis Institute, INRA, France) for the use of *Lactococcus lactis* MG1363 and SK11. This work was supported by French ANR project "Lactophages" (ANR-11-BSV8-004-01).

References

1. Delcour J, Ferain T, Deghorain M et al (1999) The biosynthesis and functionality of the cell-wall of lactic acid bacteria. In: Konings WN, Kuipers OP, In't Veld JHJH (eds) Lactic acid bacteria: genetics, metabolism and applications. Springer, Dordrecht, pp 159–184
2. Kolenbrander PE, London J (1993) Adhere today, here tomorrow: oral bacterial adherence. J Bacteriol 175:3247–3252
3. Vinogradov E, Valence F, Maes E et al (2013) Structural studies of the cell wall polysaccharides from three strains of *Lactobacillus helveticus* with different autolytic properties: DPC4571, BROI, and LH1. Carbohydr Res 379:7–12
4. Cornelissen A, Sadovskaya I, Vinogradov E et al (2016) The baseplate of *Lactobacillus delbrueckii* bacteriophage Ld17 harbors a

glycerophosphodiesterase. J Biol Chem 291:16816–16827
5. Vinogradov E, Sadovskaya I, Cornelissen A et al (2015) Structural investigation of cell wall polysaccharides of *Lactobacillus delbrueckii* subsp. *bulgaricus* 17. Carbohydr Res 413:93–99
6. Chapot-Chartier M-P, Vinogradov E, Sadovskaya I et al (2010) Cell surface of *Lactococcus lactis* is covered by a protective polysaccharide pellicle. J Biol Chem 285:10464–10471
7. Sadovskaya I, Vinogradov E, Courtin P et al (2017) Another brick in the wall: a rhamnan polysaccharide trapped inside peptidoglycan of *Lactococcus lactis*. mBio 8:e01303–e01317
8. Ainsworth S, Sadovskaya I, Vinogradov E et al (2014) Differences in lactococcal cell wall polysaccharide structure are major determining factors in bacteriophage sensitivity. mBio 5: e00880–14
9. Farenc C, Spinelli S, Vinogradov E et al (2014) Molecular insights on the recognition of a *Lactococcus lactis* cell wall pellicle by the phage 1358 receptor binding protein. J Virol 88:7005–7015
10. Chapot-Chartier M-P, Kulakauskas S (2014) Cell wall structure and function in lactic acid bacteria. Microb Cell Factories 13:S9
11. Prakobphol A, Linzer R (1980) Purification and immunological characterization of a rhamnose-glucose antigen from Streptococcus mutans 6517-T2 (serotype g). Infect Immun 30:140–146
12. Vinogradov E, Sadovskaya I, Grard T et al (2016) Structural studies of the rhamnose-rich cell wall polysaccharide of *Lactobacillus casei* BL23. Carbohydr Res 435:156–161
13. Li W, Lee REB, Lee RE, Li J (2005) Methods for acquisition and assignment of multidimensional high-resolution magic angle spinning NMR of whole cell bacteria. Anal Chem 77:5785–5792
14. Maes E, Mille C, Trivelli X et al (2009) Molecular phenotyping of mannosyltransferases-deficient *Candida albicans* cells by high-resolution magic angle spinning NMR. J Biochem (Tokyo) 145:413–419
15. Candela T, Maes E, Garénaux E et al (2011) Environmental and biofilm-dependent changes in a Bacillus cereus secondary cell wall polysaccharide. J Biol Chem 286:31250–31262
16. Dubois M, Gilles K, Hamilton JK et al (1951) A colorimetric method for the determination of sugars. Nature 168:167
17. Signoretto C, Lleò MM, Tafi MC et al (2000) Cell wall chemical composition of *Enterococcus faecalis* in the viable but nonculturable state. Appl Environ Microbiol 66:1953–1959
18. Brauge T, Sadovskaya I, Faille C et al (2016) Teichoic acid is the major polysaccharide present in the *Listeria monocytogenes* biofilm matrix. FEMS Microbiol Lett 363:fnv229
19. Hanoulle X, Wieruszeski J-M, Rousselot-Pailley P et al (2005) Monitoring of the ethionamide pro-drug activation in mycobacteria by (1)H high resolution magic angle spinning NMR. Biochem Biophys Res Commun 331:452–458

Chapter 5

Extraction and Purification of Wall-Bound Polymers of Gram-Positive Bacteria

Alice Chateau, Olaf Schneewind, and Dominique Missiakas

Abstract

The envelope of gram-positive bacteria encompasses the cell wall, a rigid exoskeleton comprised of peptidoglycan that provides protection against lysis and governs bacterial cell shapes. Peptidoglycan also serves as the site of attachment for proteins and nonproteinaceous polymers that interact with the bacterial environment. Nonproteinaceous molecules include teichoic acids, capsular polysaccharides, and secondary cell wall polysaccharides (SCWP). Treatment of gram-positive bacterial cells with proteases, nucleases, and detergents results in the isolation of "murein sacculi" (i.e., peptidoglycan with bound carbohydrate polymers). Incubation of sacculi with acid or base releases carbohydrate polymers that can be purified for further biochemical characterization. This protocol describes the hydrofluoric acid extraction and purification of the secondary cell wall polymer of *Bacillus anthracis* that is also found in the envelope of the other members of the *Bacillus cereus* sensu lato group of bacteria.

Key words Secondary cell wall polysaccharide, *Bacillus cereus*, Murein sacculus, Peptidoglycan, Hydrofluoric acid, Phosphodiester bond

1 Introduction

In gram-positive bacteria, which lack an outer membrane, the peptidoglycan cell wall represents the microbe's most outer layer and surface organelle. Peptidoglycan consists of glycan strands comprising 5–30 repeating disaccharide units, *N*-acetylmuramic acid-(β1-4)-*N*-acetylglucosamine (MurNAc-GlcNAc). MurNAc residues are amide-linked to short wall peptides that incorporate both L- and D-forms of amino acids that make the cell wall impervious to proteolytic activity. Peptides of neighboring glycan strands are cross-linked via the crossbridge, that is, the amino group derived from position three of the wall peptide (L-lysine, meso-diaminopimelic acid, or peptide extensions appended to either one of these amino acids), generating a three-dimensional molecular network that surrounds the cell akin to an exoskeleton (reviewed by [1]). Peptidoglycan also functions as a scaffold for the

Inka Brockhausen (ed.), *Bacterial Polysaccharides: Methods and Protocols*, Methods in Molecular Biology, vol. 1954,
https://doi.org/10.1007/978-1-4939-9154-9_5, © Springer Science+Business Media, LLC, part of Springer Nature 2019

immobilization of proteins [2], capsular polysaccharides [3, 4], wall teichoic acids (WTAs) [5], and secondary cell wall polysaccharides (SCWP) [6]. Cell wall-anchored proteins are covalently bound by the enzyme sortase to the peptide moiety of peptidoglycan (reviewed by [7]) while most other polymers are transferred onto the glycan strands of peptidoglycan by the LytR-CpsA-Psr (LCP) family of enzymes [4, 8–11]. Precursor polymers bound to the lipid carrier undecaprenol-$(PO_4)_2$ serve as substrates of LCPs, which attach the polymers to the C-6 hydroxyl of MurNAc via phosphodiester linkage and release undecaprenol-(PO_4) [8]. Phosphodiester bonds are sensitive to acid and base treatments [5]. Thus, peptidoglycan purified at neutral pH is decorated with covalently linked polymers that can be released by altering the pH of suspended peptidoglycan [12, 13]. Members of the *Bacillus cereus* sensu lato group, which includes *B. anthracis*, synthesize a SCWP with the trisaccharide repeat structure [→4)-β-ManNAc-(1→4)-β-GlcNAc (O3-α-Gal)-(1→6)-α-GlcNAc(O3-α-Gal, O4-β-Gal)-(1→]$_{6\text{–}12}$ [14]. The reducing end of the terminal repeat unit of SCWP is uniquely modified with ketal-pyruvyl to serve as a ligand for the S-layer homology (SLH) domains that are found in a subset of secreted proteins [15]. These include S-layer proteins, which assemble into a paracrystalline S-layer, as well as S-layer-associated proteins (BSLs) that fulfill unique functions such as host cell adhesion, nutrient transport, cell separation and chain length determination (reviewed by [16]). Unlike other gram-positive bacteria, *B. anthracis* lacks WTA and only the SCWP is released upon acid treatment of peptidoglycan preparations [17]. Figure 1 shows fluorescent micrographs of *B. anthracis* cells, revealing the surface display of SCWP.

Here we describe a simple method for the extraction and purification of SCWP that can be adapted to other peptidoglycan-bound polymers. Such preparations can be used for the determination of structure and composition using size exclusion chromatography, MALDI-TOF mass spectrometry, and NMR. The protocol may be useful to compare the composition of SCWP from different bacterial strains [14, 18–20] to study the enzymes that are implicated in SCWP synthesis [21], and to characterize ligands of the SCWP such phage endolysin [22] as well as SLH domain-containing proteins [15].

2 Materials

All solutions and buffers should be prepared with deionized water and analytical grade reagents. Unless indicated, reagents, solutions, and buffers are stored at room temperature.

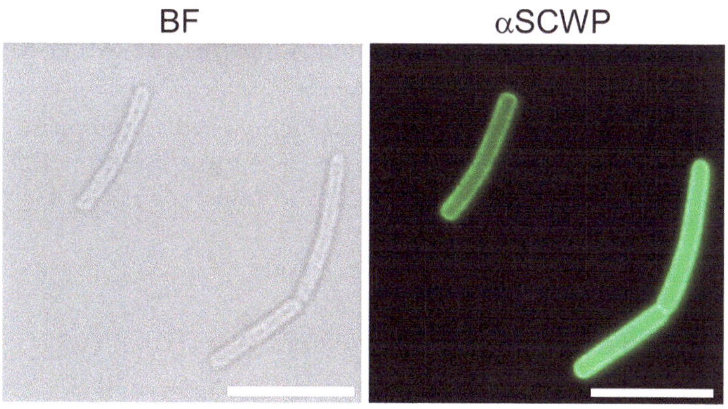

Fig. 1 Micrographs of *B. anthracis* strain Sterne cells revealing the surface display of SCWP. Cells were fixed in formalin, stained with anti-SCWP antibodies (αSCSWP) and observed by bright-field microscopy (BF, left panel) or fluorescence microscopy (right panel). Scale bar, 10 μm

2.1 Purification of Murein Sacculi (See Note 1)

1. Brain–heart infusion (BHI) broth: Dissolve 37 g of BHI in 1 L of water. Autoclave at 121 °C for 15 min.
2. BHI agar plates: add 15 g of agar to 1 L of BHI. Autoclave at 121 °C for 15 min.
3. *B. anthracis* Sterne strain (or any *B. cereus* group bacteria).
4. 250 mL conical flask.
5. Sodium dodecyl sulfate (SDS) 20%: dissolve 10 g of SDS in 500 mL of water using a water bath at 55 °C to help dissolution.
6. 50-mL conical centrifuge tubes.
7. 500-mL heat- and acid-resistant centrifuge bottles.
8. Boiling water bath.
9. 15-mL conical centrifuge tubes.
10. 0.1-mM glass beads.
11. Bead beater.
12. 100 mM Tris–HCl pH 7.5: Dissolve 12.1 g of Tris in 900 mL of water. Adjust the pH with 1 M HCl and complete 1 L with water.
13. 1 M $MgSO_4$: Dissolve 6.02 g of $MgSO_4$ in 50 mL of water.
14. RNase 10 mg/mL: Dissolve 10 mg of RNase in 1 mL of water.
15. Dnase 10 mg/mL: Dissolve 10 mg of DNase in 1 mL of water.
16. Trypsin 10 mg/mL: Dissolve 10 mg of trypsin in 1 mL of water.
17. 1 M $CaCl_2$: Dissolve 5.55 g of $CaCl_2$ in 50 mL of water.

18. 100 mM Tris–HCl, pH 8.0: Dissolve 12.1 g of Tris in 900 mL of water. Adjust the pH with 1 M HCl and complete 1 L with water.
19. 100 mM ethylenediaminetetraacetic acid (EDTA), pH 8.0: Dissolve 37.2 g of disodium EDTA (Na_2EDTA) in 800 mL of water and adjust the pH to 8 with NaOH (*see* **Note 2**). Bring volume to 1 L.
20. Acetone 100%.

2.2 Extraction of SCWP

1. Hydrofluoric acid (HF) 48%.
2. Ethanol 100% kept at −20 °C.
3. Centrifugal evaporator such as a Speed-Vac.

2.3 Reverse-Phase High-Performance Liquid Chromatography (rpHPLC)

1. One 4.6 × 250 mM Hypersil ODS C18 column (particle size 3 μm) and one 3-μm-particle-size guard column set between injector and C18 column.
2. HPLC with a photodiode array (PDA) detector.
3. HPLC grade water.
4. HPLC grade acetonitrile.
5. HPLC grade trifluoroacetic acid (TFA).
6. Buffer A: water–0.1% TFA. Add 1 mL of TFA in 1 L of water (HPLC grade).
7. Buffer B: acetonitrile–0.1% TFA. Add 1 mL of TFA in 1 L of acetonitrile (HPLC grade).
8. 1.5-mL Eppendorf tubes for sample collection.

2.4 Size Exclusion Chromatography (SEC)

1. One 7.8 × 300 mM BioBasic SEC300 column.
2. HPLC with PDA detector.
3. 50 mM sodium phosphate buffer pH 7.5. This buffer is prepared by mixing: 40 mL of 0.2 M NaH_2PO_4 H_2O (weigh 27.6 g in 1 L water), 210 mL of 0.2 M Na_2HPO_4 $7H_2O$ (weigh 53.65 g in 1 L water) and 750 mL of water. Use HPLC grade water to prepare solutions and buffer.
4. 1.5-mL Eppendorf tubes for sample collection.

3 Methods

When manipulating Risk Group 2 (RG2) live bacteria such as *B. anthracis* strain Sterne, use sterile conditions, work in a Biosafety cabinet and use proper protective equipment as stipulated by your Institutional Biosafety Committee. Discard all live and contaminated reagents and materials in a biosafety hazard container and insure that the materials are appropriately destroyed.

3.1 Purification of Murein Sacculi

1. Inoculate a single colony of *B. anthracis* into 20 mL of BHI in a 250-mL conical flask and incubate at 37 °C with aeration (220 revolution per min) until the culture reaches stationary growth or up to 12 h.
2. Transfer and spread culture aliquots of 300 μL onto 50 BHI plates. Incubate plates overnight at 37 °C or up to 16 h.
3. Scrape cells from plates and transfer to a 250-mL conical flask filled with 25 mL of water.
4. Transfer suspension to a 50-mL conical centrifuge tube and harvest cells by centrifugation at 6000 × *g* for 10 min at room temperature.
5. Discard supernatant and resuspend cells in 10 mL of water, transfer the suspension to a 500-mL centrifuge bottle. Add 270 mL water and 70 mL of 20% SDS to yield a final concentration of ~4% SDS.
6. Close bottle and securely place in a boiling water bath for 30 min.
7. Cool to room temperature, and harvest cells by centrifugation at 6000 × *g* for 10 min at room temperature. This treatment will effectively kill bacterial cells and samples may be considered safe at this point.
8. Wash materials in pellet at minimum five times with water to eliminate all traces of SDS (*see* **Notes 3** and **4**).
9. Suspend pellet in 40 mL of water, and distribute sample roughly into six 15-mL conical centrifuge tubes prefilled with roughly 3 mL of 0.1 mM glass beads (use tubes with graduation to facilitate this step).
10. Securely fasten tubes to the bead beater and subject cells to mechanical lysis by performing 10 pulses of 1 min. Transfer tubes on ice for 5 min between each pulse to avoid overheating of samples. If using the MP Biomedicals™ FastPrep −24™ Classic Instrument, set the instrument on TN 12 × 15, speed 6 m/s.
11. After the final pulse, allow beads to sediment by gravity and carefully pipette the lysates and transfer to a clean 50-mL conical centrifuge tube.
12. Refill once more the 15-mL conical centrifuge tube with water as in **step 9**, and pulse one more time as described in **steps 10** and **11**. Transfer all materials that do not sediment into the same 50-mL conical centrifuge tube.
13. Allow any leftover glass beads to sediment by gravity and carefully transfer lysate to a new 50-mL conical centrifuge tube.
14. Centrifuge tube at 6000 × *g* for 10 min at room temperature. Wash pellet once with water.

15. Repeat **step 14**, discard all fluids without perturbing pellet.
16. Suspend materials in pellet with 25 mL 100 mM Tris–HCl, pH 7.5. Add 25 μL of RNAse solution (10 mg/mL), 25 μL of DNAse solution (10 mg/mL), and 250 μL of 1 M $MgSO_4$. Incubate tube for 4 h at 37 °C with gentle agitation if possible (use a tube rotator if available).
17. Add to the same tube, 25 μL of trypsin solution (10 mg/mL) and 250 μL of 1 M $CaCl_2$. Place tube back at 37 °C for 16 h.
18. Harvest insoluble material by centrifugation at 6000 × *g* for 10 min at room temperature, wash once with water.
19. Repeat **steps 5** through **8** (*see* **Note 5**).
20. Perform one wash with 25 mL of 100 mM Tris–HCl, pH 8.0. Centrifuge tube at 6000 × *g* for 10 min at room temperature, discard supernatant.
21. Repeat **step 20** with water.
22. Repeat **step 20** with 25 mL of 100 mM EDTA, pH 8.0.
23. Repeat **step 20** with water.
24. Repeat **step 20** with 25 mL acetone but centrifuge tube at 3000 × *g* (*see* **Note 6**).
25. Repeat **step 20** twice with water (*see* **Note** 7).
26. The pellet containing cell wall fragments (murein sacculi) can be stored at −20 °C until further use.

3.2 Extraction of SCWP

1. Resuspend pellet containing cell wall fragments (as prepared above) in 5 mL of water in a 50-mL conical tube and add 25 mL of 48% HF (*see* **Note 8**).
2. Incubate for 18 h shaking gently at 4 °C.
3. Centrifuge tube at 8000 × *g* for 15 min at 4 °C.
4. Transfer the supernatant to a 500-mL centrifuge bottle and discard the HF-laden pellet in a dedicated chemical waste container.
5. Add 5 volumes of cold ethanol (−20 °C) to the bottle, close and keep overnight at −20 °C.
6. Centrifuge bottle at 10,000 × *g* for 15 min at 4 °C to harvest precipitated SCWP. Discard the HF-laden supernatant in a dedicated chemical waste container.
7. Wash pellet five times with ethanol, gently resuspending between each wash and centrifuging bottle at 10,000 × *g* for 15 min at 4 °C.
8. After the last wash, gently suspend pellet with ethanol and transfer suspension to an empty Eppendorf tube previously weighed empty.

9. Transfer tube containing sample to a Speed-Vac, and dry to completion before weighing tube once more to evaluate the yield of purified SCWP (*see* **Note 9**).
10. Store tube with dried SCWP at −20 °C.

This material can be used for *rp*HPLC and SEC purification and further analyzed by MALDI-TOF mass spectrometry or NMR.

3.3 rpHPLC Purification

1. Resuspend dried SCWP obtained in Subheading 3.2 in water to a final concentration of 100 mg/mL.
2. Using HPLC, load 100 μL of SCWP over the Hypersil ODS C18 column using the guard column set between injector and C18 column.
3. Load and elute sample with a flow rate of 0.5 mL/min and use the following separation gradient of Buffers A and B:
   (a) 10 min: 0% Buffer B (100% Buffer A).
   (b) 5 min: 0–10% linear gradient of Buffer B.
   (c) 20 min: 10–20% linear gradient of Buffer B.
   (d) 65 min: 20–100% linear gradient of Buffer B.
   (e) 10 min: 100% Buffer B.
4. Collect 0.5-mL fractions and record chromatographic UV absorption profile with PDA by monitoring absorbance at 206 nm (A_{206}). A typical elution profile for HF purified SCWP is shown in Fig. 2. The majority of the material elutes at ~25 min with 15% acetonitrile.

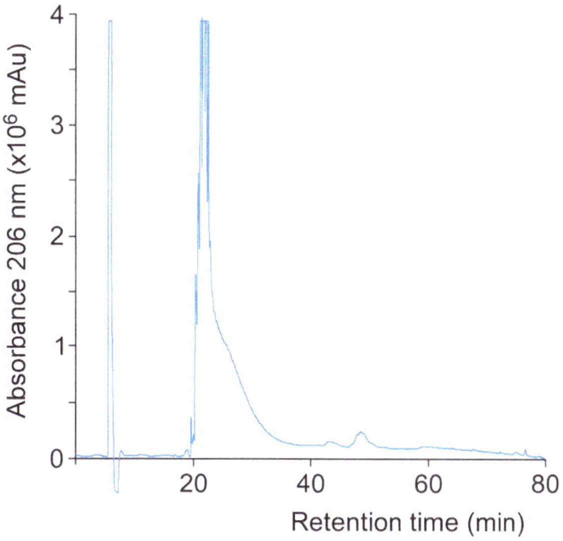

Fig. 2 *rp*HPLC chromatograph of HF purified SCWP from *B. anthracis* Sterne strain. Milli-Absorbance units (mAu) were recorded at 206 nm

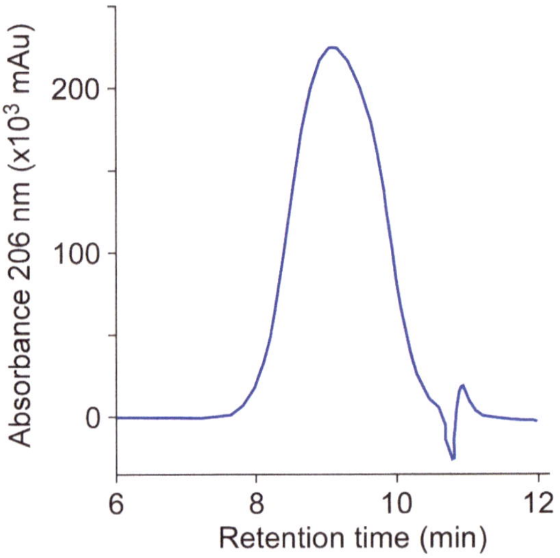

Fig. 3 Size exclusion chromatography of SCWP from *B. anthracis* Sterne strain. Absorbance was measured in milli-Absorbance units (mAu) at 206 nm

5. Fractions containing SCWP can be collected and stored at −20 °C for structural characterization and biochemical assays.

3.4 SEC

1. Resuspend dried SCWP obtained in Subheading 3.2 in 50 mM sodium phosphate pH 7.5 to a final concentration of 100 mg/mL.
2. Using HPLC, load 100 μL of SCWP over the BioBasic SEC300 column equilibrated with 50 mM sodium phosphate buffer pH 7.5.
3. Load and elute sample with a flow rate of 1 mL/min in 50 mM sodium phosphate buffer pH 7.5.
4. Collect 0.5-mL fractions and record chromatographic UV absorption profile with PDA by monitoring absorbance at 206 nm (A_{206}). The typical retention time for HF purified SCWP (~ 12 kDa) is shown in Fig. 3. The column can be calibrated using the same settings and buffer with known molecular weight markers.
5. Fractions containing SCWP can be collected and stored at −20 °C for structural characterization and biochemical assays (*see* **Note 10**).

4 Notes

1. We use the term murein sacculi to refer to peptidoglycan with bound polymers and use a bead beater during the extraction procedure to mechanically break the exoskeletal network of

peptidoglycan. This procedure results largely in murein fragments with bound polymers rather than intact sacculi.

2. The disodium salt of EDTA will not dissolve until the pH of the solution is adjusted to 8.0 with NaOH.
3. Eliminating SDS: to wash sample and eliminate all SDS, add 10–12 mL of water to the decanted cells in the 500-mL centrifuge bottle. Carefully resuspend cells before adding another 350 mL of water. Make sure to completely resuspend the cells. Repeat as needed. SDS is a detergent that is responsible for the formation of bubbles. The absence of bubbles indicates that all SDS has been washed away. Incomplete removal of SDS will interfere with the bead beating process.
4. Following removal of SDS, pellets can be stored at −20 °C before tackling the next steps. Make sure to remove all the SDS otherwise it will precipitate at the lower temperatures.
5. Second SDS boiling: this is done to heat-inactivate the enzymes added as a result of the procedure.
6. Sample solubility is decreased following acetone wash; thus, a smaller centrifugal (*g*) force is applied to the samples to facilitate resuspension of pellets.
7. Following acetone and first water washes, pellets will appear more compact. Pellets will recover a more normal and fluffy appearance with subsequent water washes.
8. HF is extremely toxic and should be used with extreme caution! A 25-square inch burn with HF can be lethal. HF should always be handled inside a chemical fume hood. HF and HF-waste products should be placed in clearly marked vials and containers. Other laboratory members should be made aware of the presence and use of HF. Use chemical gloves that cover the hands, wrists, and forearms when using HF. Call appropriate Environmental Health & Safety personnel for the immediate removal of waste material and unused HF.
9. The typical yield for SCWP purification is between 10 to 40 mg. Purity of the preparation can be rapidly validated by performing a UV–visible scan using a spectrophotometer. SCWP absorbs light with a maximum absorption at 206 nm.
10. Most purification procedures use *rp*HPLC and the use of SEC is offered as an alternative. One advantage of SEC is the ability to compare sizes of polysaccharides between related species or between isogenic mutants that might be impaired in the synthesis of SCWP.

Acknowledgments

We thank current and past members of the laboratory who helped develop and refine this protocol. Research in our laboratory is supported by grant number AI069227 from the National Institute of Allergy and Infectious Diseases, Infectious Diseases Branch.

References

1. Navarre WW, Schneewind O (1999) Surface proteins of gram-positive bacteria and the mechanisms of their targeting to the cell wall envelope. Microbiol Mol Biol Rev 63:174–229
2. Schneewind O, Fowler A, Faull KF (1995) Structure of the cell wall anchor of surface proteins in *Staphylococcus aureus*. Science 268:103–106
3. Munoz E, Ghuysen J-M, Heymann H (1967) Cell walls of *Streptococcus pyogenes* type 14. C polysaccharide-peptidoglycan and G polysaccharide-peptidoglycan complexes. Biochemistry 6:3659–3670
4. Chan YG, Kim HK, Schneewind O et al (2014) The capsular polysaccharide of *Staphylococcus aureus* is attached to peptidoglycan by the LytR-CpsA-Psr (LCP) family of enzymes. J Biol Chem 289:15680–15690
5. Coley J, Archibald AR, Baddiley J (1976) A linkage unit joining peptidoglycan to teichoic acid in *Staphylococcus aureus* H. FEBS Lett 61:240–242
6. Ries W, Hotzy C, Schocher I et al (1997) Evidence that the N-terminal part of the S-layer protein from *Bacillus stearothermophilus* PV72/p2 recognizes a secondary cell wall polymer. J Bacteriol 179:3892–3898
7. Schneewind O, Missiakas D (2014) Sec-secretion and sortase-mediated anchoring of proteins in gram-positive bacteria. Biochim Biophys Acta 1843:1687–1697
8. Kawai Y, Marles-Wright J, Cleverley RM et al (2011) A widespread family of bacterial cell wall assembly proteins. EMBO J 30:4931–4941
9. Eberhardt A, Hoyland CN, Vollmer D et al (2012) Attachment of capsular polysaccharide to the cell wall in *Streptococcus pneumoniae*. Microb Drug Resist 18:240–255
10. Chan YGY, Frankel MB, Dengler V et al (2013) *Staphylococcus aureus* mutants lacking the LytR-CpsA-Psr (LCP) family of enzymes release wall teichoic acids into the extracellular medium. J Bacteriol 195:4650–4659
11. Liszewski Zilla M, Chan YG, Lunderberg JM et al (2015) LytR-CpsA-Psr enzymes as determinants of *Bacillus anthracis* secondary cell wall polysaccharide assembly. J Bacteriol 197:343–353
12. Kojima N, Arakai Y, Ito E (1985) Structure of the linkage units between ribitol teichoic acids and peptidoglycan. J Bacteriol 161:299–306
13. de Jonge BL, Chang YS, Gage D et al (1992) Peptidoglycan composition of a highly methicillin-resistant *Staphylococcus aureus* strain. The role of penicillin binding protein 2A. J Biol Chem 267(16):11248–11254
14. Choudhury B, Leoff C, Saile E et al (2006) The structure of the major cell wall polysaccharide of *Bacillus anthracis* is species specific. J Biol Chem 281:27932–27941
15. Mesnage S, Fontaine T, Mignot T et al (2000) Bacterial SLH domain proteins are non-covalently anchored to the cell surface via a conserved mechanism involving wall polysaccharide pyruvylation. EMBO J 19:4473–4484
16. Missiakas D, Schneewind O (2017) Assembly and function of the *Bacillus anthracis* S-layer. Annu Rev Microbiol 71:79–98
17. Molnár J, Prágai B (1971) Attempts to detect the presence of teichoic acid in *Bacillus anthracis*. Acta Microbiol Acad Sci Hung 18:105–108
18. Leoff C, Saile E, Sue D et al (2008) Cell wall carbohydrate compositions of strains from *Bacillus cereus* group of species correlate with phylogenetic relatedness. J Bacteriol 190:112–121
19. Forsberg LS, Choudhury B, Leoff C et al (2011) Secondary cell wall polysaccharides from *Bacillus cereus* strains G9241, 03BB87 and 03BB102 causing fatal pneumonia share similar glycosyl structures with the polysaccharides from *Bacillus anthracis*. Glycobiology 21:934–948
20. Forsberg LS, Abshire TG, Friedlander A et al (2012) Localization and structural analysis of a conserved pyruvylated epitope in *Bacillus anthracis* secondary cell wall polysaccharides and characterization of the galactose deficient wall polysaccharide from avirulent *B. anthracis* CDC 684. Glycobiology 22:1103–1117

21. Chateau A, Lunderberg JM, Oh SY et al (2018) Galactosylation of the secondary cell wall polysaccharide of *Bacillus anthracis* and its contribution to anthrax pathogenesis. J Bacteriol 200 (5). https://doi.org/10.1128/JB.00562-17

22. Mo KF, Li X, Li H et al (2012) Endolysins of *Bacillus anthracis* bacteriophages recognize unique carbohydrate epitopes of vegetative cell wall polysaccharides with high affinity and selectivity. J Am Chem Soc 134:15556–15562

Chapter 6

Purification and Analysis of Mycobacterial Phosphatidylinositol Mannosides, Lipomannan, and Lipoarabinomannan

Kathryn C. Rahlwes, Julia Puffal, and Yasu S. Morita

Abstract

Mycobacteria and related bacteria in the Actinobacteria phylum are unusual in that they produce phosphatidylinositol (PI) as a major phospholipid species. PI can be further modified by glycan polymers, leading to the synthesis of PI mannosides (PIMs), lipomannan (LM), and lipoarabinomannan (LAM). Small lipids such as PI and PIMs are extracted with a mixture of chloroform, methanol, and water and analyzed by thin layer chromatography. For larger glycolipids, such as LM and LAM, more hydrophilic solvent is needed for the extraction, and SDS-PAGE is better suited for the analysis. For LM, further structural characterization can be performed by MALDI-TOF mass spectrometry. Precise quantification of PIMs, LM, and LAM can be performed by quantification of glycan staining using analytical software. The metabolic radiolabeling protocol is also described.

Key words Glycolipids, Glycan staining, Lipomannan, Lipoarabinomannan, Phosphatidylinositol mannosides

1 Introduction

The glycolipid components of mycobacteria are major host immune modulators and frequently vital to the integrity of the cell envelope [1–3]. Among them, phosphatidylinositol (PI)-anchored glycolipids, such as PI mannosides (PIMs), lipomannan (LM) and lipoarabinomannan (LAM), are emerging as major pathogen-associated molecular patterns [3, 4]. Furthermore, LAM is released into the urine of active tuberculosis patients, and the released LAM is being pursued as a diagnostic tool for hospitalized HIV-infected patients who do not produce sputum needed for more standard testing [5–8]. These recent research developments in immunology and medicine signify the importance of understanding the basic molecular mechanisms behind the biosynthesis, trafficking, and release of these PI-linked glycans. Having the initial discovery of PIMs dating back more than half a century

Inka Brockhausen (ed.), *Bacterial Polysaccharides: Methods and Protocols*, Methods in Molecular Biology, vol. 1954, https://doi.org/10.1007/978-1-4939-9154-9_6,

ago [9, 10], extensive literature is now available for the structure and biosynthesis of PIMs/LM/LAM (*see* reviews [11–15]), and yet our understanding of these glycolipids is far from complete with particularly scarce information available on the regulation of biosynthesis, trafficking across the cell envelope, and shedding to the extracellular milieu. Although we cannot review the history of method developments in detail for this short protocol chapter, we acknowledge that many laboratories around the world have contributed to the development of tools to analyze PIMs/LM/LAM. Here, we describe the basic methods used in our laboratory to purify and analyze PIMs/LM/LAM from *Mycobacterium smegmatis* and *Mycobacterium tuberculosis.*

2 Materials

Prepare all solutions using ultrapure water. Organic solvents such as chloroform, methanol, and propanol should be stored in borosilicate glass bottles with polytetrafluoroethylene (PTFE)-lined caps in a flammable cabinet. When handling organic solvents, glass beakers and graduated cylinders should be used. Avoiding plastics for organic solvent-based lipid extraction is particularly important when preparing samples for mass spectroscopic analysis. For other experiments, polypropylene plastics, which are relatively resistant to organic solvents, can be used as organic solvent containers. Polystyrene plastics should be avoided for any procedures involving organic solvents. Volatile organic solvents should be handled in a fume hood. All materials can be stored at room temperature unless otherwise indicated.

2.1 Extraction of PIMs/LM/LAM from Cultured Mycobacterial Cells

1. Middlebrook 7H9 medium (BD Biosciences). Follow the manufacturer's guidelines, except that Middlebrook ADC Growth Supplement (BD Biosciences) is replaced with the dextrose-sodium chloride (DC) supplement for the growth of *M. smegmatis* strain mc^{2}155 (*see* below).
2. DC supplement: 4.25 g sodium chloride and 10 g glucose, bring to a total volume of 500 mL. Autoclave for 15 min at 121 °C.
3. 20% Tween 80: carefully pour Tween 80 to the 10-mL graduation mark on a 50-mL conical tube, and add water to the 50-mL graduation mark. Gently mix the conical tube overnight at room temperature on a rotator. Filter-sterilize.
4. Chloroform.
5. Chloroform–methanol (2:1): mix 240 mL chloroform and 120 mL methanol.

6. Chloroform–methanol–water (1:2:0.8): mix 100 mL chloroform, 200 mL methanol, and 80 mL water.
7. Phenol, saturated solution (pH 6.6), (e.g., MP Biomedicals), store at 4 °C (*see* **Note 1**).
8. 10 mg/mL Proteinase K: dissolve 10 mg of Proteinase K (Thermo Fisher) in 1 mL of 10 mM Tris–HCl, pH 7.5, 20 mM calcium chloride, and 50% glycerol. Store at −20 °C.
9. Water bath sonicator (e.g., Branson Ultrasonics).
10. Speed-vac concentrator (e.g., Eppendorf Vacufuge).
11. Centrifuge (e.g., Eppendorf 5810R).
12. Microcentrifuge (e.g., Eppendorf 5418R).

2.2 Metabolic Radiolabeling of PIMs/LM/LAM

1. En^3Hance™ Autoradiography Enhancer (PerkinElmer).
2. En^3Hance™ Spray Surface Autoradiography Enhancer (PerkinElmer).
3. [Methyl-^{14}C] Methylated, Protein Molecular Weight Markers, (PerkinElmer).
4. Gel dryer (e.g., Bio-Rad, Model 583 Gel Drier).
5. BioMAX MR film (Kodak) and an X-ray film cassette.
6. Coomassie Blue.
7. [^{3}H]mannose or [^{3}H]inositol (1 mCi/mL, PerkinElmer).
8. 1 M mannose.
9. 1 M inositol.
10. Whatman paper.
11. Thin layer chromatography (TLC) developing tanks.
12. High performance TLC (HPTLC) silica gel 60 (Millipore 5547).
13. 13 NH_3: 87.8 mL of ammonium hydroxide (28% NH_3, 14.8 M). Add water to 100 mL.
14. 1 M ammonium acetate.
15. TLC developing solvent (chloroform–methanol–13 M NH_3–1 M NH_4Ac–water, 180:140:9:9:23): mix 180 mL chloroform, 140 mL methanol, 9 mL 13 M NH_3, 9 mL 1 M ammonium acetate, and 23 mL water.

2.3 Octyl-Sepharose Column Purification

1. Octyl-Sepharose (GE Healthcare), store at 4 °C.
2. 1-Propanol.
3. Lyophilizer (e.g., LABCONO).
4. Reacti-Vap PTFE-coated needles, Reacti-Vap evaporators, and single-block Reacti-Therm modules (Thermo Fisher).
5. Poly-Prep® chromatography columns (Bio-Rad).

6. 1 M ammonium acetate: dissolve 7.71 g ammonium acetate in ~80 mL water, bring to a total volume of 100 mL with water. Store at −20 °C.
7. 5% propanol in 0.1 M ammonium acetate: mix 5 mL 1-propanol and 50 mL water. Add 10 mL 1 M ammonium acetate, and bring the volume to 100 mL using water.

2.4 LM/LAM Analysis

1. 1.5 M Tris–HCl, pH 8.8: dissolve 81.7 g Tris in ~300 mL of water. Adjust pH to 8.8 with 6 M HCl. Then bring to a final volume of 450 mL with water. Store at 4 °C.
2. 0.5 M Tris–HCl, pH 6.8: dissolve 24.2 g Tris in ~250 mL of water. Adjust pH to 6.8 with 6 M HCl and bring to a final volume of 400 mL with water. Store at 4 °C.
3. 10% sodium dodecyl sulfate (SDS): dissolve 10 g of SDS in ~80 mL of water and bring to a final volume of 100 mL.
4. 40% acrylamide–bis-acrylamide (37.5:1) (e.g., Alfa Aesar).
5. 10% ammonium persulfate (APS): 20 mg APS, 200 μL water (*see* **Note 2**).
6. Tetramethylethylenediamine (TEMED) (MP Biomedicals).
7. Dithiothreitol (DTT).
8. Bromophenol Blue.
9. Glycerol.
10. Precision Plus Protein Kaleidoscope Standards (Bio-Rad).
11. Mini-PROTEAN short plates (Bio-Rad).
12. Mini-PROTEAN spacer plates 0.75 mm spacer (Bio-Rad).
13. Mini-PROTEAN tetra electrode assembly (Bio-Rad).
14. Mini-PROTEAN tetra vertical electrophoresis buffer tank and lid (Bio-Rad).
15. MINI-PROTEAN tetra stand and clamps (Bio-Rad).
16. 15% separating gel: mix 4.31 mL water, 3 mL 1.5 M Tris–HCl, pH 8.8, 120 μL 10% SDS, and 4.50 mL 40% acrylamide–bis-acrylamide. Use immediately upon the addition of 60 μL 10% APS and 6 μL TEMED.
17. 4% stacking gel: mix 3.17 mL water, 1.25 mL 0.5 M Tris–HCl, pH 6.8, 50 μL 10% SDS, and 500 μL 40% acrylamide–bis-acrylamide. Use immediately upon the addition of 25 μL 10% APS and 5 μL TEMED.
18. SDS-PAGE gel: Assemble glass plates on a gel casting stand and load 4.5 mL of 15% separating gel. Carefully apply 500 μL of 0.1% SDS on top of the 15% gel and solidify for 45 min. Discard the top 0.1% SDS layer and add ~1 mL 4% stacking gel. Place an appropriate comb (*see* **Note 3**).

19. 4× Reducing Loading Buffer: 4.96 mL 0.5 M Tris–HCl, pH 6.8, 5.04 mL glycerol, 308 mg DTT, 800 mg SDS, 4 mg bromophenol blue, bring to a total volume of 10 mL with water.
20. Pro-Q™ Emerald 488 Glycoprotein Gel and Blot Stain Kit (Invitrogen) contains: CandyCane molecular weight standards, oxidizing solution, ProQ Emerald 488 Staining solution, and ProQ Emerald 488 reagent.
21. Glacial acetic acid.
22. ProQ Fixative Solution: 200 mL methanol, 20 mL glacial acetic acid, 180 mL water.
23. ProQ Wash Solution: 30 mL acetic acid, 970 mL water.
24. ImageQuant LAS 4000 mini Biomolecular Imager (GE Healthcare) with a green fluorescence filter (Cy2 Y515).
25. Standard LM and LAM from *M. tuberculosis* (Biodefense and Emerging Infections Research Resources Repository).
26. ImageQuant TL 8.1 (GE Healthcare).
27. MicroFlex MALDI-TOF (Bruker).
28. MSP 96 target polished steel BC plate (Bruker).
29. Sinapinic acid (Millipore Sigma).
30. Acetonitrile (Thermo Fisher).
31. Trifluoroacetic acid (TFA) (Thermo Fisher).
32. Matrix: 20 mg/mL sinapinic acid, 30% acetonitrile, and 1% TFA in water.

2.5 Analysis of PIMs

1. Water-saturated butanol: Mix 100 mL 1-butanol with 100 mL water by vigorous shaking. Once the solution is settled, the upper and lower phases are water-saturated butanol and butanol-saturated water, respectively.
2. TLC developing tanks.
3. HPTLC silica gel 60 (Millipore 5547).
4. TLC developing solvent (chloroform–methanol–13 M NH_3–1 M NH_4Ac–water, 180:140:9:9:23).
5. Orcinol spray reagent: dissolve 180 mg orcinol in 5 mL water. Once dissolved, add 75 mL ethanol. Place solution on ice and slowly add 10 mL of concentrated (~18 M) H_2SO_4. Store at 4 °C in the dark.
6. 50 mL Kimble™ Reagent Sprayers with Screw Thread Ground Joint.

2.6 Measuring TLC Band Intensity

1. Fiji (ImageJ, version 2.0.0-rc-43/1.51d) [16].

3 Methods

3.1 Extraction of PIMs/LM/LAM from Cultured Mycobacterial Cells

1. Using a heated inoculation loop, inoculate a loopful of a frozen stock of *Mycobacterium smegmatis* strain mc^{2}155 in 18 mL of Middlebrook 7H9 supplemented with 2 mL of DC, 50 μL of 20% Tween 80. Incubate at 30 °C with shaking at 135–180 rpm for 3–4 days (*see* **Note 4**).
2. Inoculate 1 mL of the starter culture from the **step 1** into 100 mL of complete Middlebrook 7H9 (90 mL 7H9 supplemented with 10 mL of DC and 250 μL of 20% Tween 80), and grow at 30 °C with shaking at 135–180 rpm until an OD_{600} between 0.6 and 1.0.
3. Harvest 50 OD units of cells (*see* **Note 5**) to a preweighed 50 mL conical tube, and centrifuge at 3220 × *g* on a tabletop centrifuge (e.g., Eppendorf 5810R) for 10 min at room temperature.
4. Discard the supernatant and aspirate the residual fluid as much as possible. Weigh the tube and calculate the pellet weight by subtracting the weight of the empty tube (*see* **Note 6**).
5. Add 20 volumes of chloroform–methanol (2:1) (*see* **Note 7**). Sonicate in a water bath sonicator for ~30 s or longer until the pellet is broken and the suspension is completely homogeneous (*see* **Note 8**). Incubate for 1–2 h at room temperature (or longer at −20 °C) and centrifuge as above. Decant the supernatant into a new 50 mL conical tube (*see* **Note 9**).
6. Resuspend the pellet in ten volumes of chloroform–methanol (2:1). Sonicate as above to break the pellet completely. Incubate for 1–2 h and centrifuge as above, decant supernatant into the same 50 mL tube as before.
7. Resuspend the pellet in ten volumes of chloroform–methanol–water (1:2:0.8). Sonicate for ~30 s to break the pellet completely. Incubate for 1–2 h and centrifuge as above, decant supernatant into the same 50 mL tube as before. This combined supernatant is used for further purification and analysis of PIMs (*see* Subheading 3.5).
8. Resuspend the pellet in four volumes of water, sonicate to break the pellets, and add four volumes of phenol (saturated solution, pH 6.6) (*see* **Note 10**). Incubate at 55 °C in a water bath for 2 h.
9. Vortex briefly and centrifuge for 5 min at 3220 × *g*. Transfer the entire sample to a 2-mL tube and centrifuge 1 min at 16,800 × *g* using a microcentrifuge (*see* **Note 11**). Transfer the upper aqueous phase to a new 2-mL tube.

10. Add four volumes of chloroform to the aqueous phase, vortex for 30 s, and centrifuge 1 min at 16,800 × *g*.
11. Transfer the upper aqueous phase to a 1.5-mL tube. Dry on a speed vac concentrator at 45 °C until the volume becomes just under 100 μL (*see* **Note 12**).
12. Adjust the volume to 2 volumes using water, sonicate briefly, and vortex about 10 s. The final concentration of LM/LAM is 0.5 g wet pellet equivalent/mL.
13. Add 10 mg/mL Proteinase K to the extracted LM/LAM to achieve a final concentration of 0.1 mg/mL. Incubate at 50 °C for 1 h. Save 9 μL as a starting material for analysis by SDS-PAGE and ProQ Emerald staining.

3.2 Metabolic Radiolabeling of PIMs/LM/LAM

Metabolic labeling is used to examine the biosynthetic activity of PIMs/LM/LAM. Both [^{3}H]mannose and [^{3}H]inositol are used for the detection of PIMs/LM/LAM biosynthesis [17, 18].

1. Using a heated inoculation loop, inoculate a loopful of a frozen stock of *Mycobacterium smegmatis* strain mc^{2}155 in 18 mL of Middlebrook 7H9 supplemented with 2 mL of DC, 50 μL of 20% Tween 80. Incubate at 30 °C with shaking at 135–180 rpm for 3–4 days.
2. Inoculate 5 mL of the starter culture from the **step 1** into 500 mL of complete Middlebrook 7H9 (450 mL 7H9 supplemented with 50 mL of DC and 1.25 mL of 20% Tween 80), and grow at 30 °C with shaking at 135–180 rpm until an OD_{600} between 0.6 and 1.0.
3. After 12–16 h, transfer 50 mL of culture (roughly 4 mg wet pellet per mL at OD_{600} = 1.0) into a preweighed 50-mL conical tube and spin at 3220 × *g* for 15 min at 30 °C.
4. Resuspend the pellet in 10 mL of 30 °C 7H9 with gentle pipetting. Spin again at the same condition. Decant supernatant, and weigh the pellet.
5. Resuspend the pellet at 0.2 g/mL in 7H9, and incubate for 15 min at 30 °C shaking at 135–180 rpm.
6. Transfer 500 μL (= 100 mg wet pellet) to a tube containing 25 μL of 1 mCi/mL [^{3}H]mannose or [^{3}H]inositol (final concentration 50 mCi/mL), and further incubate with shaking for 15 min.
7. For a pulse-chase experiment, add 5 mL of ice-cold 7H9 and spin at 3220 × *g* for 10 min. Decant the supernatant.
8. Add 5 mL of ice-cold 7H9 to resuspend the pellet and spin at the same condition again. Decant the supernatant.
9. Resuspend the pellet in 500 μL of ice-cold 7H9, immediately take 100 μL (= 20 mg wet pellet) of the sample and snap-

freeze. Transfer 380 μL into 19 mL of 7H9 prewarmed at 30 °C.

10. Supplement with 1 M mannose or 1 M inositol to achieve a final concentration of 1 mM and 5 mM respectively, and start the incubation at 30 °C.
11. At 0.5, 2, and 6 h, take 5 mL (= 20 mg wet pellet) of the sample from each culture and snap freeze.
12. Thaw the samples and proceed with lipid extraction described in Subheading 3.1.
13. For LM/LAM analysis: take 4 μL of the final LM/LAM extract for scintillation count and mix 6 μL with 2 μL of 4× reducing loading buffer, incubate at 100 °C for 5 min (*see* **Note 13**).
14. Incubate 0.5 μL of [methyl-^{14}C] methylated protein molecular weight markers mixed with 20 μL of 1× reducing loading buffer at 100 °C for 5 min.
15. Separate by SDS-PAGE as described in Subheading 3.5. Stain with Coomassie Blue and destain. Soak the gel in En^{3}Hance™ Autoradiography Enhancer solution for 30 min, and dry gel on a piece of Whatman paper at 70 °C for 2 h using a gel dryer. Expose the dried gel to a BioMAX MR film at −80 °C (*see* **Note 14**).
16. For PIM analysis: Take 10 μL for scintillation count, dry an appropriate amount of the butanol phase on speed-vac concentrator, and resuspend in 10 μL of chloroform–methanol–water (10:10:3). Vortex to resuspend pellet and centrifuge briefly to collect the lipid extract to the bottom of the tube.
17. Equilibrate a TLC tank with ~50 mL of TLC developing solvent.
18. On a 10-cm height HPTLC plate (Merck 5547), mark a straight line 1.5 cm from the bottom edge using a soft pencil. Then, on the straight line, mark lanes that are 0.5 cm wide with a 0.5 cm space between them.
19. Spot 10 μL of samples (1.5 μL per application, dry completely after each application) on a HPTLC plate and develop for ~45 min until the solvent front reaches to the top end of the HPTLC plate.
20. Dry the HPTLC plate completely (longer than 10 min), apply En^{3}Hance™ Spray Surface Autoradiography Enhancer and expose to a BioMAX MR film in an X ray film cassette at −80 °C.

3.3 Octyl-Sepharose Column Purification

Crude nonradiolabeled LM/LAM extracted above (Subheading 3.1) can be analyzed by SDS-PAGE and visualized by the Pro-Q™ Emerald 488 Glycoprotein Gel Stain Kit. However, further purification by hydrophobic column chromatography is recommended for cleaner results.

1. Briefly rinse with water the appropriate number of PolyPrep chromatography columns and place in the column stand.
2. Gently swirl the bottle of octyl-Sepharose CL-4B resin. Cap the bottom of the column and lay ~1 mL resin bed. Drain excess fluid by removing the cap.
3. Apply 1 mL of 50% propanol along the side of the column to the resin. Repeat this wash **step 4** times. Be sure to continue adding 50% propanol before the column starts to dry (*see* **Note 15**).
4. Apply 1 mL of 5% propanol in 0.1 M NH_4Ac along the side of the column to the resin. Repeat this **step 4** times (*see* **Note 16**).
5. Place a 15-mL conical tube beneath the column. Apply 2 mL of LM/LAM extract and collect flow-through (*see* **Note 17**).
6. Apply 1 mL 5% propanol in 0.1 M NH_4Ac for 4 times to the column, collecting the flow-through in the same conical tube. Save the flow-through at −20 °C for later analysis.
7. Place a new 15-mL conical tube beneath the column. Elute with 1 mL of 30% propanol twice.
8. Elute with 1 mL of 40% propanol twice, collecting the eluate in the same conical tube (*see* **Note 18**).
9. Elute with 1 mL of 50% propanol twice, collecting the eluate in the same conical tube.
10. Dry the combined eluate under a flow of nitrogen gas stream to 1–2 mL. Periodically during the evaporation, mix the eluate gently to facilitate homogeneous drying. Freeze the samples at −80 °C overnight (*see* **Note 19**).
11. Take off the cap of the conical tube containing samples and cover the opening of the tube with aluminum foil, and tighten the aluminum foil cover by a rubber band. Poke a few small holes into aluminum using a needle. Avoid thawing of the samples.
12. Following the manufacturer's instruction, freeze-dry the samples overnight. Resuspend in 100 μL of water.
13. Octyl-Sepharose columns are reusable. To regenerate the column, wash the resin by adding 4 mL of 50% propanol. Re-equilibrate the resin with 2 mL of 5% propanol in 0.1 M NH_4Ac. Firmly place the cap on the bottom of the column. Add enough (5–10 mL) 5% propanol in 0.1 M NH_4Ac so that the resin will not dry during storage. Place the cap on the top of the column, and tightly wrap both top and bottom caps with a piece of Parafilm.

3.4 LM/LAM Analysis

For LM/LAM visualization, we use Pro-Q™ Emerald 488 Glycoprotein Gel and Blot Stain Kit. We follow the manufacturer's protocol, which is only briefly described below.

1. Aliquot 9 μL of LM/LAM extracts, mix with 3 μL of 4× Reducing Loading Buffer, and boil for 5 min.
2. For a standard curve, use increasing concentrations of standard LM or LAM generally ranging between 1 and 10 μg (*see* **Note 20**).
3. CandyCane molecular weight standards: Dilute 2 μL of the CandyCane standards with 4 μL of water and 2 μL of 4× Reducing Loading Buffer. Boil for 5 min.
4. Aliquot 10 μL of Precision Plus Protein Kaleidoscope Standards into a microtube, and boil for 5 min (*see* **Note 21**).
5. Load samples onto 15% SDS-PAGE and run at constant 110 V until the dye front is 1 cm above the bottom of the gel (~80 min) (*see* **Note 22**).
6. Fix the gel in ~100 mL of fixative solution in a clean plastic container with gentle agitation for 1 h.
7. Replace with ~100 mL of fresh fixative solution in the same plastic container and continue the incubation with gentle agitation overnight (*see* **Note 23**).
8. Wash twice in ~100 mL of Wash Solution for 20 min each with gentle agitation.
9. Incubate the gel in 20 mL of oxidizing solution with gentle agitation for precisely 20 min (*see* **Note 24**).
10. Wash four times in ~100 mL of Wash Solution for 20 min each with gentle agitation. Bring one vial of Pro-Q Emerald 488 reagent to room temperature before opening. Add 0.5 mL of DMSO to the vial and mix thoroughly. Always use the stock solution within a few hours.
11. Dilute 0.5 mL of stock solution into 25 mL of the Pro-Q Emerald 488 staining buffer. Incubate gels in 25 mL of Pro-Q Emerald 488 Staining Solution for 2 h (or up to overnight) in the dark at room temperature.
12. Wash three times with 100 mL of Wash Solution at room temperature for 20 min with gentle agitation.
13. Visualize using the ImageQuant LAS 4000 mini Biomolecular Imager with the fluorescence mode using a Cy2 Y515 filter.
14. ImageQuant analysis software can be used to quantify the band intensity. Use the empty area of the gel for the background subtraction. The standard curve can be made using LM/LAM standards, and used for the determination of LM/LAM concentrations in the samples (Fig. 1) (*see* **Note 25**).

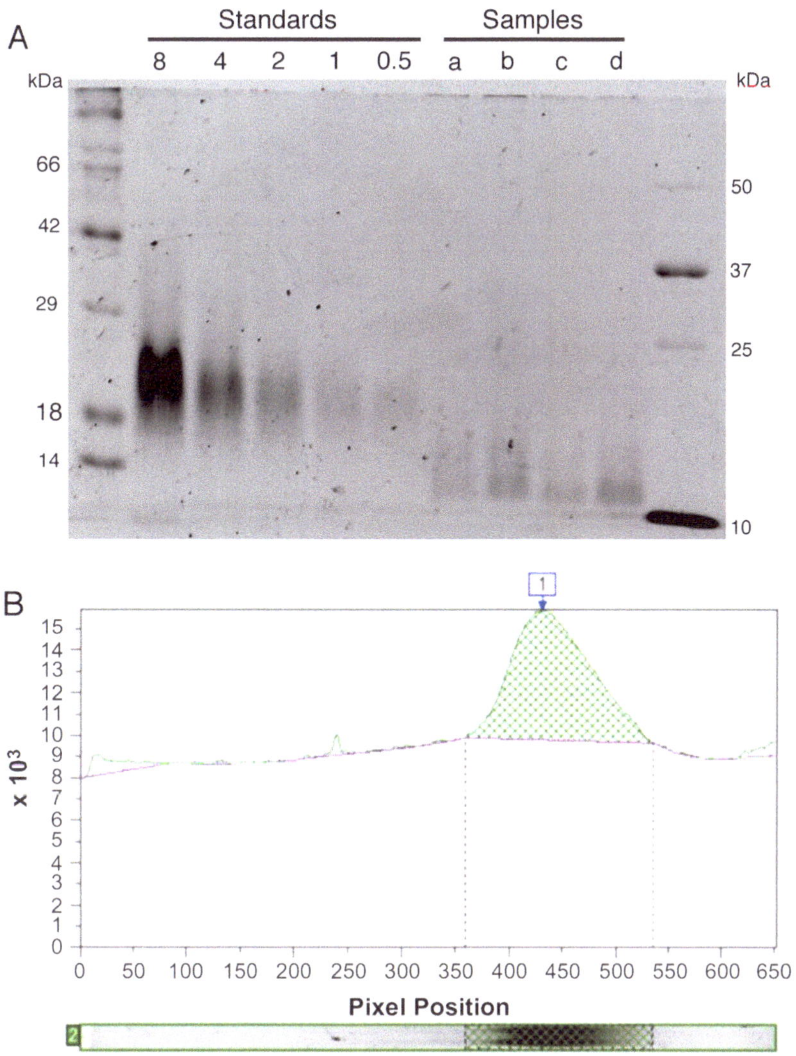

Fig. 1 Calibration of a truncated lipomannan intermediate accumulating in the *mptA* depletion strain. (**a**) Concentrations (μg) of *M. tuberculosis* LM are indicated above. Samples a, b, c, and d are truncated LM purified from the *mptA* depletion strain. The molecular weight standard on the left of the gel is CandyCane and the standard on the right of the gel is Precision Plus Protein Kaleidoscope prestained protein standards. (**b**) ImageQuant analysis of the intensity of the band in lane 2 (i.e., 8 μg of *Mycobacterium tuberculosis* LM)

3.5 Analysis of PIMs

1. Dry 800 μL of the combined organic solvents (~20 mg pellet equivalent) in a microtube under a N_2 stream until the sample is completely dry.
2. Resuspend in 100 μL water and 200 μL of water-saturated butanol. Vortex for 15 s and centrifuge for 1.5 min at 16,800 × *g* at room temperature using a microcentrifuge, transfer the upper butanol phase to a new tube (*see* **Note 26**).

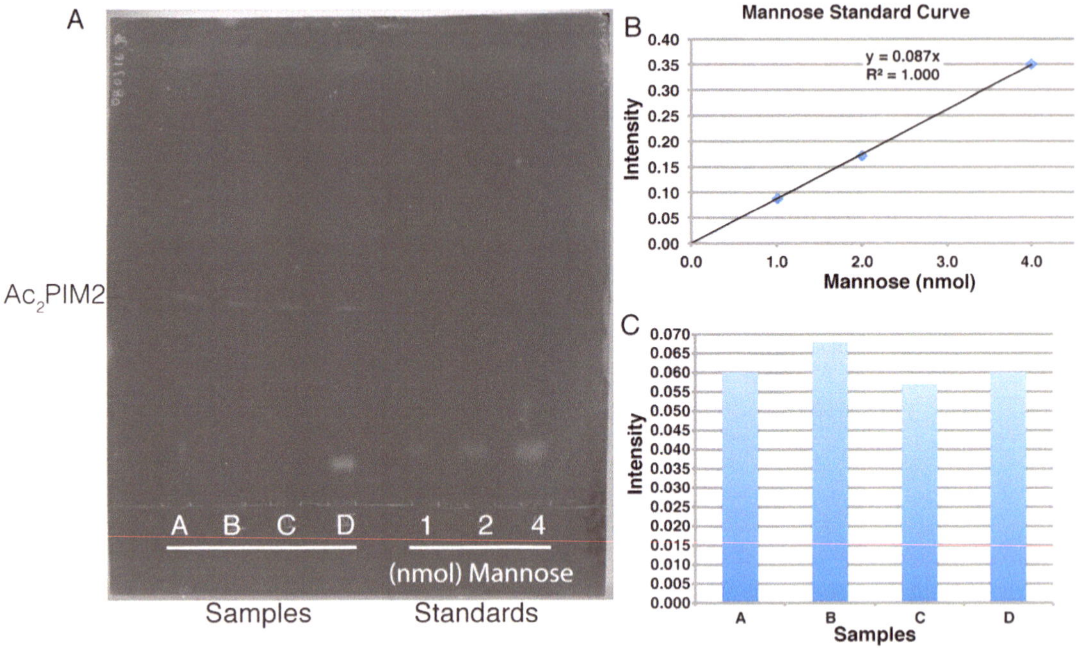

Fig. 2 Quantification of PIMs. (**a**) Inverted grayscale image of an orcinol staining of PIMs from various lipid extracts, A, B, C, and D and mannose standards at 1, 2, and 4 nmol. (**b**) Standard curve of mannose calculated from Fiji. The slope is calculated and used to determine the concentration of Ac_2PIM2 from the intensity. (**c**) The intensity of Ac_2PIM2 of IMD samples A, B, C, and D calculated from Fiji

3. Mix 100 μL of water-saturated butanol with the remaining lower aqueous phase. Vortex for ~15 s, centrifuge as before, transfer the upper butanol phase to the same tube.
4. As an additional cleanup step, back-extract the combined butanol phase with 100 μL of water. Vortex and centrifuge as before, and transfer the upper butanol phase to a new tube (*see* **Note 27**).
5. To concentrate the sample, dry the upper butanol phase using a speed-vac concentrator, and resuspend in 20 μL of water-saturated butanol (1 mg pellet equivalent per μL). Briefly sonicate and vortex to resuspend the pellet.
6. Equilibrate a TLC tank with 50 mL of TLC developing solvent.
7. Prepare a 10-cm height HPTLC plate as described in Subheading 3.2 (also *see* Fig. 2 for the format).
8. Spot 1.5 μL of sample over a 0.5-cm lane on the plate, let the plate completely dry and then spot an additional 1.5 μL of sample over the first. This should be equivalent to 3.0 mg wet pellet of samples. Let the plate completely dry and then develop for ~45 min or until the solvent front reaches the top of the plate.

9. A known standard can be run to determine the concentration of each species, such as mannose at concentrations ranging between 0.5 and 10 μg.
10. Completely dry the plate (longer than 10 min).
11. Spray with the orcinol reagent four times, turning the plate 90° between each spray (*see* **Note 28**).
12. Let the plate completely dry and then bake at 100 °C for 5 min or until bands develop. Scan the plate, save as a tiff file with the resolution of 300 dpi, and analyze by Fiji.

3.6 Measuring TLC Band Intensity

1. Open the scanned tiff image with Fiji.
2. Convert image to grayscale and invert:
   (a) "Image" → "Type" → "16-bit."
   (b) "Edit" → "Invert."
3. Select a band using the rectangular selection tool in the Main Window and open a new "ROI Manager" window by pressing the keyboard shortcut "T."
4. Check "Show all" and "Labels" in the "ROI Manager" window.
5. Click on the first selection in the ROI Manager window and press "Add [t]"—another rectangular selection box should appear; if not, click "Update" and "Add [t]" again.
6. Drag the new rectangular selection box to a new location for background measurement.
7. Repeat "Add [t]" to select all bands and corresponding background areas. For creating a standard curve, all selection boxes should be the same size.
8. Select: "Analyze" → "Set Measurements" → select "Area" and "Integrated density."
9. Hold the "Shift" or "Command" key down, and select all ROIs listed on the "ROI Manager" window. Click "Measure"—a new "Results" window will pop up with the measurement values.
10. Copy and paste "IntDen" values into a Microsoft Excel spreadsheet to create a standard curve.
11. After selecting and measuring the mannose standards, move on to selecting your bands of interest (Fig. 2).

4 Notes

1. The organic phenol is beneath the top aqueous buffer layer, which protects phenol from oxidation.

2. Make APS fresh.
3. Gels may be stored at 4 °C in wet paper towels in a plastic bag for a few days.
4. Culture growth of pathogenic *M. tuberculosis* must be conducted in a Biosafety Level 3 facility following a protocol approved by the Institutional Biosafety Committee. The growth medium is supplemented with Middlebrook ADC or OADC Growth Supplement instead of the DC supplement, and the cells are grown at 37 °C.
5. To calculate OD_{600} units, multiply the OD_{600} value of the culture by the volume of the culture. For example, 62.5 mL of a culture with the OD_{600} value of 0.80 would correspond to 50 OD_{600} units. Note that some mutants of mycobacteria form clumps, which prevent accurate OD measurements. While OD_{600} units is a convenient and reproducible reference for culture sampling, we rely on the wet cell pellet weight for the subsequence lipid extraction procedures.
6. Excess supernatant will make the calculation of the wet pellet weight inaccurate. It is therefore important to carefully remove as much supernatant as possible.
7. Here and also for the rest of the steps, solvent additions are based on the wet pellet weight. If the pellet weight is 0.2 g, then 20 volumes of chloroform–methanol (2:1) would be 4 mL, assuming the wet pellet density of 1 g/mL.
8. Manually position the tube to hotspots of sonic vibration in the water bath so that the sonication is maximally applied to the cell pellet inside the tube. Wear an ear protector during sonication.
9. If the pellet is loose, centrifuge for an extra 10 min. Additionally, if the pellet is accidentally transferred to the supernatant collection tube, centrifuge the collection tube to remove the residual pellet. The residual pellet can be combined back to the original pellet for the next lipid extraction step.
10. The pellet at this step is often difficult to resuspend. Be sure to resuspend the pellet in water first before adding phenol.
11. The separation of the aqueous phase from the phenol phase is technically difficult in a 50 mL conical tube. Therefore, first transfer the entire content of the tube to either a 2 mL tube or, if the volume is greater than 2 mL, a 15 mL conical tube. Increasing the centrifugation time or speed will help reducing the thickness of the interface, and may help collecting more upper aqueous phase. Avoid the contamination of the interphase as much as possible. However, a small amount of the contaminating lower phase or the interphase can be cleaned by the subsequent chloroform wash. Therefore, attempt to collect as much upper phase as possible.

12. This drying step may take 2 h or longer. Avoid drying LM/LAM completely, as it becomes more difficult to resuspend. We recommend to dry until the volume is reduced to 50–100 μL, measure the volume of the remaining fluid using a Pipetman, and adjust the volume to two volumes of the wet pellet weight.
13. We usually do not purify radiolabeled LM/LAM.
14. The exposure may vary from a few days to a few months depending on the radioactivity count.
15. Do not let the column run dry. A dry column will interfere with elution of the sample.
16. Be sure not to disrupt the settled resin. Add slowly along the side of the column.
17. To enhance LM/LAM binding to the column, stop the flow of the column and let the sample sit for 5 min.
18. The increase of propanol concentration decreases the flow rate.
19. The sample must be completely frozen.
20. This is the total amount per lane loaded onto the gel. Adjust the loading volume to be between 10 and 15 μL for the consistency of loading.
21. Precision Plus Kaleidoscope standard bands at 10, 25, 37, and 50 kDa can be visualize by ProQ Emerald staining.
22. The percentage of the gel can be increased to 18% for better resolution of smaller bands and the runtime can take up to 4 h at 100 V.
23. In our experience, this step can be extended to 2 days.
24. The timing on this step is important. Follow the manufacturer's guideline for details.
25. To determine the precise size distribution of LM, purified LM/LAM extract can be analyzed using MALDI-TOF mass spectroscopy. For the matrix, prepare a solution containing 20 mg/mL sinapinic acid, 30% acetonitrile, 1% TFA in water. Mix 1 μL of 1 μg/μL LM solution with 1 μL of the matrix and load onto the target plate. Let the sample dry completely. Analyze using MicroFlex MALDI-TOF instrument with linear/positive mode, m/z range from 5000 to 20,000, an accelerating voltage of 20,000 V, grid voltage at 95%, guide wire voltage at 0.1%, laser intensity between 70 and 85%, and delay time 500–800 ns. Analysis of mass spectrum is performed using the Flex Analysis software. Changing the ratio of matrix to sample, concentration of TFA, or acetonitrile may enhance the likelihood of detection. Under this condition, sinapinic acid may form an adduct with LM. Increasing the delay time and setting the guide wire voltage higher may improve

resolution of peaks for glycolipids. For truncated LM, we set the laser intensity to 74%.

26. Avoid the contamination of the lower aqueous phase. We generally leave ~1 mm thickness of butanol phase above the interface.
27. For the maximal recovery of hydrophilic glycolipids, we repeat the extraction from the same aqueous phase up to four times, and avoid the water back-extraction.
28. Place the TLC plate into a cardboard box and spray gently from 30 to 40 cm away. Spray in a circular motion, or up and down across the plate to achieve an even spraying.

Acknowledgments

This work was supported by a Biomedical Research Grant (RG-414805) from the American Lung Association and a Research Grant from the Pittsfield Anti-Tuberculosis Association to YSM.

References

1. Mishra AK, Driessen NN, Appelmelk BJ et al (2011) Lipoarabinomannan and related glycoconjugates: structure, biogenesis and role in *Mycobacterium tuberculosis* physiology and host-pathogen interaction. FEMS Microbiol Rev 35:1126–1157
2. Oldenburg R, Demangel C (2017) Pathogenic and immunosuppressive properties of mycobacterial phenolic glycolipids. Biochimie 141:3–8
3. Ishikawa E, Mori D, Yamasaki S (2017) Recognition of mycobacterial lipids by immune receptors. Trends Immunol 38:66–76
4. Källenius G, Correia-Neves M, Buteme H et al (2016) Lipoarabinomannan, and its related glycolipids, induce divergent and opposing immune responses to *Mycobacterium tuberculosis* depending on structural diversity and experimental variations. Tuberculosis 96:120–130
5. Achkar JM, Lawn SD, Moosa MY et al (2011) Adjunctive tests for diagnosis of tuberculosis: serology, ELISPOT for site-specific lymphocytes, urinary lipoarabinomannan, string test, and fine needle aspiration. J Infect Dis 204 (Suppl 4):S1130–S1141
6. Sabur NF, Esmail A, Brar MS et al (2017) Diagnosing tuberculosis in hospitalized HIV-infected individuals who cannot produce sputum: is urine lipoarabinomannan testing the answer? BMC Infect Dis 17:803
7. LaCourse SM, Cranmer LM, Njuguna IN et al (2018) Urine tuberculosis lipoarabinomannan predicts mortality in hospitalized human immunodeficiency virus-infected children. Clin Infect Dis 66:1798–1801
8. Shah M, Hanrahan C, Wang ZY et al (2016) Lateral flow urine lipoarabinomannan assay for detecting active tuberculosis in HIV-positive adults. Cochrane Database Syst Rev 19: CD011420
9. Ballou CE, Vilkas E, Lederer E (1963) Structural studies on the myo-inositol phospholipids of *Mycobacterium tuberculosis* (var. bovis, strain BCG). J Biol Chem 238:69–76
10. Lee YC, Ballou CE (1964) Structural studies on the *myo*-inositol mannodides from the glycolipids of *Mycobacterium tuberculosis* and *Mycobacterium phlei*. J Biol Chem 239:1316–1327
11. Morita YS, Fukuda T, Sena CBC et al (2011) Inositol lipid metabolism in mycobacteria: biosynthesis and regulatory mechanisms. Biochim Biophys Acta 1810:630–641
12. Sancho-Vaello E, Albesa-Jové D, Rodrigo-Unzueta A et al (2017) Structural basis of phosphatidyl-*myo*-inositol mannosides biosynthesis in mycobacteria. Biochim Biophys Acta 1862:1355–1367
13. Jankute M, Cox JAG, Harrison J et al (2015) Assembly of the mycobacterial cell wall. Annu Rev Microbiol 69:405–423

14. Daffé M, Crick DC, Jackson M (2014) Genetics of capsular polysaccharides and cell envelope (glyco)lipids. Microbiol Spectr 2. MGM2–0021–2013
15. Angala SK, Belardinelli JM, Huc-Claustre E et al (2014) The cell envelope glycoconjugates of *Mycobacterium tuberculosis*. Crit Rev Biochem Mol Biol:1–39
16. Schindelin J, Arganda-Carreras I, Frise E et al (2012) Fiji: an open-source platform for biological-image analysis. Nat Methods 9:676–682
17. Patterson JH, Waller RF, Jeevarajah D et al (2003) Mannose metabolism is required for mycobacterial growth. Biochem J 372:77–86
18. Haites RE, Morita YS, McConville MJ et al (2005) Function of phosphatidylinositol in mycobacteria. J Biol Chem 280:10981–10987

Chapter 7

Conjugation of Synthetic Oligosaccharides to Proteins by Squaric Acid Chemistry

Hélène B. Pfister, Xiaowei Lu, Sameh E. Soliman, and Pavol Kováč

Abstract

Oligosaccharides equipped with amine-containing linkers can be conjugated to carrier proteins using squaric acid chemistry. In a two-step process, a squarate derivative of such oligosaccharide is formed first, which is followed by its reaction with a protein carrier. Monitoring of the conjugation reaction is achieved by SELDI-TOF-MS or MALDI-TOF-MS. This experimentally simple procedure yields desired glycoconjugates in high yields and with reproducible hapten–protein ratios.

Key words Oligosaccharides, Neoglycoconjugates, Squaric acid chemistry, Conjugation, Vaccines

1 Introduction

The coupling of two amines by squaric acid chemistry was introduced by Tietze's group in 1991 [1] and was since successfully used in glycoconjugate vaccine development [2–6]. The method relies on the sequential reaction of squaric acid diesters **1** with amines (Fig. 1). At pH 7, the squaric acid monoester monoamide **2** is formed. The second amidation occurs only at higher pH (e.g., pH 9) to form the squaric acid diamide **3**. When the first amine is an amine-equipped carbohydrate and the second amine is a protein, a glycoprotein (glycoconjugate) is obtained. This conjugation method takes place at relatively mild conditions (no oxidative or reductive conditions are necessary) and is selective for amino groups (alcohols and phenols do not react). The conjugation method based on squaric acid chemistry is reliable to obtain glycoconjugates with a predetermined and reproducible hapten–protein ratio [7, 8].

Using squaric acid chemistry, glycoconjugates can be prepared step-wise, with purification of the intermediate **2** [7, 8], or in a one-pot manner [9, 10]. Herein, as an example, we describe the step-wise conjugation of hexasaccharide **4** (Fig. 2) [11], the

Inka Brockhausen (ed.), *Bacterial Polysaccharides: Methods and Protocols*, Methods in Molecular Biology, vol. 1954,
https://doi.org/10.1007/978-1-4939-9154-9_7,

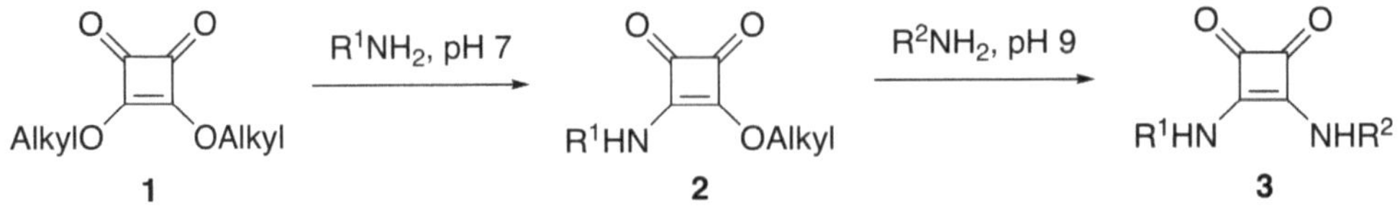

Fig. 1 General scheme for the sequential amidation of squarates. Reaction of squaric diester **1** with an amine at pH 7 gives the squaric monoester monoamide **2**. The second amidation at pH 9 gives the squaric acid diamide **3**

complete [12, 13] O-specific polysaccharide of *Vibrio cholerae* O139, with Bovine Serum Albumin (BSA). First, reaction of hexasaccharide **4** with squaric acid dimethyl ester **5** gave squarate **6**. Reaction of the latter with BSA at pH 9 gave the desired neoglycoconjugate with a predictable carbohydrate–protein ratio (loading).

Historically, the main obstacle to the use of glycoconjugates in the life sciences was the unavailability of methods for monitoring the progress of the conjugation reaction. Our finding [14] that the progress of conjugation can be monitored in near-real time by surface-enhanced laser desorption/ionization-TOF-MS (SELDI-TOF-MS) brought the conjugation of carbohydrates to proteins to a new, higher level of sophistication. Making glycoconjugate had ceased to be a hit or miss situation. In the experiment described below, the predetermined loading was ~5 mol of hapten per BSA. It was achieved without monitoring the reaction because the revised protocol applied herein resulted from detailed studies [2, 8] which included frequent monitoring of the conjugation process. The cited authors showed that, with oligosaccharides, efficiency of conjugation is ~80%, and that the conjugation reaction is virtually complete after ~16 h. While SELDI-TOF-MS is still an invaluable tool for those who have access to such equipment, the SELDI-TOF instrument is no longer an indispensable tool. With the invention of reversed-phase chromatography in a pipette tip for desalting [15], access to the costly SELDI-TOF instrument, while still immensely helpful, is not absolutely necessary, and the same results can be obtained with less expense using a more conventional MALDI-TOF instrument.

2 Materials

2.1 Buffers

1. 0.50 M Phosphate Buffer, pH 7.0: Weigh 13.6 g potassium phosphate monobasic and 21.0 g sodium phosphate dibasic. Transfer to a 500 mL graduated cylinder. Add water to a volume of 450 mL. Adjust pH with solid KOH to pH 7. Make up to 500 mL, transfer the solution to a screw-cap bottle, and store it at room temperature.

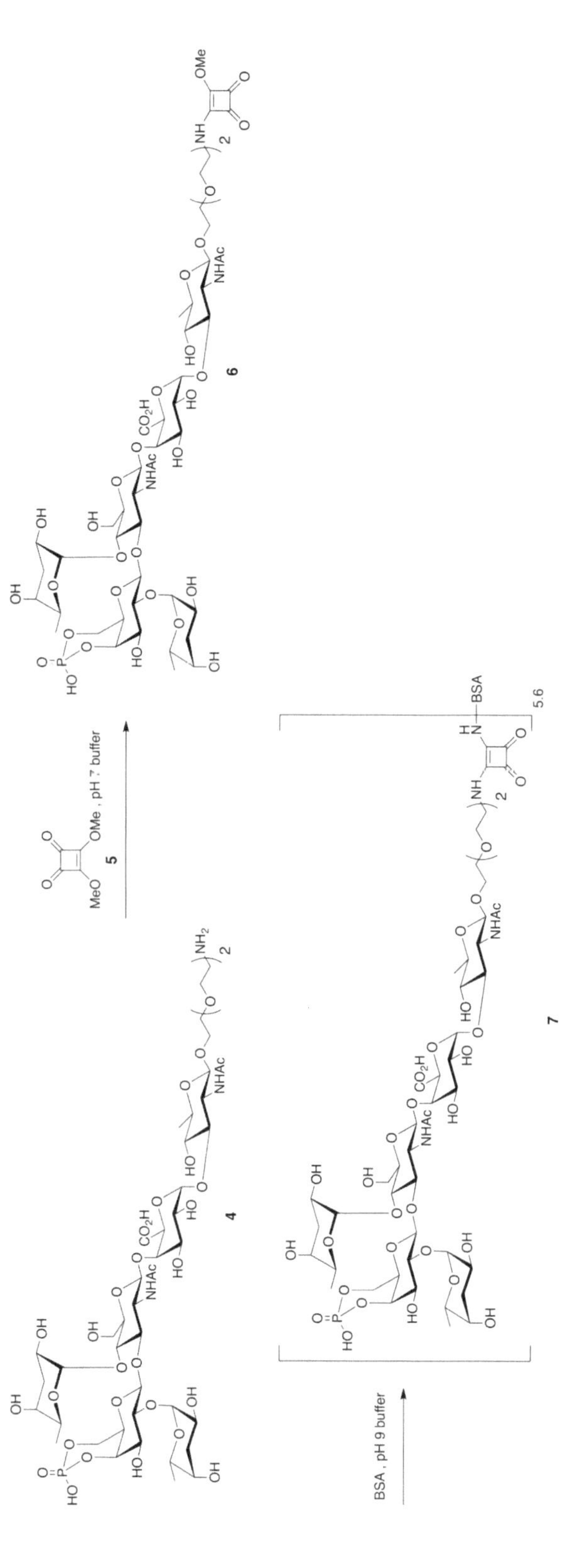

Fig. 2 Conjugation of hexasaccharide **4** to BSA by squaric acid chemistry. Reaction of the amine **4** with squaric dimethyl ester **5** gives squarate **6**. Conjugation with BSA yields conjugate **7**

2. Borax Buffer pH 9.0: 0.50 M boric acid, 0.35 M KCl, and 0.15 M KOH. Weigh 30.9 g boric acid and 26.1 g KCl. Transfer to a 1 L graduated cylinder. Add water to a volume of 900 mL. Weigh 8.42 g KOH and add portion-wise to the solution while stirring. Adjust pH to 9.0 using solid KOH. Make up to 1 L with water. Transfer to a screw-cap bottle and store at room temperature.
3. 10 mM aq ammonium carbonate: Dissolve 960 mg of ammonium carbonate in 1 L water. Store at room temperature.
4. 10% aq TFA solution: Add 1.0 mL TFA to 9.0 mL water in a vial. Cap the vial using a Teflon-lined cap and store at room temperature.

2.2 Reagents

1. 3,4-Dimethoxy-3-cyclobutene-1,2-dione (squaric acid dimethyl ester) **5**. Store at 4 °C
2. Bovine Serum Albumin (BSA, Sigma-Aldrich). Store at 4 °C.
3. Sinapinic acid. Store at 4 °C.
4. Trifluoroacetic acid (TFA).
5. Acetonitrile (ACN).

2.3 Chromatography

1. Silica gel 60-coated TLC plates with inorganic binder and fluorescent indicator.
2. Solution for charring TLC plates: H_2SO_4 in EtOH (5% v/v). To a graduated beaker in an ice bath, add EtOH to a volume of 950 mL. Slowly add with stirring 50 mL H_2SO_4. Transfer to a Teflon-lined screw-cap bottle and store at room temperature.
3. P-2 size exclusion chromatography: P-2 polyacrylamide gel, fine (Bio-Rad). Column size: 10 × 850 mm. Bed volume ~20 mL.

2.4 SELDI-TOF

1. An appropriate mass spectrometer such as the Bio-Rad Protein Chip SELDI System.
2. NP-20 chip arrays.
3. 1% aq TFA solution: Add slowly 100 μL TFA to cold water (10 mL). Transfer to a Teflon-lined, screw-cap bottle and store at room temperature.
4. Sinapinic acid solution: Add 5 mg sinapinic acid to ACN–1% aq TFA (1:1, 200 μL). Vortex the suspension for 1 min and allow to settle for 5 min. Use the supernatant. Make fresh daily and store at room temperature.

2.5 MALDI-TOF

1. An appropriate mass spectrometer such as the Bruker Autoflex III.

2. Stainless steel MALDI-TOF MS plate appropriate for the instrument.
3. C4 ZipTips (such as Merck Millipore).
4. ZipTip activation solution: ACN–H_2O–TFA 50:50:0.1, v/v. Mix 500 μL ACN, 490 μL water, and 10 μL 10% aq TFA. Make fresh daily and store at room temperature.
5. ZipTip washing solution: 0.1% aq TFA. Add and 10 μL 10% aq TFA to 990 μL water. Make fresh daily and store at room temperature.
6. ZipTip elution solution: ACN–H_2O–TFA 75:25:0.1, v/v. Mix 750 μL ACN, 240 μL water, and 10 μL 10% aq TFA. Make fresh daily and store at room temperature.
7. Sample reconstitution solution: ACN–H_2O–TFA 5:95:0.1, v/v. Mix 50 μL ACN, 940 μL water, and 10 μL 10% aq TFA. Make fresh daily and store at room temperature.
8. Sinapinic acid solution: as in Subheading 2.4, **item 4**.

2.6 Ultrafiltration

1. An appropriate centrifuge capable of handling standard centrifugal devices at 7500 rcf such as Eppendorff centrifuge 5804 R.
2. Centrifugal filtering device 30 kDa cut-off. Volume: 4 mL (such as Merck Millipore Amicon Ultra 4).

3 Methods

Carry out all operations at room temperature, unless otherwise specified.

3.1 Squarate Formation

1. Weigh the amine-equipped oligosaccharide **4** (Fig. 2) (5.0 mg, 4.1 μmol) into a 2 mL V-shaped vial, equipped with a V-shaped, Teflon-coated magnetic stir bar. Weigh 3,4-dimethoxy-3-cyclobutene-1,2-dione **5** (5.9 mg, 41 μmol, 10 equiv., *see* **Note 1**) and transfer crystals into the 2 mL V-shaped oligosaccharide-containing vial. Add 1.0 mL of 0.5 M phosphate buffer, pH 7, with slow stirring. The dissolution of squarate **5** is slow. Cap the vial and stir overnight (14 h, *see* **Note 2**).
2. The next morning, check the reaction progress by TLC, using an appropriate solvent system (*see* **Note 3**). When the reaction is complete, proceed to **step 3**. If the conversion of the amine is not complete, *see* **Note 4**).
3. Directly load the reaction mixture onto the P-2 size exclusion column. Rinse the vial twice (≈0.5 mL each) and load onto the P-2 column (Flow rate: 12 mL/h. 3 mL fractions collected). Check for the presence of carbohydrate in the eluate by

spotting a small amount of each fraction onto a TLC plate and charring. Pool fractions containing the carbohydrate, concentrate to a small volume and freeze-dry (*see* **Note 5**).

4. The identity and purity of squarate **6** (4.90 mg, 3.7 μmol, 90%) can be checked by ^{1}H and/or ^{13}C NMR spectroscopy, and MS (*see* **Note 6**).

3.2 Conjugation

1. In a 0.5 mL V-shaped vial (*see* **Note** 7), weigh squarate **6** (0.60 mg, 0.46 μmol, 6.0 equiv., *see* **Notes 8–10**) and BSA (5.10 mg, 0.077 μmol, 1.0 equiv). Add 115 μL of borax buffer, pH 9.0. After stirring for a few minutes, a clear solution is obtained. Cap the vial. Stir the reaction mixture overnight (14 h).

3.3 Verify the Progress of the Reaction Using SELDI-TOF-MS

1. Apply 1 μL of the reaction mixture onto one of the empty spots of the NP-20 chip (*see* **Note 11**) and air-dry for 5 min.
2. Wash the spot by pulsing 10 times 5 μL water on the edge of the spot. Air-dry after last wash for 5 min (*see* **Note 12**).
3. Apply 2 μL of the sinapinic acid solution (matrix) onto the dry sample. Gently stir the matrix drop with the pipette tip and, when crystals start forming, air-dry for 5 min (*see* **Note 13**).
4. Record the SELDI spectra using parameters recommended by the manufacturer (*see* **Notes 14** and **15**). If the desired loading is achieved, go to Subheading 3.5. If not, *see* **Note 16**.

3.4 Verify the Progress of the Reaction Using MALDI-TOF-MS

1. Prepare a vial for waste and two small Eppendorf tubes. The first one should contain 14 μL of reconstitution solution. The second should be empty and available for addition of 1.6 μL of elution solution, just before elution from the ZipTip (**step 5**). Add 1 μL of the reaction mixture to the first Eppendorf tube containing 14 μL of reconstitution solution (*see* **Note 11**).
2. Apply a C4 ZipTip to an air displacement pipette set to 10 μL (*see* **Note 17**). Prepare another air displacement pipette set to 1.6 μL. Activate the ZipTip: Pipet two times 10 μL of activation solution, then discard (*see* **Note 18**). Pipet two times 10 μL of washing solution, then discard.
3. Load the sample onto the Ziptip: Slowly pulse 15 times 10 μL of the sample in the reconstitution solution (from **step 1**), aspirate then expel back into the Eppendorf tube. After the last pulse, the remaining liquid should be just slightly above the surface of the C4 layer of the tip.
4. Wash the Ziptip: Pipet and discard 10 μL of the washing solution. Repeat eight times. Leave the last wash solution in the ZipTip until the elution solution is prepared (next step).

5. Eluting the sample: Place 1.6 μL of the elution solution in a small Eppendorf tube (*see* **Note 19**). Discard the last wash solution from the ZipTip. Pulse the elution solution through the ZipTip eight times. Be careful to not introduce air bubbles. After the last pulse, expel all the liquid into the small Eppendorf tube.
6. Apply the elution solution onto the stainless-steel plate. Air-dry for 5 min. Apply 2 μL of the sinapinic acid solution (matrix) onto the dry sample. Gently stir the matrix drop with the pipette tip and, when crystals start forming, air-dry for 5 min (*see* **Note 13**).
7. Record the MALDI spectra using parameters recommended by the manufacturer (*see* **Notes 14** and **15**). If the desired loading is achieved, go to Subheading 3.5. If not, *see* **Note 16**.

3.5 Purification of the Conjugate

1. Purification of the conjugate by ultrafiltration: Wet the membrane of the centrifugal filtering device with 1 mL of 10 mM aq ammonium carbonate. Transfer the conjugation mixture from the reaction vial into the centrifugal tube.
2. Wash the vial with 10 mM aq ammonium carbonate and transfer washings to the centrifugal tube. Make-up the volume of the top chamber of the centrifugal tube to 4 mL.
3. Centrifuge at 7500 rcf for 7 min at 4 °C to bring the volume of the retentate to ≈200 μL. Discard the permeate and make up the volume in the top chamber to 4 mL with 10 mM aq ammonium carbonate. Repeat the operation seven times.
4. Transfer the retentate into a vial for subsequent freeze-drying. Wash the membrane three times by adding water (1 mL each) and gentle vortexing. Combine the washings with the bulk and freeze-dry.
5. Run a MALDI-TOF or SELDI-TOF analysis for the freeze-dried material as follows: dissolve a very small portion of conjugate 7 (Fig. 2) in a small amount of water (e.g., a barely visible speck of the conjugate dissolved in ~5 μL water).

   Apply 1 μL of the conjugate solution onto the MALDI-TOF plate or SELDI chip (*see* **Note 11**). Air-dry for 5 min.
6. Apply on the plate or chip 2 μL of the sinapinic acid solution (matrix). Gently stir the matrix drop with the pipette tip until crystals start forming. Air-dry for 5 min (*see* **Notes 13** and **20**).
7. Record the MALDI or SELDI spectra using parameters recommended by the manufacturer (Fig. 3).

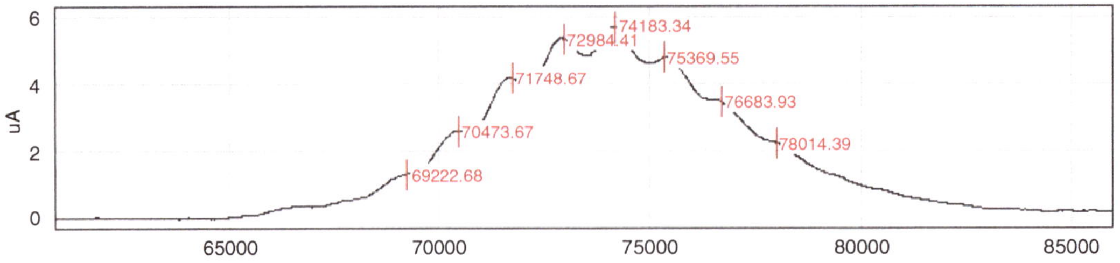

Fig. 3 SELDI-TOF spectra of the purified glycoconjugate **7**

4 Notes

1. Although other squaric acid dialkyl ester reagents are commercially available, we find reagent **5** especially convenient to handle because of its crystalline nature [8, 9].
2. When working with shorter oligosaccharides (disaccharides and trisaccharides), the reaction times might decrease (e.g., to 5–6 h).
3. The squaric acid dimethyl ester **5** is visible under UV light (254 nm) but does not char with dilute sulfuric acid.
4. The squarate formation is usually finished overnight. Check the reaction periodically by TLC, and add more squarate if necessary (~2–4 equiv). When the reaction is complete, proceed to **step 3**.
5. Although slow, the hydrolysis of squaric acid monoester monoamide **6** to its corresponding acid occurs at pH 7. Therefore, reaction time for the formation of squarates should be kept to a minimum, and solutions of monoamides should be concentrated without delay.
6. Figure 4 shows a comparison of the 1H NMR spectra of amine **4** and squarate **6**. In the ^{1}H and ^{13}C NMR spectra, the peaks corresponding to the atoms of, or close to, the squarate ring are split due to their vinylic nature [1].
7. V-shaped vials and stir bars are optimal for small reaction volumes, in order to minimize the loss of reagents on the walls of the vial.
8. Because of the small quantity of squarate **6** and BSA, the latter being electrostatic, we recommend use of an antistatic gun and weighing both solids directly into the reaction vial.
9. To avoid weighing errors when working with small quantities, it is preferred to prepare a concentrated stock solution of the squarate **6**, and add an aliquot to the solution of BSA in buffer, while maintaining the intended final hapten concentration.

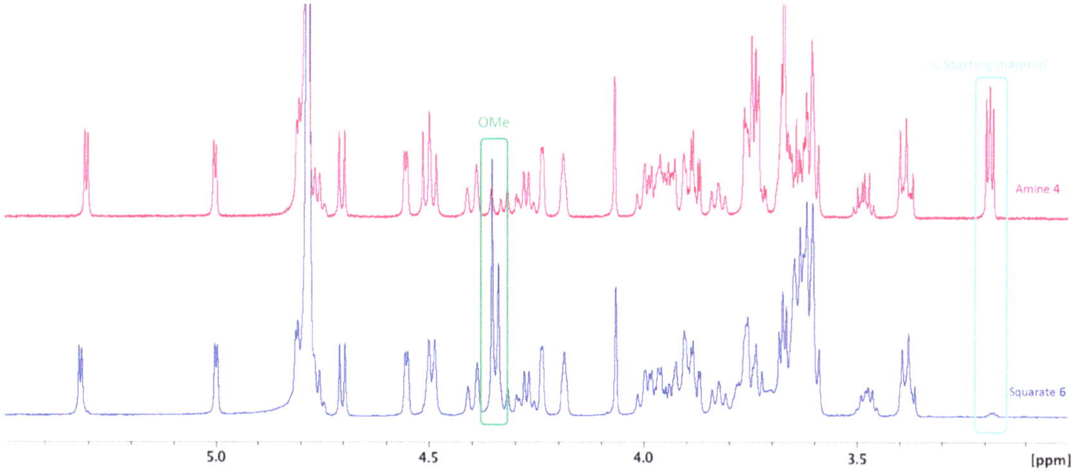

Fig. 4 ^{1}H NMR of amine **4** (in red) and squarate **6** (in blue)

The excess stock solution should be freeze-dried immediately after use, to avoid any hydrolysis of squarate **6**.

10. To maintain a reasonable reaction rate, conjugation reaction with oligosaccharides are usually conducted with a small excess (20%) of the squarate reagent.
11. The intensity of the MALDI-TOF and SELDI-TOF MS signal decreases when either too much or too little conjugate is applied on the chip. If the intensity of the spectra is too low, prepare two samples, one with larger and one with smaller quantity of the sample applied.
12. This operation does not remove completely the salts, which normally interfere with MALDI/SELDI-TOF measurements, but the amount of salts left behind is negligible. On the other hand, ample amount of the glycoprotein, the molecular mass we want to determine, remains on the chip after this incomplete wash.
13. Gently stirring the matrix solution induces crystallization and even distribution of sinapinic acid on the chip. This results in higher intensity of the recorded spectrum.
14. When conjugating large oligosaccharides (trisaccharides and higher), the spectrogram shows a dented curve wherein each peak corresponds to a glycoconjugate with a different hapten--protein ratio (Fig. 5a). For smaller haptens (monosaccharides and disaccharides), the mass difference between glycoconjugates with increasing hapten–protein ratio is close to the limit of the resolution of the spectrometer, resulting in a smooth bell-shaped curve (Fig. 5b).
15. In the case of the dented spectrogram curve (Fig. 5b), each peak maximum does not corresponds exactly to the calculated

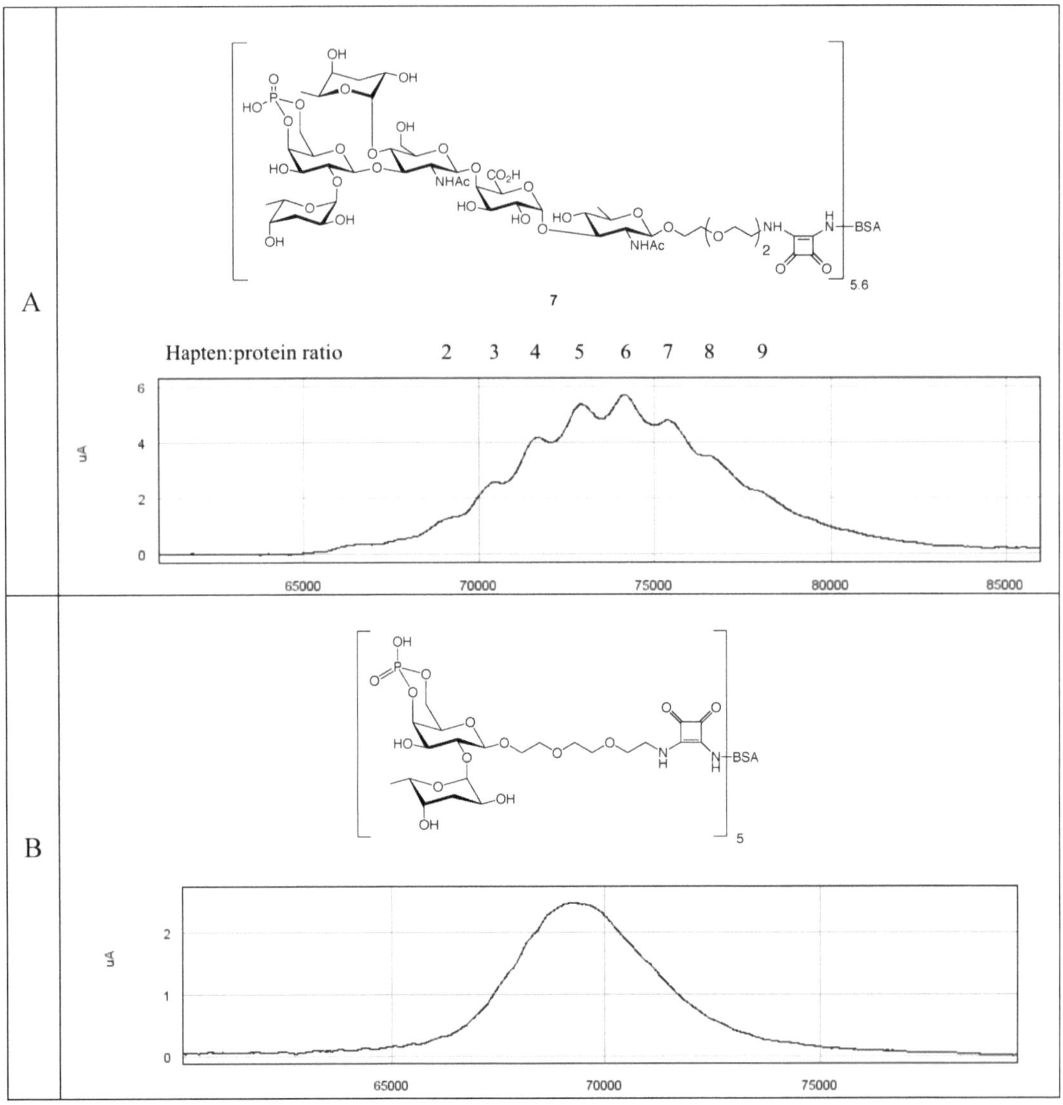

Fig. 5 Structures of glycoconjugates and SELDI spectrograms thereof. (**a**) Structure of a glycoconjugate with a hexasaccharide hapten and its corresponding SELDI spectrogram; (**b**) Structure of a glycoconjugate with a disaccharide hapten and its corresponding SELDI spectrogram

molecular weight of the conjugate for this hapten–protein ratio: the accuracy of the instrument is ~0.3%.

16. If the desired loading is not reached, the reaction should not be terminated. Although the hydrolysis of squarate **6** in pH 9 buffer is faster than in pH 7 buffer, there is still squarate present for up to 24 h [9]. After that time, if the desired hapten–protein ratio is still not reached, more squarate **6** must be added.
17. It is critical to avoid introduction of air into the C4 gel of the ZipTip during pipette manipulations. Avoid aspiration of air

while drawing solution into the pipette. During all wash and elution steps, the liquid should only be expelled to reach the surface of the gel in the ZipTip.

18. To avoid contamination when discarding liquid from the ZipTip, eject the liquid from the tip to the surface of the C4 gel, and discard the last droplet by touching the inner wall of the waste tube, taking care not to contaminate the tip itself.
19. The elution buffer should be added to the air displacement pipette immediately prior to elution of the Ziptip [16].
20. At that stage, the washing prior to the MALDI/SELDI-TOF MS measurement is not necessary, as salts would have been removed during the ultrafiltration step.

References

1. Tietze LF, Arlt M, Beller M et al (1991) Anticancer agents, 15. Squaric acid diethyl ester: A new coupling reagent for the formation of drug biopolymer conjugates. Synthesis of squaric acid ester amides and diamides. Chem Ber 124:1215–1221
2. Kováč P, Xu P (2017) Controlled and highly efficient preparation of carbohydrate-based vaccines: squaric acid chemistry is the way to go. In: Pilar Rautner A, Lindhorst T, Queneau Y (eds) Carbohydrate chemistry, The royal society of chemistry, vol 42, pp 83–115
3. Wurm FR, Klok H-A (2013) Be squared: expanding the horizon of squaric acid-mediated conjugations. Chem Soc Rev 42:8220–8236
4. Palitzsch B, Hartmann S, Stergiou N et al (2014) A fully synthetic four-component antitumor vaccine consisting of a mucin glycopeptide antigen combined with three different T-helper-cell epitopes. Angew Chem Int Ed 53:14245–14249
5. Fallarini S, Buzzi B, Giovarruscio S et al (2015) A synthetic disaccharide analogue from *Neisseria meningitidis* A capsular polysaccharide stimulates immune cell responses and induces immunoglobulin G (IgG) production in mice when protein-conjugated. ACS Infect Dis 1:487–496
6. Sayeed MA, Bufano MK, Xu P et al (2015) A cholera conjugate vaccine containing O-specific polysaccharide (OSP) of V. cholerae O1 Inaba and recombinant fragment of tetanus toxin Heavy Chain (OSP: rTTHc) induces serum, memory and lamina proprial responses against OSP and is protective in mice. PLoS Neglected Trop Dis 9: e0003881
7. Xu P, Kelly M, Vann WF et al (2017) Conjugate vaccines from bacterial antigens by squaric acid chemistry: A closer look. Chembiochem 18:799–815
8. Hou S-J, Saksena R, Kováč P (2008) Preparation of glycoconjugates by dialkyl squarate chemistry revisited. Carbohydr Res 343:196–210
9. Xu P, Trinh MN, Kováč P (2018) Conjugation of carbohydrates to proteins using di(triethylene glycol monomethyl ether) squaric acid ester revisited. Carbohydr Res 456:24–29
10. Saksena R, Ma X, Kováč P (2003) One-pot preparation of a series of glycoconjugates with predetermined antigen–carrier ratio from oligosaccharides that mimic the O-PS of *Vibrio cholerae* O:1, serotype Ogawa. Carbohydr Res 338:2591–2603
11. Soliman SE, Kováč P (2016) Total synthesis of the complete protective antigen of *Vibrio cholerae* O139. Angew Chem Int Ed 55:12850–12853
12. Knirel YA, Widmalm G, Senchenkova SN et al (1997) Structural studies on the short-chain lipopolysaccharide of *Vibrio cholerae* O139 Bengal. Eur J Biochem 247:402–410
13. Knirel YA, Senchenkova SN, Jansson P-E et al (1996) Structure of the O-specific polysaccharide of an *Aeromonas Trota* strain cross-reactive with *Vibrio cholerae* O139 Bengal. Eur J Biochem 238:160–165
14. Chernyak A, Karavanov A, Ogawa Y et al (2001) Conjugating oligosaccharides to proteins by squaric acid diester chemistry: Rapid monitoring of the progress of conjugation, and recovery of the unused ligand. Carbohydr Res 330:479–486

15. Pluskal MG (2000) Microscale sample preparation. Nat Biotechnol 18:104
16. Zhang Y, Sinaiko AR, Nelsestuen GL (2012) Glycoproteins and glycosylation: apolipoprotein C3 glycoforms by top-down MALDI-TOF mass spectrometry. In: Josic D, Hixson DC (eds) Liver proteomics: methods and protocols, Methods in molecular biology, vol 909. Springer Science, New York, pp 141–150

Chapter 8

Direct Conjugation of Bacterial Polysaccharides to Proteins by Squaric Acid Chemistry

Peng Xu and Pavol Kováč

Abstract

Bacterial polysaccharides that contain one amino group can be conjugated using squaric acid chemistry directly to a protein carrier. The conjugation is a two-step process consisting of labeling the polysaccharide with a squarate group and a reaction of the squarate formed with protein. The intermediate squarate derivative and the product glycoconjugate can be easily purified using centrifugal filtration devices. This method is experimentally simple and affords glycoconjugates with predictable carbohydrate–protein ratio (carbohydrate content), high conjugation efficiency, and excellent yield.

Key words Bacterial polysaccharides, O antigens, O-specific polysaccharides, Squaric acid chemistry, Conjugation, Glycoconjugate

1 Introduction

Glycoconjugates from bacterial polysaccharides are important tools in the life sciences. They can be made according to two different conjugation models, that is, multiple- and single-point attachment models [1]. According to the former, multiple reactive groups are either introduced into the polysaccharide or advantage is taken of reactive groups present in the polysaccharide, and conjugation between these groups and functional groups present in the protein carrier is effected by a suitable coupling reagent. The product thus formed is a highly cross-linked large molecule, which is difficult to characterize and with poor batch-to-batch reproducibility. Fine work in the life sciences requires well-defined glycoconjugates, which result from conjugation according to the single-point conjugation model. This requires that only one reactive group per molecule be present in the polysaccharide. The product glycoconjugate obtained by single-point conjugation is a star-shaped and better chemically defined construct.

Application of squaric acid chemistry to conjugation of carbohydrates was introduced by Tietze [2]. The method lies in the

Inka Brockhausen (ed.), *Bacterial Polysaccharides: Methods and Protocols*, Methods in Molecular Biology, vol. 1954,
https://doi.org/10.1007/978-1-4939-9154-9_8,

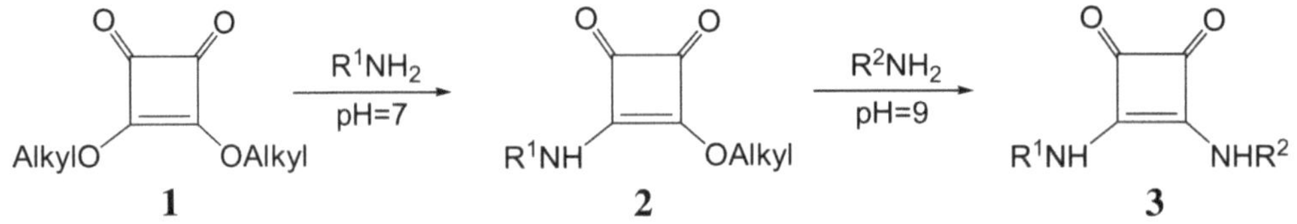

Fig. 1 Conjugating amines through squaric acid diester chemistry.

selective reaction of squaric acid diesters (Fig. 1) with amines. At pH 7, one of the ester groups in **1** reacts with an amine to form a monoamide (**2**). At more basic conditions, further amidation takes place to form a diamide (**3**). When the first amine is an amino group-containing carbohydrate and the second amine is anything else, a glycoconjugate is formed.

During the years following its inception, the method was revised based on various studies on conjugation of monosaccharides and oligosaccharides [3, 4] and, more recently, on conjugation of bacterial polysaccharides [5]. Unlike with small molecules, conversion of a bacterial polysaccharide to the corresponding squarate derivative cannot be monitored by thin-layer chromatography (TLC) because of the heterogeneity of the polysaccharide. However, the assay of primary amines with fluorescamine is a very valuable method for the follow-up of the reaction [6]. Secondly, the progress of the reaction of the formed squarate with a carrier protein can be monitored by SELDI-TOF [7, 8] or MALDI-TOF mass spectrometry [9, 10].

As a typical example of the squaric acid chemistry for conjugation of a bacterial polysaccharide, described below is conjugation of a single amino group-containing O-specific polysaccharide—core fragment of *Vibrio cholerae* O1, serotype Inaba **4** (average molecular weight ~6000 Da, Fig. 2) with bovine serum albumin (BSA, molecular weight 66,430 Da).

Other, single amino group-containing, water-soluble polysaccharides can be conjugated in a similar way. The protocol described herein includes the more recent additions and improvements made [11]. The targeted carbohydrate–protein ratio (loading) in the product glycoconjugate **6** was 5, and the observed loading in the product conjugate was 5.1 and the efficiency of conjugation 51%.

2 Materials

High-quality solvents and reagents should be used.

2.1 Reagents

1. Sinapinic acid.
2. Bovine serum albumin (BSA), Sigma-Aldrich; the most homogeneous protein available should be used.
3. 3,4-Dimethoxy-3-cyclobutene-1,2-dione.

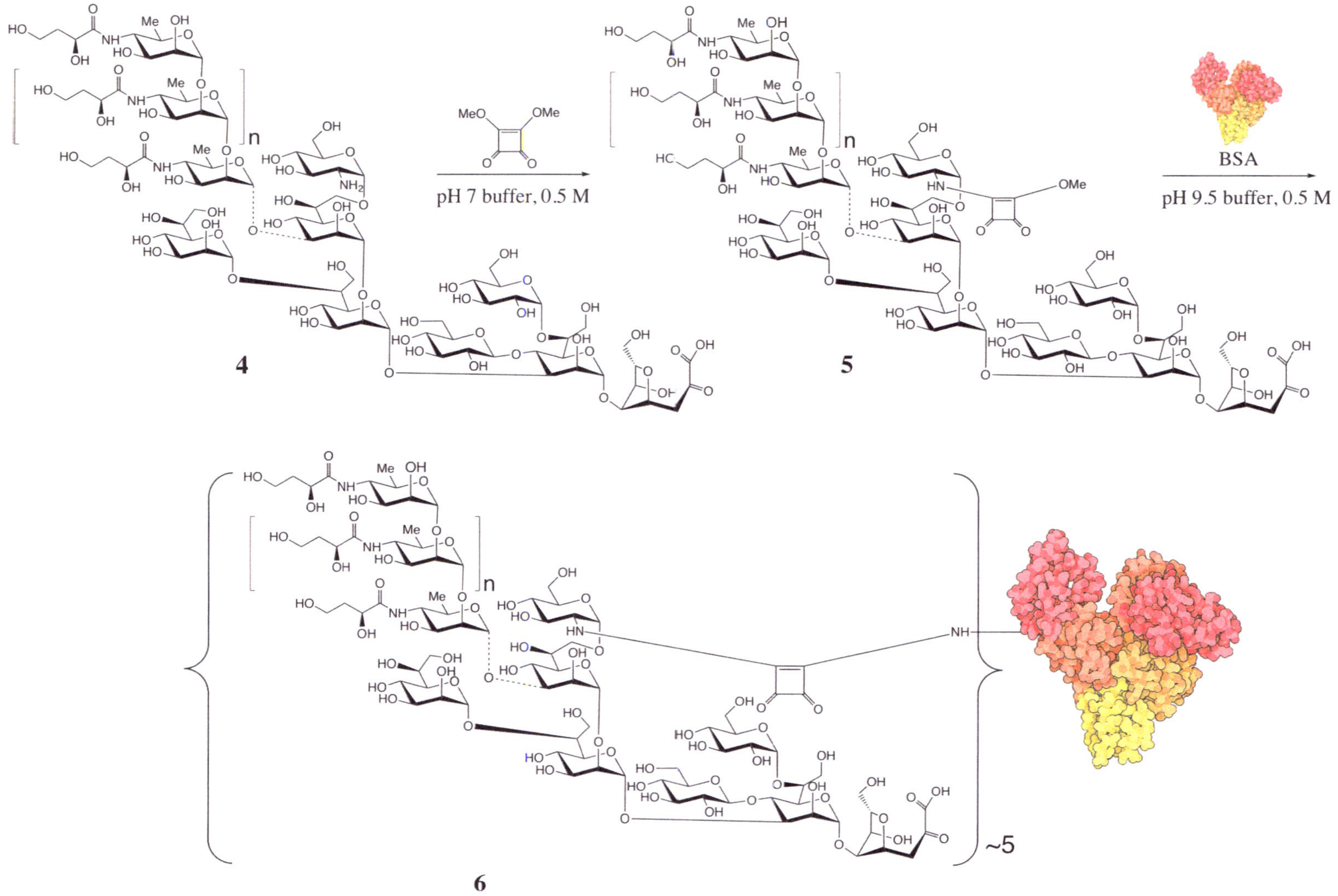

Fig. 2 Conjugation of *V.c.* Inaba O-SPcNH$_2$ to BSA using squaric acid chemistry: Treatment of O-SPcNH$_2$ (**4**) with dimethyl squarate, followed by conjugation of product **5** to BSA gives glycoconjugate (glycoprotein) **6**.

4. Fluorescamine (Sigma-Aldrich).
5. Trifluoroacetic acid (TFA).

2.2 Buffers and Solutions

1. Phosphate buffer, pH 7.0 (0.5 M): Weigh potassium phosphate monobasic (13.6 g) and sodium phosphate dibasic (21.0 g). Transfer to a 500 mL graduated cylinder. Add water to bring the volume to 450 mL. Adjust the pH with solid KOH to pH 7.0. Bring volume to 500 mL, transfer the solution to a screw-cap bottle, and store it at room temperature.
2. Borate buffer, pH 9.5 (0.5 M): In a bottle containing 1 L of water, add boric acid, (30.9 g), potassium chloride (26.1 g), and potassium hydroxide (8.42 g). After dissolution, add small portions of KOH to bring pH to 9.5.
3. PBS buffer (1×, pH 7.2, KD Medical).
4. Fluorescamine solution: In a glass vial protected from light (e.g., aluminum foil), dissolve fluorescamine in DMSO to make a 3 mg/mL clear solution (this solution should be made fresh on the day of the assay).
5. ZipTip sample reconstitution solution (ACN–H_2O–TFA 5:95:0.1, v/v): Mix 50 μL ACN, 940 μL water, and 10 μL 10% aq TFA. Make fresh daily and store at room temperature.
6. ZipTip activation solution (ACN–H_2O–TFA 50:50:0.1, v/v). Mix 500 μL ACN, 490 μL water, and 10 μL 10% aq TFA. Make fresh daily and store at room temperature.
7. ZipTip washing solution (0.1% aq TFA). Add 10 μL 10% aq TFA to 990 μL water. Make fresh daily and store at room temperature.
8. ZipTip elution solution (ACN–H_2O–TFA 75:25:0.1, v/v). Mix 750 μL ACN, 240 μL water, and 10 μL 10% aq TFA. Make fresh daily and store at room temperature.
9. Matrix solution for SELDI/MALDI-MS: In an Eppendorf tube, place 5 mg of sinapinic acid and add 100 μL acetonitrile followed by 100 μL 1% TFA aqueous solution. Mix vigorously (vortex) for 20 s. Centrifuge for 10 min at 1000 rcf and use the supernatant as the matrix solution.

2.3 Instruments

1. Centrifuge 5804R (Eppendorf) or any centrifuge accepting standard plastic centrifuge tubes capable of achieving 7500 rcf.
2. NanoDrop 3300 (Thermo Scientific) fluorospectrometer or any suitable fluorescence reader.
3. FiveEasy Plus (Mettler Toledo) pH meter with electrode: MI-410 (Microelectrodes Inc.) or similar.
4. Mass spectrometers: Protein Chip SELDI system (Bio-Rad) with NP-20 chip arrays. MALDI micro MX Micromass system (Waters).

2.4 Miscellaneous

1. V-Vial reaction vessel (Wheaton, Millville, NJ, USA)
2. V-shaped stirring bar.
3. Millipore C4 ZipTip.
4. Corning plastic microspatula.

3 Methods

3.1 Squarate Preparation

1. Place *V. cholerae* O1 Inaba antigen **4** (39.0 mg, 6.5 μmol, 1 equiv) [5] into a 5 mL V-vial reaction vessel equipped with a V-shaped stirring bar, then add 0.5 M pH 7.0 phosphate buffer (1.63 mL) to make a 4 mM solution with respect to **4** (*see* **Note 1**).
2. Withdraw 2 μL of the above reaction mixture into a 2.0 mL HPLC glass sample vial, dilute with 46 μL of pH 7.2 PBS (1×) buffer to make a 1 mg/mL solution with respect to **4**. Cover the vial with aluminum foil, add fluorescamine/DMSO solution (one-third of the volume of the sample solution, 16 μL) into the vial and shake to mix. Keep the solution at room temperature for 15 min before analyzing by fluorospectrometer. For each analysis, transfer 2 μL of the analyte solution onto the optical measurement surface of the NanoDrop 3300 spectrometer and take the RFU (relative fluorescence units) reading. Measure 5 replicates for each sample solution using a fresh 2 μL aliquot with a fresh pipette tip for each measurement. The average of the five measurements is used as the initial readout of the fluorescamine assay.
3. Add 3,4-dimethoxy-3-cyclobutene-1,2-dione (18.5 mg, 0.13 mmol, 20 equiv) into the reaction vessel and stir for a few minutes to form a clear solution. Keep stirring at room temperature and check the progress of the reaction using the fluorescamine assay at 1, 3, 6, and 21 h (*see* **Note 2**).
4. Once the reaction is completed, indicated by a plateau in fluoresecence reading, the reaction mixture is transferred into a Millipore Amicon Ultra-15 tube (3 kDa cutoff). Ultrafiltration is done at 4 °C against pure water (5000 rcf, eight times, for detailed protocol, follow manufacturer's instruction) (*see* **Note 3**).
5. After the last ultrafiltration run, the retentate is transferred into a sample vial and the membrane is washed three times by adding ultrapure water (2 mL) and vortexing for ~15 s. Combine the washes with the main product and lyophilize to give a fluffy solid product (*see* **Note 4**). Characterization can be done by MS (a 110 Da increase of molecular weight should be observed).

3.2 Conjugation

1. Place BSA (12.2 mg, 0.183 μmol, 1 equiv) into a 2 mL V-vial equipped with a V-shaped stir-bar (*see* **Note 5**) and add 0.5 M pH 9.5 buffer (366 μL), to form a 5 mM solution of BSA with respect to the amount of antigen to be added (*see* **Note 6**).
2. Add **5** (11.0 mg, 1.83 μmol, 10 equiv) into the reaction vessel (*see* **Note 7**).

3.3 Monitoring Progress of Conjugation by SELDI-TOF-MS

The progress of the conjugation can be monitored by SELDI-TOF-MS every 24 h until the antigen–carrier ratio levels off. The targeted antigen–carrier ratio in this protocol is 5.

1. Apply 1 μL of the reaction mixture onto one of the eight spots of the NP-20 SELDI chip. Dry with a stream of cold air for 5 min.
2. Wash the dried sample spot by pulsing 5 μL of water on the edge of the spot for 10 times. The last pulsing after aspiration, the washing solution is discarded. Dry for another 5 min by stream of cold air.
3. Apply 2 μL of the sinapinic acid matrix solution and gently stir the droplet on the chip using the pipette tip to induce crystallization. Dry the sample for 5 min by stream of cold air (*see* **Note 8**).
4. Analyze by SELDI-TOF-MS in positive mode.

3.4 Monitoring Progress of Conjugation by MALDI-TOF-MS (See Note 9)

1. Transfer 1 μL of the reaction mixture into an Eppendorf tube containing 14 μL of ZipTip sample reconstitution solution.
2. Attach a C4 ZipTip to a single channel pipette and set the volume to 10 μL. Activate the C4 ZipTip by pipetting 10 μL of the activation buffer and discard the buffer into a waste tube (*see* **Note 10**), repeat once. Wash the tip with 10 μL of washing buffer and discard the buffer, repeat once.
3. Slowly pulse 15 times 10 μL of the sample in the reconstitution solution (from **step 1**), aspirate then expel back into the Eppendorf tube. After the last pulse, the remaining liquid should be just slightly above the surface of the C4 layer of the tip.
4. Pipet and discard 10 μL of the washing solution. Repeat eight times. Leave the last wash solution in the ZipTip until the elution solution is prepared.
5. Place 1.6 μL of the elution solution in a small Eppendorf tube. Discard the last washing solution from the ZipTip (from **step 4**). Pulse the elution solution through the ZipTip eight times. Be careful to not introduce air bubbles. After the last pulse, expel all the liquid into the small Eppendorf tube, creating thereby the sample solution.

6. Apply the sample solution onto MALDI chip and allow it to air-dry.
7. Add 2 μL of sinapinic acid matrix solution to the sample on the MALDI chip and gently stir the droplet using pipette tip to induce crystallization. Allow the sample spot to air-dry for 5 min.
8. Analyze by MALDI-TOF-MS in linear positive mode.

3.5 Ultrafiltration

1. Once the desired loading is reached, work up by transferring the reaction mixture into a Millipore Amicon Ultra-15 tube (30 kDa cutoff). Ultrafiltration is done at 4 °C against 10 mM ammonium carbonate (7500 rcf, eight times, for detailed protocol follow manufacturer's instruction).
2. After the last ultrafiltration run, the retentate is transferred into a sample vial and the membrane is washed three times by adding 10 mM ammonium carbonate (2 mL) and gentle vortexing for ~15 s. Combine the washes with the main product and lyophilize to give a fluffy solid product (yield >95%, efficiency 51%, *see* **Notes 11** and **12**).

4 Notes

1. Unlike the preparation of squarates from synthetic oligosaccharides, squarate formation from **4** cannot be monitored by conventional methods used in small molecular organic synthesis, such as TLC, HPLC, NMR and MS because **4** is a heterogenic polymer. Fluorescamine assay [6] is a convenient way for monitoring the progress of squarate formation from such materials as it measures consumption of the amine in the reaction mixture. Therefore, the initial fluorescamine reading has to be taken before the squarate reagent is added. The progress of the reaction is then indicated by decrease (in %) of the fluorescamine reading.
2. The amount of the squarate reagent, 3,4-dimethoxy-3-cyclobutene-1,2-dione, has to be judiciously decided upon. Larger excess of the reagent results in faster reaction rate. However, due to the relatively fast hydrolysis of the reagent (half-life in water <3 h [12]), when larger excess of the reagent is needed, the strength/capacity of the reaction buffer has to be sufficient to keep the pH from dropping too much during the course of the reaction (preferred pH > 6.5 [11]). When working with small volume of the reaction mixture, the pH can be monitored by a pH meter equipped with a microelectrode that allows pH measurements in very small volume.

3. The size of the ultrafiltration tube to be used depends on the scale of the reaction. When working with <1 mg of antigen, the smallest tube (such as Millipore Amicon Ultra-0.5) is recommended. Follow the manufacturer's recommendation for choosing the membrane cutoff. With Millipore centrifugal filtration devices, the membrane safely retains molecules two times larger than the cutoff value. When working with polysaccharides of smaller molecular weight, size exclusion chromatography (for example P-2 gel) can be used, instead of ultrafiltration, to purify the squarate derivatives of the antigens if no proper cutoff membrane is available.
4. Squarate **5** slowly hydrolyzes in aqueous solutions, even under neutral condition. Even though the process is quite slow (<3% in 24 h), unnecessarily long contact with water during purification should be avoided. The sample should be freeze-dried immediately after purification.
5. Because BSA can be fluffy and electrostatic, care has to be taken during the operation. Transfer BSA by using a sterilized plastic spatula, such as a Corning microspatula, and use an antistatic gun.
6. The buffer should be added before the squarate, to minimize squarate hydrolysis because squarate **5** is prone to hydrolysis under conjugation condition. Based on our experience, conjugation is more efficient under higher pH. However, the efficiency varies with different proteins and different antigens. We also achieved satisfactory conjugation with a pH 9.0 buffer.
7. Because of the difference in reactivity of different proteins and antigens, the molar amount of squarate derivatives of antigens needed at the onset of the reaction has to be optimized case by case when other protein–antigen combinations are used.
8. As shown in Fig. 3, at the early stages of the reaction, the peaks and valleys in the SELDI profile are more prominent (4 h, peaks at 66434.70, 72781.97, and 78737.52 represent loading 0, 1, and 2 respectively) compared to the late stages of the reaction (72 h, peaks at 86065.33, 92134.75, 97722.03, and 104008.50 represent loading 3, 4, 5, and 6 respectively). This is caused by the heterogeneous nature of the antigen, manifested by broadened peaks. As the loading increases, the SELDI profile becomes smoother.
9. Considering the unavailability of SELDI-TOF-MS in many laboratories, we have confirmed that MALDI-TOF-MS is a suitable alternative for monitoring the progress of the conjugation. This involves pretreatment of the sample with C4 ZipTip [9, 10]. The MALDI-TOF results obtained are in close agreement with the SELDI-TOF results.

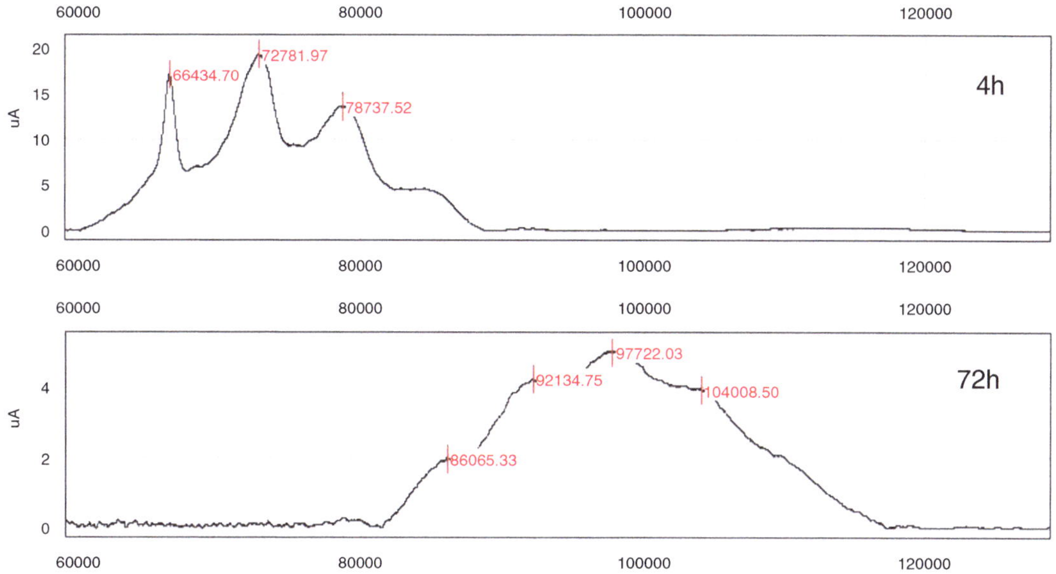

Fig. 3 SELDI-TOF-MS result of **5**–BSA conjugation at 4 h and 72 h

10. Avoiding introduction of air into the C4 gel of the ZipTip during pipette manipulations is critical. Avoid aspiration of air while drawing solution into the pipet. During all wash and elution steps, the liquid should only be expelled to reach the surface of the gel in the ZipTip.
11. The yield of the conjugation reaction is relative to the protein. It is the molar ratio of the conjugate obtained to the carrier protein used at the onset of the conjugation. Conversely, conjugation efficiency is based on the antigen loading. It is the molar ratio of the squarate derivative attached to the carrier to the squarate derivative used at the onset of the conjugation. The higher the conjugation efficiency, the better the conjugation method.
12. Bacterial polysaccharides and their squarate derivatives are expensive materials. During conjugation, most of the squarate that is not attached onto the protein will be hydrolyzed to form the corresponding acid. After purification of the conjugate, the hydrolyzed squarate is collected as "waste" in the down chamber of the ultrafiltration device. When conjugation efficiency is low, it may be beneficial to recover the hydrolyzed squarate and convert it back to an active squarate so that it can be used for a second round of conjugation. Such recovery/reactivation protocol has been developed and reported [11].

References

1. Dick JW, Beurret M (1989) Glycoconjugates of bacterial carbohydrate antigens. In: Cruse JM, Lewis JRE (eds) Conjugate vaccines, Contributions to microbiology and immunology, vol 10. Karger, Switzerland, pp 48–114
2. Tietze LF, Schroter C, Gabius S et al (1991) Conjugation of p-aminophenyl glycosides with squaric acid diester to a carrier protein and the use of neoglycoprotein in the histochemical detection of lectins. Bioconjug Chem 2:148–153
3. Astronomo RD, Lee HK, Scanlan CN et al (2008) A glycoconjugate antigen based on the recognition motif of a broadly neutralizing human immunodeficiency virus antibody, 2g12, is immunogenic but elicits antibodies unable to bind to the self glycans of gp120. J Virol 82:6359–6368
4. Hou S-J, Saksena R, Kováč P (2008) Preparation of glycoconjugates by dialkyl squarate chemistry revisited. Carbohydr Res 343:196–210
5. Xu P, Alam MM, Kalsy A et al (2011) Simple, direct conjugation of bacterial o-sp-core antigens to proteins: development of cholera conjugate vaccines. Bioconjug Chem 22:2179–2185
6. Udenfriend S, Stein S, Bohlen P, Dairman W, Leimgruber W, Weigele M (1972) Fluorescamine: a reagent for assay of amino acids, peptides, proteins, and primary amines in the picomole range. Science 178:871–872
7. Chernyak A, Ogawa Y, Kováč P (2001) Conjugating oligosaccharides to proteins by squaric acid diester chemistry: rapid monitoring of the progress of conjugation, and recovery of the unused ligand. Carbohydr Res 330:479–486
8. Saksena R, Chernyak A, Karavanov A et al (2003) Conjugating low molecular mass carbohydrates to proteins 1. Monitoring the progress of conjugation. Methods Enzymol 362:125–139
9. Pluskal MG (2000) Microscale sample preparation. Nat Biotechnol 18:104–105
10. Zhang Y, Sinaiko AR, Nelsestuen GL (2012) Glycoproteins and glycosylation: Apolipoprotein c3 glycoforms by top-down MALDI-TOF mass spectrometry. Methods Mol Biol 909:141–150
11. Xu P, Kelly M, Vann WF et al (2017) Conjugate vaccines from bacterial antigens by squaric acid chemistry: a closer look. Chembiochem 18:799–815
12. Cohen S, Cohen S (1966) Preparation and reactions of derivatives of squaric acid. Alkoxy-, hydroxy-, and aminocyclobutenediones. J Am Chem Soc 88:1533–1536

Chapter 9

Chemical Modification of the N-Acetyl Moieties of Hyaluronic Acid from *Streptococcus equi* for Studies in Cytokine Production

Siziwe Bebe and Tassos Anastassiades

Abstract

Partial N-deacetylation and certain N-reacylations of low-molecular-weight hyaluronic acid (hyaluronan) abate its proinflammatory properties in mammalian systems. Here, we describe the treatment of bacterial hyaluronic acid by hydrazine or NaOH to yield smaller partially deacetylated polymers. These N-deacetylated polymers can be reacylated with acyl anhydrides to yield substituted hyaluronic acid derivatives of equivalent size and equimolar N-acyl substitutions.

Key words Hyaluronic acid, Hyaluronan, Chemical N-acylation, *Streptococcus equi*, Proinflammatory cytokines

1 Introduction

Anionic glycoconjugates, mainly glycosaminoglycans, are widely distributed in mammalian, bacterial [1–3] and algae systems [4–7] and have multifaceted applications such as in bone and cartilage biochemistry. Low molecular mass hyaluronans are known to induce inflammation [8]. The key role of N-acetyl groups of low molecular mass hyaluronan in stimulating the production of proinflammatory cytokines can be demonstrated by the partial N-deacetylation, which is carried out by chemical hydrolysis using two different methods, described in the synthesis protocols. For the conditions described, the hydrolysis of the hyaluronan chain results in a reduction of the molecular mass to 30–214 kDa, (Fig. 1) with a percentage of total NH_2 and N-acetyl moieties in the reacetylated hyaluronan of 0% and $98.7 \pm 1.5\%$, respectively (Figs. 2 and 3) [9]. For partially butyrylated hyaluronan, the total NH_2, N-acetyl, and N-butyryl moieties are 0, 82.2 ± 4.6, and $22.7 \pm 3.8\%$, respectively, based on ^{1}H-NMR. The method can be applied to comparative studies of chemically N-acylated low molecular weight

Inka Brockhausen (ed.), *Bacterial Polysaccharides: Methods and Protocols*, Methods in Molecular Biology, vol. 1954,
https://doi.org/10.1007/978-1-4939-9154-9_9,

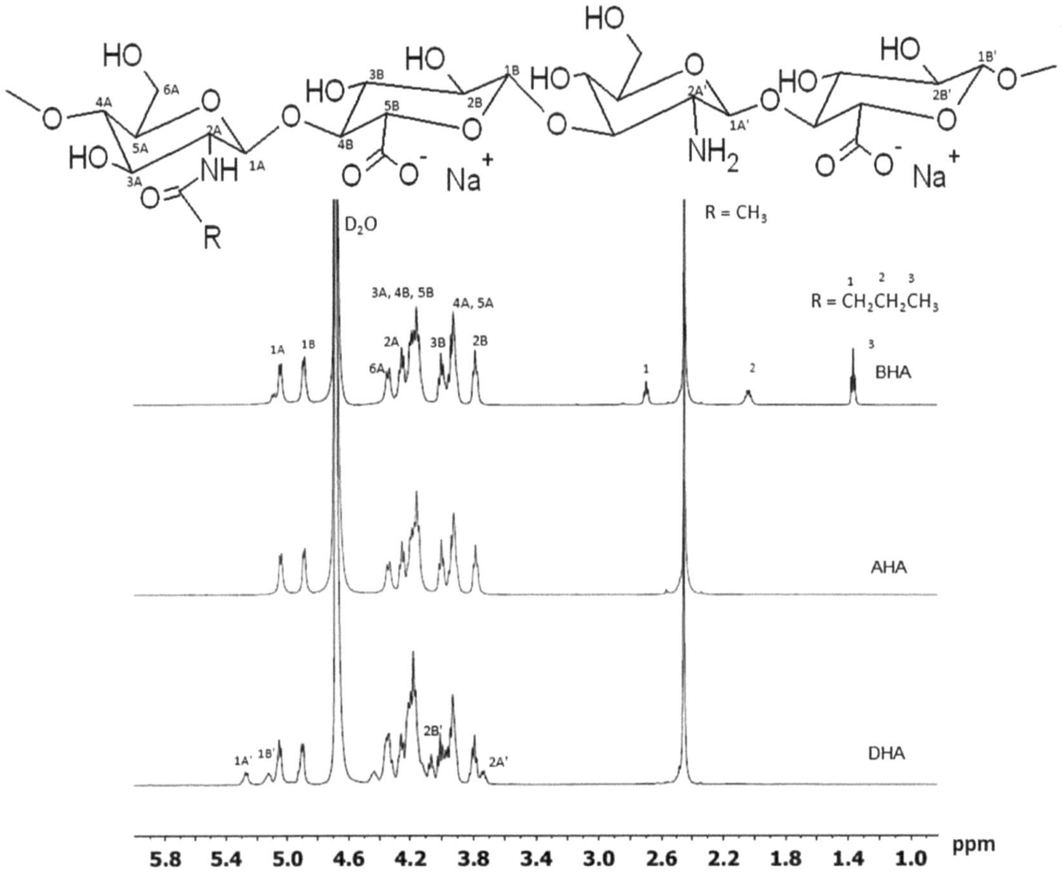

Fig. 1 ^{1}H NMR of DHA, AHA, and BHA in D_2O at 349 K. 25 mg/mL of the polymers dissolved in deuterated water.^{1}H NMR spectra of the polymers are recorded at 348 K in D_2O at 600 MHz. The resulting peaks are compared with the solvent peaks relative to tetramethylsilane as reference, because in the native polymer repeat unit, there are three methyl protons (acetyl group) for every two anomeric protons, and the ratio of the signal for these protons is 1.5 [1]. The method for estimating the degrees of deacetylation at the various time points is described in the legend to Fig. 3c [20]

hyaluronans of equivalent mass and extent of N-acyl substitution, utilizing relatively less expensive bacterial hyaluronan as the parent compound.

2 Materials

Prepare all solutions using ultrapure water (prepared by purifying deionized water, to attain a sensitivity of 18 MΩ-cm at 25 °C) and analytical grade reagents. Prepare and store all reagents at room temperature (unless indicated otherwise). Diligently follow all waste disposal regulations when disposing of waste materials.

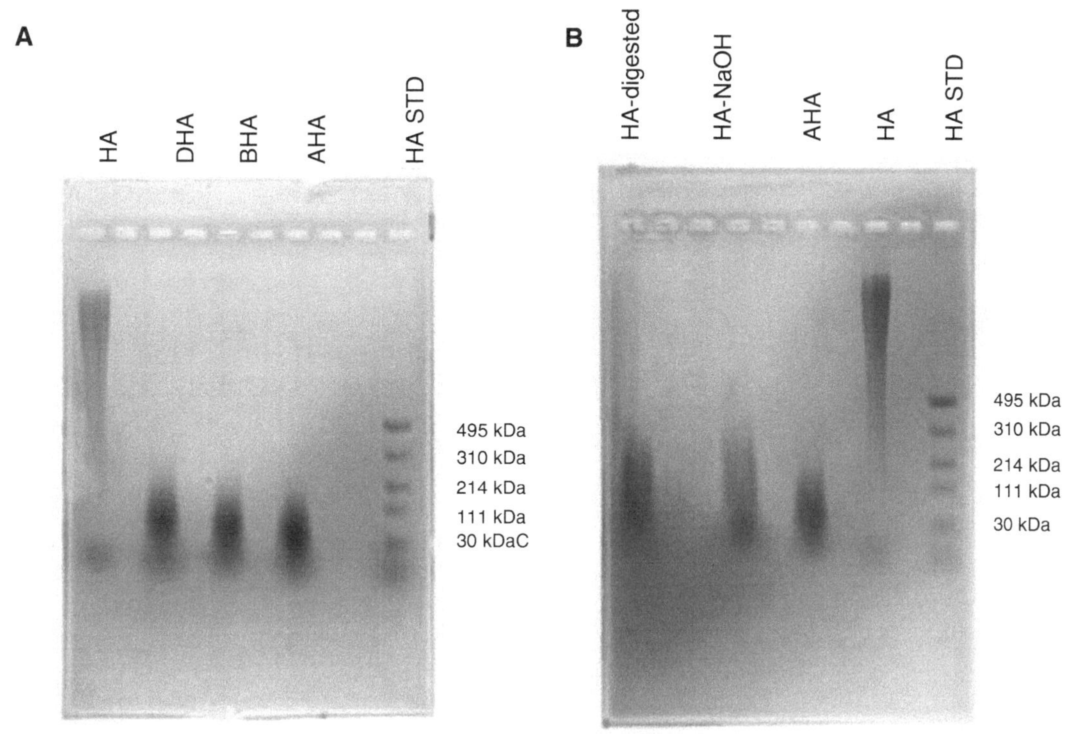

Fig. 2 Electrophoretic separation of HA derivatives. (**a**) HA, DHA, BHA, AHA, and Select-HA LoLadder (HA STD) were run on a 0.75% (w/v) agarose gel in TAE running buffer at a constant voltage of 100 V for 1 h and then visualized with 0.005% (w/v) Stains-All. (**b**) HA-digested, AHA-NaOH, AHA, HA, and Select-HA LoLadder were run on a 0.75% (w/v) agarose gel in TAE running buffer, at a constant voltage of 100 V for 1.5 h and then visualized with 0.005% (w/v) Stains-All [20]

2.1 Hyaluronic Acid

1. Sodium salt of hyaluronic acid (HA) from *Streptococcus equi* (1500–1800 kDa) (Sigma).
2. Acetic anhydride, propanoic anhydride, butyric anhydride, valeric anhydride, isovaleric anhydride (Sigma).
3. Select-HA LoLadder (Sigma).
4. Dialysis membrane with molecular mass cutoff of 6000–8000 (Spectrum Laboratories, Inc.).

2.2 Agarose Gel

1. TAE Buffer: weigh out 242 g of Tris base and transfer to 2 L beaker/conical flask.

   Add 750 mL deionized/Milli-Q water and mix until all Tris base has dissolved completely. Add 100 mL of 0.5 M EDTA solution and 57.1 mL glacial acetic acid. Mix the solution again. Adjust pH to 8.3, if required. Adjust the solution volume to 1000 mL with deionized/Milli-Q water. Mix the solution again. Sterilize the solution by autoclaving (20 min at 15 lb/sq.in. (psi) from 121 to 124 °C on liquid cycle) (*see* **Note 1**).

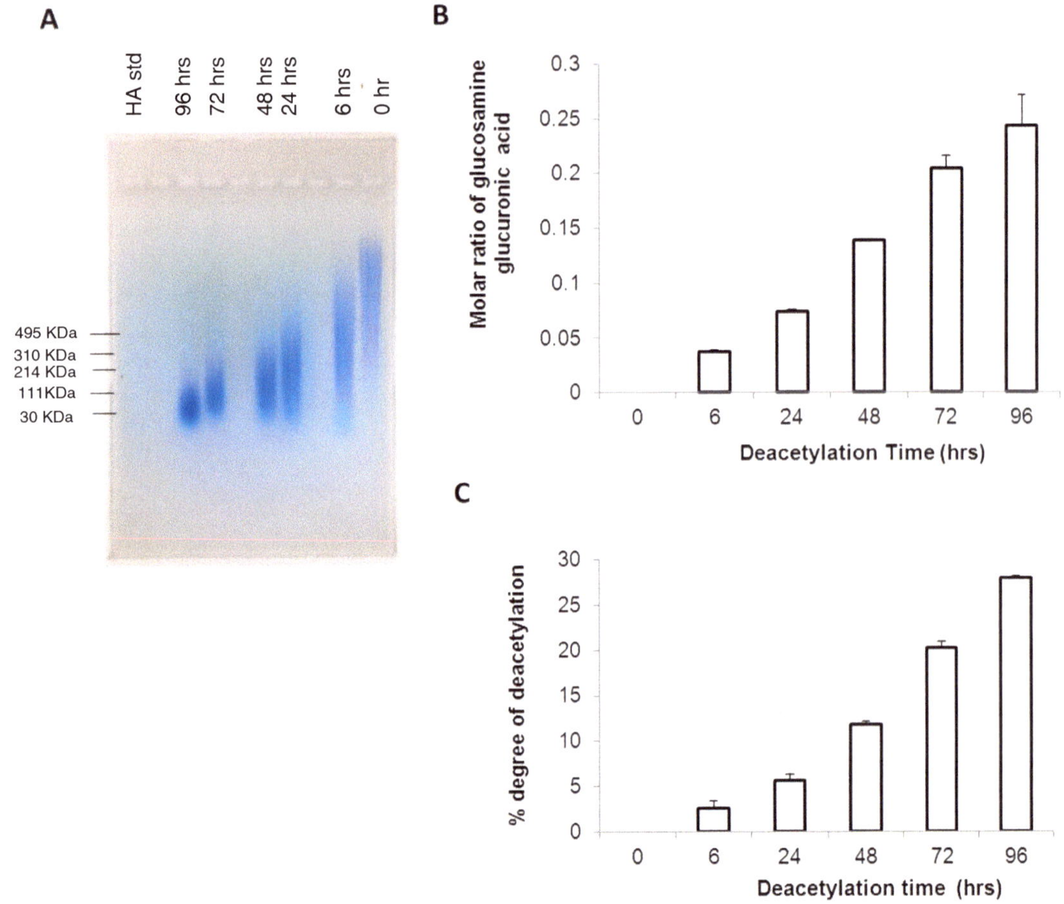

Fig. 3 Characterization of HA and HA deacetylated for 6, 24, 48, 72, and 96 h. (**a**) HA and HA deacetylated for 6, 24, 48, 72, and 96 h and Select-HA LoLadder were run on a 0.75% w/v agarose gel in TAE running buffer, at a constant voltage of 100 V for 90 min and then visualized with 0.005% w/v "Stains-All." (**b**) The molar ratio of glucosamine to glucuronic acid for HA and HA deacetylated for 6, 24, 48, 72, and 96 h. This was obtained by the colorimetrically using the *o*-phthalaldehyde and Carbazole colorimetric assays, respectively. (**c**) The degree of deacetylation of polymers deacetylated for 6, 24, 48, 72, and 96 h is obtained from the ^{1}H-NMR peaks recorded at 348 K in D_2O at 600 MHz. The resulting peaks were compared to the solvent peaks relative to tetramethylsilane as reference, since in the native polymer repeat unit there are three methyl protons (acetyl group) for every two anomeric protons, and the ratio of the signal for these protons is 1.5 [9]. The degree of deacetylation in our sample DHA was estimated by using the Equation ($DD = 1 - (X/Y)/1.5) \times 100$ [9]. X is the integration of the acetyl group signal in DHA at 2.4–2.5 ppm, and Y is the integration of the two anomeric protons' signals at 4.9–5.3 ppm. Their ratio is approximately 1.24 ± 0.05. Therefore, the % of glucosamine was estimated to be $19.7 \pm 3.5\%$ [20]

2. Agarose gel: 0.75% w/v agarose gel in TAE running buffer. Weigh 0.75 g of agarose and resuspend in 100 mL TAE buffer. Add buffer (at room temperature) to a flask that is 2.5–4 times the volume of gel solution. Add a Teflon-coated stir bar, then add biotechnology grade agarose powder while stirring vigorously so the agarose is dispersed uniformly.

Stir for 2 min to hydrate the agarose. Tare the flask and solution.

3. Place in a microwave oven and heat at 100% power using 20–60 s intervals. Swirl gently between intervals to resuspend the agarose. Continue the cycle of heating and swirling until the agarose is completely dissolved (no visible particles are present). Add distilled water to return the solution to its initial weight and mix.
4. Cool the solution to 50–60 °C before pouring the gel.
5. Pour the solution into the mold so as to dispense the entire amount in 30–60 s, without generating bubbles (*see* **Note 2**).

2.3 Analysis of HA and HA Derivatives

1. AR2000 TM rheometer (TA Instruments, New Castle, DE).
2. Mass spectrometry: Q STARXL hybrid quadrupole/TOF tandem mass spectrometer (Applied Biosystems/MDSSCIEX).
3. NMR: Bruker AMX 600 operating at 600.13 MHz.
4. Viscosity: AR2000 TM rheometer.
5. Filtration: micron centrifugal filter device (Millipore Corporation).

2.4 Cytokine Responses to HA and HA Derivatives

1. Microplate spectrometer (Quant; Bio-Tek Instruments, Inc.).
2. Enzyme-linked immunosorbent assay kits (Cedarlane).
3. Inverted tissue culture microscope.

3 Methods

3.1 Partial Deacetylation of HA—Hydrazinolysis

To form the partially deacetylated HA (DHA) use the hydrazinolysis reaction.

1. Weigh out 3 g of HA and 1.5 g of hydrazine sulfate on an analytical balance.
2. Dissolve compounds in 150 mL of hydrazine monohydrate and at 55 °C for 6, 24, 48, 72 and 96 h [10, 11] (*see* **Note 3**).
3. Precipitate the polymeric product with ethanol (*see* **Note 4**).
4. Wash the precipitated product with ethanol.
5. Dry the product from **step 4** in an oven at 25 °C for 24 h.
6. Mix 5% (v/v) acetic acid and 30 mL of 0.5 M iodic acid.
7. Dissolve product from **step 5** in the 50 mL mixture from **step 6** and keep in a water bath at 4 °C for at least 1 h [10, 11].
8. Stir 8.75 mL of aqueous 57% (v/v) CH_3I into the mixture from **step 7** for 15 min to see a deep violet color.
9. Remove the deep violet color of the solution via liquid–liquid extraction with ethyl ether:

Transfer deep violet liquid solution to a separatory funnel. Add 150 mL of diethyl ether.
Stir vigorously for a few min. Recover the aqueous layer. Add 150 mL of diethyl ether and stir vigorously. Repeat until complete discoloration.

10. Adjust the pH of the aqueous layer to 7.0–7.5.
11. Precipitate out the deacetylated HA with ethanol.
12. Dissolve the precipitate from **step 11** in water and dialyze against double-distilled H_2O (*see* **Note 5**).
13. Freeze-dry the resultant solution: transfer the solution into a 50 mL tube and store in the freezer at −80 °C overnight. Put the frozen tube in a lyophilizer flask and freeze-dry.

3.2 Partial Deacetylation of HA—NAOH Method

HA is also deacetylated using NaOH as the deacetylating agent (DHA-NaOH) according to the method used by Wada et al. [12].

1. Weigh out 2 g of HA.
2. Dissolve it in 200 mL of 2.5 N NaOH.
3. Stir the solution at room temperature for 72 h.
4. Add 140 mL of water to the reaction mixture and stir in an ice bath.
5. Adjust the pH of the solution to 10.0 with concentrated HCl.
6. Pour the solution into 800 mL of a 7:1 acetone–H_2O mixture at 0 °C.
7. Recover the white fibrous precipitate formed by filtration.
8. Dissolve the precipitate in water.
9. Adjust the pH to 7.0 using NaOH.
10. Evaporate acetone using a rotary evaporator (*see* **Note 6**).
11. Freeze-dry the solution.
12. Redissolve the polymer in water.
13. Remove dissolved salt by dialysis.
14. Freeze-dry/relyophilize the sample.

3.3 Reacylation of DHA or DHA-NaOH

Reacylation of DHA and DHA-NaOH to form AHA or AHA-NaOH is done according to the method used for N-propionylation of the glucosamine monomer [13], and recently used by Zhang et al. [14], with some modifications.

1. Dissolve 0.1 g of DHA or DHA-NaOH in 30 mL of distilled water.
2. Add 6 mL of saturated $NaHCO_3$.
3. Add acetic, butyric or other acyl anhydride to absolute alcohol at concentration of 0.05, 0.5, 2.5, 5.0% v/v (*see* **Note 7**).

4. Add solutions at each concentration (10.0% v/v) to the reaction mixture.
5. Stir the reaction mixture for 10 min.
6. Quench in a boiling water bath for 5 min.
7. Evaporate residual ethanol by using the rotary evaporator.
8. Freeze-dry/lyophilize the sample to remove the water.
9. Dialyze the lyophilized sample against distilled water in a 6000 Da molecular mass cutoff dialysis tubing to ensure that the LMHA preparations do not contain oligosaccharides.
10. Dissolve the sample and relyophilize.

3.4 Digestion of HA

As control, HA is digested with hyaluronidase from bovine testes to obtain low molecular mass HA with similar molecular mass with the DHA and AHA.

1. Digest 3 mg/mL HA with 10 units/mL hyaluronidase from bovine testes in PBS (pH 7.2) at 37 °C for 30 min (HA-digested) [15].
2. Stop the reaction by boiling for 5 min.
3. Lyophilize the sample.
4. Remove dissolved salt and oligosaccharides by dialysis (6000-Da molecular mass cutoff).
5. Relyophilize the sample.
6. Separate digested HA into fractions smaller and larger than 30 kDa using a polyethersulfone gel filtration spin column.

3.5 ^{1}H-NMR Analysis

To confirm the structure and purity of the polymers and to provide preliminary measurements of the degree of deacetylation, ^{1}H-NMR spectra of the polymers are recorded at 348 K in D_2O at 600 MHz.

1. Dissolve sample in D_2O (99.9 atom % D purity.
2. Perform ^{1}H-NMR at 348 K on Bruker AMX 600 operating at 600.13 MHz.
3. Suppress HOD signal using soft signal presaturation.
4. Perform at 2D experiments: ^{1}H–^{1}H correlation spectroscopy (COSY) and total correlation spectroscopy (TOCSY).
5. Perform ^{13}C-NMR at 348 K on Bruker AMX 600 operating at 150.9 MHz.
6. Perform at 2D experiments: ^{1}H–^{13}C heteronuclear multiple-bond coherence spectroscopy (HMBC) and heteronuclear single-quantum coherence spectroscopy (HSQC).
7. Obtain assignment by heteronuclear multiple quantum coherence spectroscopy (HMQC). Use tetramethyl silane as

reference to relatively compare the resulting peaks with the solvent peaks. The degree of deacetylation is estimated from the ^{1}H NMR spectrum according to the method used by Crescenzi et al. [10, 11]. In the native polymer repeating unit, there are three methyl protons in the GlcNAc unit for every two anomeric protons from the GlcNAc and GlcUA unit. The ratio of the signal for these protons is 1.5 [10, 11]. From the ^{1}H NMR spectra of the DHA obtained, the ratio of the integration of the peaks (Y) for three methyl protons in the GlcNAc unit observed at 2.4–2.5 ppm and the anomeric protons of the GlcNAc, GlcUA, and glucosamine units observed at 4.9–5.3 ppm is calculated. The deacetylation degree, defined as the molar ratio of glucosamine to glucosamine + GlcNAc unit, is calculated according to the equation, Deacetylation degree $DA = (1.0 - (Y/1.5))100)$ [1, 2].

3.6 Colorimetric Assays

To further confirm the ^{1}H NMR data, colorimetric quantification of the glucosamine and glucuronic content of the polymer is carried out. The glucosamine content is quantified according to the method used previously, for chitosan [16], which is based on the derivatization reaction of its primary amino group with *o*-phthalaldehyde and the thiol group of N-acetyl-L-cysteine. The glucuronic content is quantified according to the carbazole reaction method used by Cesaretti et al. [17]. The results are expressed as the molar ratios of glucosamine to glucuronic acid.

3.7 Molecular Mass Estimation

1. Cast 00.75% (w/v) agarose gel (*see* **Notes 1** and **2**) [18].
2. Run HA samples in TAE buffer (pH 8.0) at 100 V for 90 min. The bromophenol blue tracking dye migrates close to the end of the gel during this period (*see* **Note 8**).
3. Immediately after the run, place the gel in ~100 mL of solution containing 0.005% w/v Stains-All in 50% (v/v) ethanol overnight in the dark at room temperature.
4. To destain, transfer the gel to 10% (v/v) ethanol solution.

3.8 Viscosity of Polymers

Steady shear viscosity of the polymers is assessed with an AR2000 TM rheometer equipped with a cone and plate fixture consisting of a 0.5°, 4-cm-diameter stainless steel cone, in the steady shear of mode (*see* **Note 9**).

1. Add approximately 300 μL fluid sample on the plate.
2. Control the temperature at 37 °C.
3. Use a concentration of 5 mg/mL for HA and the HA derivatives.

3.9 Mass Spectrometry

1. For HA: Digest 3 mL of HA (5 mg/mL) with 1 mL of 800–2000 units/mL of bovine testes hyaluronidase by incubating at 37 °C for 24 h.
2. For HA derivatives: digest 500 μL: 500 μL of HA (30 mg/mL) with 1 mL of 800–2000 units/mL of bovine testicular hyaluronidase.
3. Precipitate the enzyme by boiling for 5 min.
4. Centrifuge at 2050 rpm (850 RCF) for 10 min.
5. Collect the supernatant and separate with a micron centrifugal filter device (Millipore Corporation) with a molecular mass cutoff of 3000 Da.
6. Lyophilize the filtrate obtained from this digestion and use for MS analyses.
7. MS and MS/MS analyses are performed on a Q STARXL hybrid quadrupole/TOF tandem mass spectrometer equipped with electro spray ionization source and is further confirmed using Thermo LTQ Orbitrap Velos Pro with heated electrospray ionization source under negative mode. For electrospray ionization analysis under negative mode, the following source conditions are used: a curtain gas setting of 25, ion spray voltage of −4000 V, Q0 declustering potential of −90 V, and focusing potential of −350 V. Nitrogen is used as the collision gas (CAD setting 5) for both TOF-MS and MS/MS scans.

3.10 Proinflammatory Cytokine Response to Reacylated HA (AHA)

The size range of the polymer and the molar ratio of its free amino groups of glucosamine to glucuronic acid are determined. The degree of AHA acetylation necessary for inflammatory activity is determined by the biological assay.

1. Prior to the study: Place the polymers into 1.5 mL Eppendorf tubes sterilized by exposing to germicidal UV source for 30 min in the laminar flow hood.
2. Maintain the human suspension monocyte cell line (THP-1) in RPMI 1640 medium (*see* **Note 10**), supplemented with 10% fetal bovine serum at 37 °C in humidified atmosphere with 5% CO_2 in tissue culture flasks.
3. To investigate the proinflammatory cytokine response to our HA compounds, seed THP-1 cells at 500,000 cells/mL in RPMI 1640, 10% fetal bovine serum, 50 μM β-mercaptoethanol, and 100 nM phorbol 12-myristate 13-acetate (PMA), at 37 °C, in humidified atmosphere, with 5% CO_2 in 24- or 96-well plates.
4. Seed THP-1 cells with 100 nM PMA [19, 20] for the induction of monocyte-macrophage differentiation.

5. After 72 h, remove nonattached cells by aspiration and washed three times with serum-free medium.
6. Replace the medium of the adherent macrophages with RPMI 1640, 1% fetal bovine serum, 100 nM PMA with either (a) HA derivatives, (b) without HA derivatives, (c) TLR agonists (*see* **Note 11**), (d) antagonists, or neutralizing antibodies.
7. After 24 h, remove media and determine the proinflammatory cytokine concentrations using enzyme-linked immunosorbent assay kits for (a) TNF-α, (b) IL-1β, (c) IL-6, and (d) IL-8.
8. Measure the absorbance on a microplate spectrometer at a single wavelength of 450 nm.
9. Construct a curve of absorbance versus log concentration of individual cytokines in the standard wells.
10. Use the curve to determine the concentrations of cytokines in the media.
11. Plot the standard curve using GraphPad Prism version 5.00 for Windows.
12. Determine a measure of the attached THP-1 macrophage cell numbers using the XTT/PMS assay [21].
13. Treat THP-1 macrophages with hyaluronan (HA) derivatives and Toll-like receptor (TLR) agonists for 24 h.
14. Assay the media for cytokines using ELISA.
15. Determine a measure of the attached cell numbers using the XTT/PMS cell proliferation assay.
16. Treat DHA with increasing concentrations of acetic anhydride in absolute ethanol.
17. Assay the inflammatory activity of these differently acetylated HA derivatives by measuring the IL-1β secretion.
18. Fractionate DHA on the basis of size using polyether sulfone gel filtration spin columns with molecular mass cutoffs of 30, 50, 100, or 300 kDa.
19. Freeze-dry the filtrates.
20. Resuspend with 0.01 N HCl 6. Determine the glucosamine and glucuronic acid concentrations using the o-phthalaldehyde and carbazole colorimetric assays, respectively [17, 18] and determine its relationship to the cytokine production (Fig. 4, *see* **Note 12**).

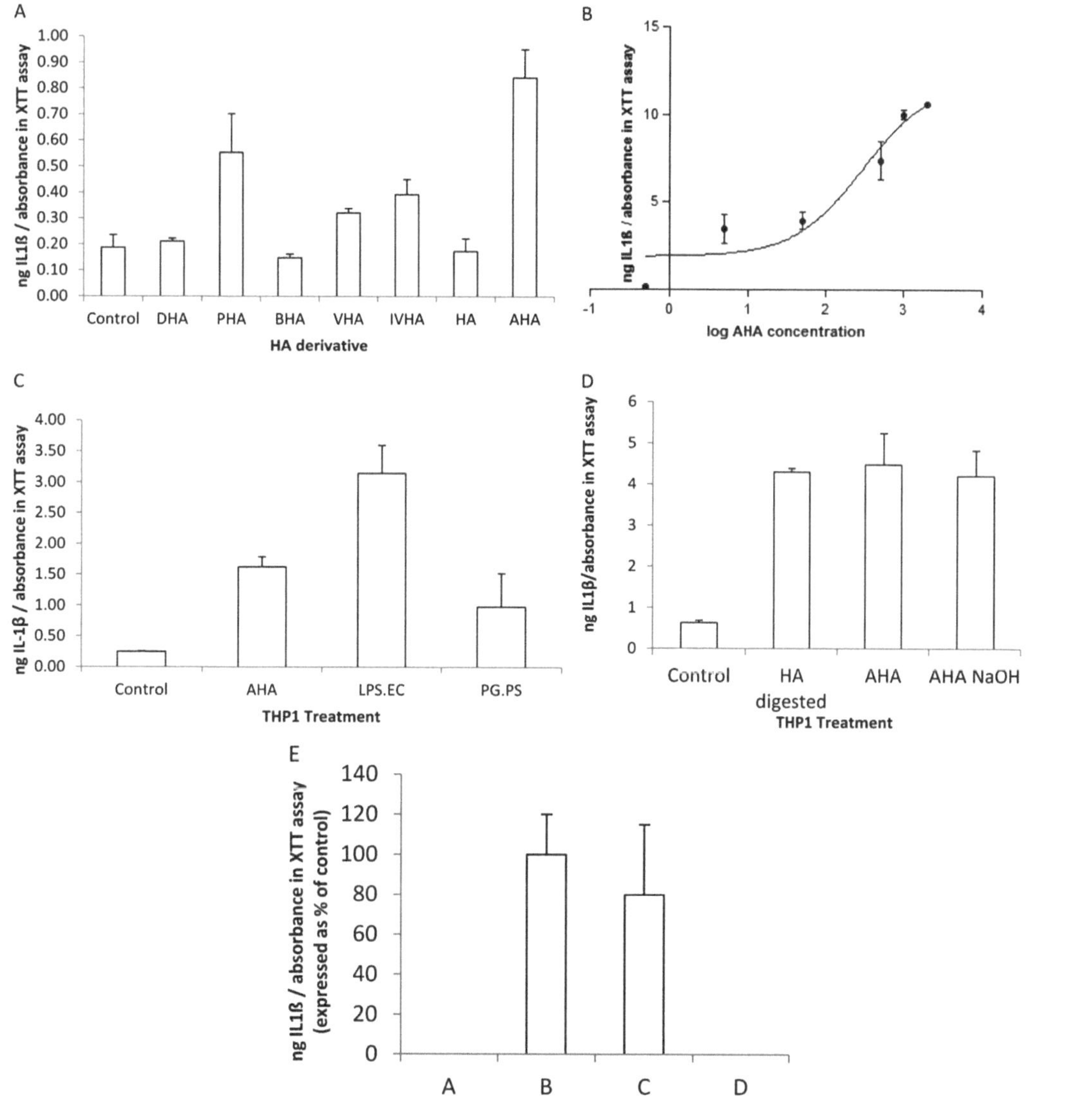

Fig. 4 Stimulation of the inflammatory response in THP-1 macrophages. In this figure, the total ng of cytokines secreted per well by THP-1 cells, determined by ELISA, is divided by the absorbance measured at 450 nm using the XTT/PMS cell proliferation assay, as an estimate of the ng of cytokines secreted per cell. The data are means $\pm$ S.D. ($n = 3$). (**a**) THP-1 IL-1*J3* secretion in the presence of different HA derivatives. THP-1 cells were treated with 500 μg/mL, for 24 h of: *DHA* deacetylated HA, *PHA* propionylated HA, *BHA* butyrylated HA, *VHA* valerylated HA, *IVHA* isovalerylated HA. Treatment with reacetylated HA (AHA) was significantly greater than the control (no HA derivative) ($p = 0.0486$, Student's *t* test). (**b**) THP-1 IL-1*J3* secretion in the presence of increasing concentrations of AHA. There was a dose-dependent response of IL-1*J3* secretion with increasing concentrations of AHA. The EC50 is 292 μg/mL. (**c**) THP-1 IL-1*J3* secretion in the presence and absence of AHA and a TLR-4 agonist (LPS.EC), or a TLR-2 agonist (PG.PS). Stimulation of IL-1*J3* by AHA (500 mg/mL), THP-1 was treated with 500 μg/mL of AHA and 250 ng/mL of TLR-4 agonist (lipopolysaccharide from *E. coli 0111:B4*, LPS.EC), or 250 ng/mL TLR-2 agonist (peptidoglycan polysaccharide from *S. pyogenes* group A, PG. PS). For AHA and LPS.EC, the increase in IL-1*J3* secretion over the untreated control is statistically significant ($p < 0.05$, Student's *t* test). (**d**) THP-1 IL-1*J3* secretion in the presence of acetylated low molecular mass

4 Notes

1. Since pH is dependent on temperature, we recommend adjusting the solution pH at room temperature (25 °C). Sterilize the solution by passing through 0.22 μ filter unit. Filter sterilization removes all suspended particles with sizes more than 0.22 μ which includes most bacteria, their spores but not mycoplasma. Autoclaving inactivates most enzymes except some (e.g., RNases) but kills most microorganisms including mycoplasma. Solutions can be stored at 15–25 °C (room temperature) for several months.
2. Use 0.5× TBE buffer or 1× TAE buffer for running and preparation of agarose gel. Always use the same running buffer in which the agarose gel is prepared. Sometimes, there is a significant loss of water during the melting process, which depends on the melting procedure. You can calculate the loss of water by weighing the flask/bottle just before and after the melting process. Loss of water is significantly high when you prepare agarose suspension in conical flask/beaker (no lid) and melt it in a microwave to melt agarose suspension. Agarose gel can be stored for few days. To store the agarose gel, we recommend not to remove comb and tape. Dip the agarose gel in the TAE buffer so that it contains moisture. Seal the gel in a plastic wrap and store in the cold room (4 °C). Before starting electrophoresis, warm to room temperature.
3. The reacylation reactions with the anhydrides are carried out in a temperature-controlled oven that has a shaker and is in a well-ventilated place. Use protective gloves made of Neoprene. Wear approved safety goggles. Wear appropriate clothing to prevent any possibility of skin contact.
4. Procedures that follow in the deacetylation process must be done in a fully functional fume hood.

Fig. 4 (continued) HA. THP-1 cells were then treated with 500 μg/mL of these acetylated low molecular mass HA derivatives for 24 h. Abbreviations of the low molecular HA are: reacetylated with acetic anhydride (*AHA*), HA partially digested with bovine testicular hyaluronidase (*HA digested*), and HA deacetylated with NaOH and reacetylated with acetic anhydride (*AHA NaOH*). Control is THP-1, that is not treated with any low molecular mass HA compound. (**e**) THP-1 IL-1*J3* secretion in the presence of size-fractionated and unfractionated hyaluronidase digests of HA (expressed as percent of control). The HA was digested for 30 min with testicular hyaluronidase to generate low molecular mass HA. THP-1 cells were treated for 24 h with either total/unfractionated low molecular mass HA or size-fractionated low molecular mass HA, and were adjusted to equivalent molar range. (*column A*) is the control THP-1 cultures that were not treated with a low molecular mass HA compound, Total/unfractionated low molecular mass HA is (*column B*) is low molecular mass HA >30 kDa (*column C*) was used at 500 μg/mL. The low molecular mass HA fraction of <30 kDa (*column D*) was used at 12.5 μg/mL [20]

5. Dialysis membrane used has a mass molecular weight cutoff at 6000 Da. Dialysis performance is determined by factors that affect the rate of diffusion of a molecule such as temperature, membrane thickness and solution concentration as well as molecular weight of the molecules in solution. Recommended temperature is room temperature and membrane thickness of between 12 and 30 μm. More concentrated solution will speed up the process.
6. It is very important to use the rotor evaporator to remove the acetone and this might remain in the sample. It is not to be assumed that lyophilization will take care of what is left in the sample at that stage. The observation has been that because the molecular weight of HA is very big, some acetone remains lodged in the molecule and acetone has been detected with samples that have not gone through the rotor evaporation process.
7. This process must be performed under a fully functional fume hood with appropriate clothing to prevent any possibility of skin contact.
8. The agarose gel electrophoresis method is based on an optimized method for HA separations [18]. Please pay particular attention to the following points. The agarose gel should be submerged in buffer, but take care not to add too much electrophoresis buffer over the agarose gel since it can cause slow run and distorted bands. Place the lid on the gel box, and connect electrodes with power supply using electric wires, supplied with the electrophoresis apparatus. The distance between electrodes in the electrophoretic apparatus determines the maximum voltage, which should not exceed 5 V/cm. Since the most commonly used electrophoresis devices have a distance of 14–20 cm between electrodes, 70–100 V is the default voltage. By detecting air bubbles from the electrodes (negative electrode), one can ensure that the electric supply is functional. Always place the lid of electrophoretic apparatus to avoid electric shock. Make sure that connections to positive (red in color) and negative poles (blue in color) to power pack are proper. If it is reversed, sample will run in the opposite direction and will come out of the gel. The direction of the run can be monitored by observing the movement of the loading dye.
9. Use 40 mm-diameter cone-and-plate geometry with a 1° cone and roughened surfaces to avoid wall slip. Before performing any experiments, preshear samples at a shear rate of 100 s^{-1} for 30 s, followed by a rest period of 30 s in order to create a controlled initial state in the samples.
10. THP-1 is a human monocytic line derived from the peripheral blood of a 1 year old patient with acute monocytic leukemia

and can be adapted to resemble macrophage-like cells for immune regulation [19]. PMA is required for the differentiation of the THP-1 cells to a macrophage phenotype, which then will adhere to plastic. The differentiated THP-1 cultures should be checked for adherence to the plastic flasks by gently disturbing the medium while observing the cells under the inverted tissue culture microscope. If adherence is lost, consider obtaining a fresh line by contacting the provider. PMA is a skin irritant and should be handled with protective gloves, as in **Note 3**.

11. The TLR-2 agonist is purified peptidoglycan polysaccharide polymer from the cell wall of *Streptococcus pyogenes* group A, D58 strain. LPS.EC (L4391) is lipopolysaccharide from *E. coli* 0111:B4. The TLR-4 antagonist, LPS-RS Ultrapure (tlrl-prslps), has been isolated from the bacterium *Rhodobacter sphaeroides*. This ultrapure grade has only TLR-4 antagonist activity, as opposed to the standard grade, which has TLR-4 antagonist and TLR-2 agonist activities.
12. If it is desired to carry out a comparison between equimolar amounts of different N-acyl derivatives of low molecular mass HA in order to determine relative degrees of either cytokine stimulation or inhibition [20], it is essential that the partial deacetylation of the high molecular mass HA be done in one large batch. Aliquots of DHA from this batch are then used for obtaining the various low molecular mass N-acyl HA derivatives, utilizing a large molar excess of the corresponding acyl anhydride, as described in Subheading 3.3.

Acknowledgments

This work was supported by National Sciences and Engineering Research Council of Canada (NSERC) Strategic Grant 381564-09.

References

1. Beaty NB, Mello RJ (1987) Extracellular mammalian polysaccharides: glycosaminoglycans and proteoglycans. J Chromatogr 418:187–222
2. Whitfield C (2006) Biosynthesis and assembly of capsular polysaccharides in *Escherichia coli*. Ann Rev Biochem 75:39–68
3. Widner B, Behr R, Von Dollen S et al (2005) Hyaluronic acid production in *Bacillus subtilis*. Appl Environ Microbiol 71:3747–3752
4. Pulz O, Gross W (2004) Valuable products from biotechnology of microalgae. Appl Microbiol Biotechnol 65:635–648
5. Wang T, Jonsdottir R, Ólafsdóttir G (2009) Total phenolic compounds, radical scavenging and metal chelation of extracts from Icelandic seaweeds. Food Chem 116:240–248
6. Wijffels RH, Barbosa MJ (2010) An outlook on microalgal biofuels. Science 329:796–799
7. Cavalier-Smith T (2007) Evolution and relationships. In: Brodie J (ed) Unravelling the algae: the past, present, and future of algal systematics. CRC Press, Boca Raton, FL USA, p 21
8. Weldon S, Mitchell S, Kelleher D et al (2004) Conjugated linoleic acid and atherosclerosis:

no effect on molecular markers of cholesterol homeostasis in THP-1 macrophages. Atherosclerosis 174:261–273

9. Scheibner KA, Lutz MA, Boodoo S et al (2006) Hyaluronan fragments act as an endogenous danger signal by engaging TLR2. J Immunol 177:1272–1281
10. Crescenzi V, Francescangeli A, Renier D et al (2002) New cross-linked and sulfated derivatives of partially deacetylated hyaluronan: synthesis and preliminary characterization. Biopolymers 64:86–94
11. Crescenzi V, Francescangeli A, Segre A et al (2002) NMR structural study of hydrogels based on partially deacetylated hyaluronan. Macromol Biosci 2:272–279
12. Wada T, Chirachanchai S, Izawa N (1994) Synthesis and properties of hyaluronic-acid conjugated nucleic-acid analogs: 1. synthesis of deacetylhyaluronan and introduction of nucleic-acid bases. J Bioact Compat Pol 9:429–447
13. Anastassiades TP (1973) Effect of a synthetic hexosamine derivative on mucopolysaccharide synthesis by human capsule and synovium. Biochem Pharmacol 22:3013–3023
14. Zhang W, Mu H, Zhang A (2013) A decrease in moisture absorption–retention capacity of N-deacetylation of hyaluronic acid. Glycoconj J 30:577–583
15. Maharjan AS, Pilling D, Gomer RH (2011) High and low molecular weight hyaluronic acid differentially regulate human fibrocyte differentiation. PLoS One 6:e26078
16. Larionova NI, Zubaerova DK, Guranda DT et al (2009) Colorimetric assay of chitosan in presence of proteins and polyelectrolytes by using o-phthalaldehyde. Carbohydr Polym 75:724–727
17. Cesaretti M, Luppi E, Maccari F et al (2003) A 96-well assay for uronic acid carbazole reaction. Carbohydr Polym 54:59–61
18. Cowman MK, Chen CC, Pandya M et al (2011) Improved agarose gel electrophoresis method and molecular mass calculation for high molecular mass hyaluronan. Anal Biochem 417:50–56
19. Chanput W, Mes JJ, Wichers HJ (2014) THP-1 cell line: an in vitro cell model for immune modulation approach. Int Immunopharmacol 23:37–45
20. Babasola O, Rees-Milton KJ, Bebe S et al (2014) Chemically modified N-acylated hyaluronan fragments modulate proinflammatory cytokine production by stimulated human macrophages. J Biol Chem 289:24779–24791
21. Roehm NW, Rodgers GH, Hatfield SM et al (1991) An improved colorimetric assay for cell-proliferation and viability utilizing the tetrazolium salt Xtt. J Immunol Methods 142:257–265

Chapter 10

Assays for the Enzymes Catalyzing the O-Acetylation of Bacterial Cell Wall Polysaccharides

Ashley S. Brott, David Sychantha, and Anthony J. Clarke

Abstract

The polysaccharides that comprise bacterial cell walls are commonly O-acetylated. This modification confers resistance to hydrolases of innate immune systems and/or controls endogenous autolytic activity. Herein, we present protocols for the compositional analysis of bacterial cell wall O-acetylation, and assays for monitoring *O*-acetyltransferases and *O*-acetylesterases. The assays are amenable for the development of high-throughput screens in search of inhibitors of the respective enzymes.

Key words O-Acetylation, *O*-acetyltransferase, *O*-acetylesterase, Peptidoglycan, Chitooligosaccharides

1 Introduction

The O-acetylation of one or more constituent carbohydrates in cell walls represents a form of preservation used by both plants and bacteria. In plants, the O-acetylation of noncellulosic cell wall polysaccharides, including xylan and the galacturonans, is very common (reviewed in [1]). Likewise, the O-acetylation of bacterial cell wall polysaccharides is a common occurrence. In addition to peptidoglycan (PG) [2], the secondary cell-wall polysaccharides of some gram-positive bacteria [3], the bacterial cellulose associated with biofilms [4], and the alginate of Pseudomonads [5] are O-acetylated. With each, the modification sterically hinders binding of respective lytic enzymes that normally target the wall component. As such, the enzymes conferring the modification have attracted the attention of researchers seeking alternative targets for antibiotic development. This chapter presents protocols that we developed for the study and characterization of the enzymes involved with PG O-acetylation [6–12], but they can be adapted for the investigation of other systems. Indeed, these assays have been applied to studies on the secondary-cell wall polysaccharides of bacilli [13], *Pseudomonas aeruginosa* alginate [14], mycobacterial

Inka Brockhausen (ed.), *Bacterial Polysaccharides: Methods and Protocols*, Methods in Molecular Biology, vol. 1954, https://doi.org/10.1007/978-1-4939-9154-9_10,

lipopolysaccharides [15], an extracellular glycoprotein involved in Sec transport [16], plant xylan [17], and mammalian sialoglycans [18].

PG is both essential and unique to bacteria. These properties have made it, and the enzymes associated with its metabolism, proven and effective targets for antibiotics. However, the issues related to antibiotic resistance have required the search and development of alternative strategies to combat problematic infections. One strategy is to develop antivirulence agents which would aid the innate immunity system of the host to more effectively manage the infecting bacterium. A promising target involved in PG metabolism that is common to both gram-positive and gram-negative pathogens is O-acetylation. Despite being ignored in textbook descriptions, PG is often modified by O-acetylation. Indeed, the PG of *Staphylococcus aureus,* the first to be have its composition characterized [19, 20] and used widely as a model for PG structure, is heavily O-acetylated [21]. The PG produced by most other gram-positive species is also O-acetylated (only a few Bacilli and nonpathogenic Staphylococci appear to be the exceptions), as is the PG supporting the walls of many gram-negative pathogens (recently reviewed in [2]).

The O-acetylation of PG occurs predominantly at the C-6 hydroxyl group of *N*-acetylmuramoyl (MurNAc) residues (Fig. 1); the O-acetylation of the C-6 hydroxyl of *N*-acetylglucosaminyl (GlcNAc) residues has been reported for some species of Lactobacilli [22]. As the C-6 hydroxyl of MurNAc represents a major contributor to recognition and then binding by lysozyme, its O-acetylation confers resistance to this first line of attack by innate immune systems against gram-positive pathogens where the PG is exposed (Fig. 1). With gram-negative bacteria, their O-acetylated PG (*O*-acetyl-PG) would also be protected from lysozyme digestion by O-acetylation if other components of the immune system succeed in permeabilizing the outer membrane. However, PG O-acetylation in these bacteria plays a major housekeeping function as it controls the autolytic activity of the complement of lytic transglycosylases (LTs) that are produced to permit PG biosynthesis, modification (for insertion of for example flagella, secretion systems), and turnover [23]. The product of LT activity is 1,6-anhydroMurNAc thus requiring a free C-6 hydroxyl; the presence of a C-6 acetyl group totally precludes the lytic activity (Fig. 1). Given these critical roles, the enzymes involved in PG O-acetylation and, in gram-negative bacteria, its de-O-acetylation, have attracted attention as potential new antibiotic/antivirulence targets.

As with the O-acetylation of other bacterial cell wall components, such as the secondary cell-wall polysaccharides produced by some gram-positive pathogens [3] and the alginate produced by *P. aeruginosa* [5], the O-acetylation of PG occurs as a post biosynthetic modification [2, 24]. Consequently, acetyl groups have to be

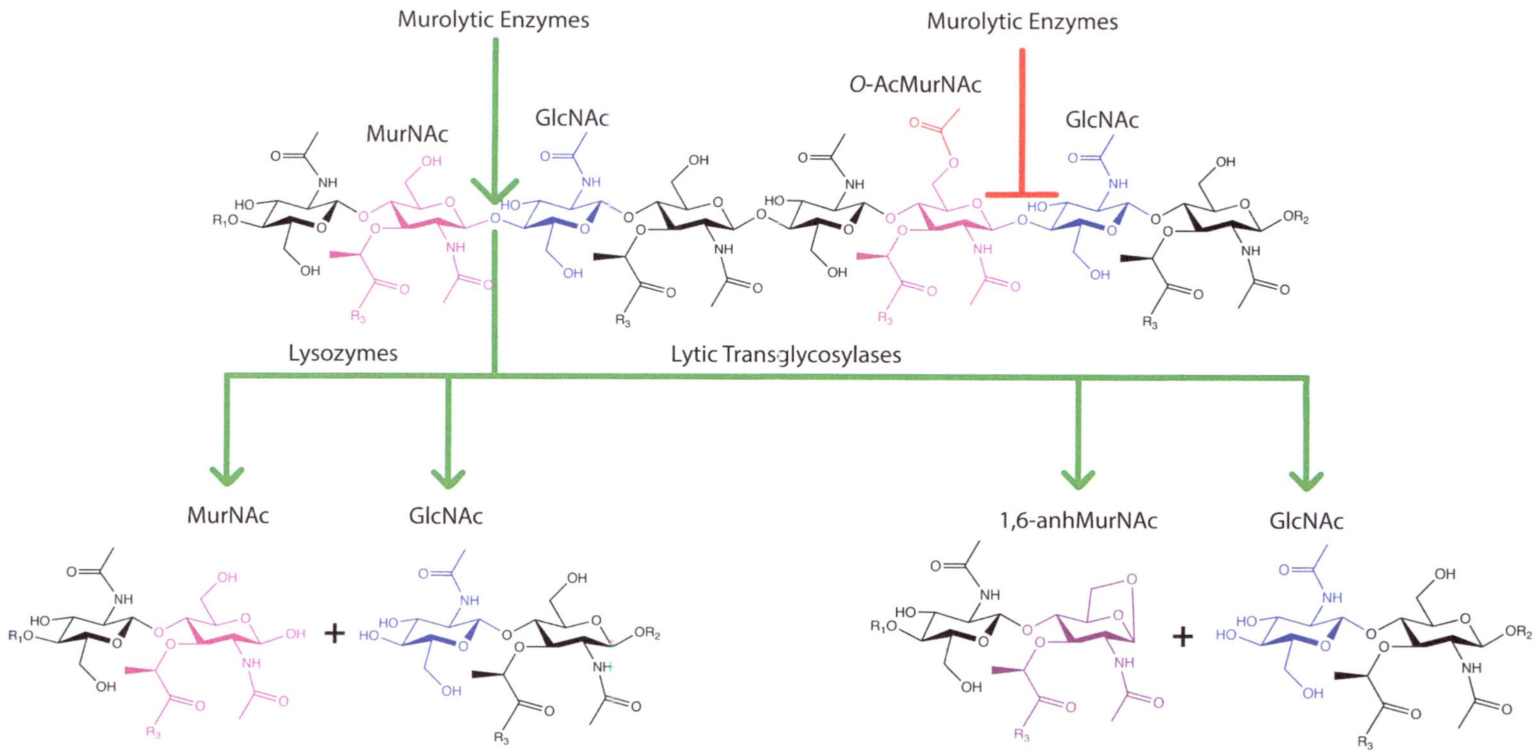

Fig. 1 PG as a substrate for lysozymes and LTs. Both lysozymes and the lytic transglycosylases (LTs) are muralytic enzymes cleaving the β-1,4 glycosidic linkage between MurNAc and GlcNAc residues in PG. However, the LTs are not hydrolases and instead of adding water across these linkages, they utilize the C-6 hydroxyl group of MurNAc residues to form a 1,6-anhydroMurNAc (1,6-anhMurNAc) reaction product. O-Acetylation of PG occurs at the C-6 hydroxyl group of MurNAc residues to preclude, and therefore control, LT activity, and sterically hinder lysozyme binding, thereby providing protection from the lytic action of innate immune systems

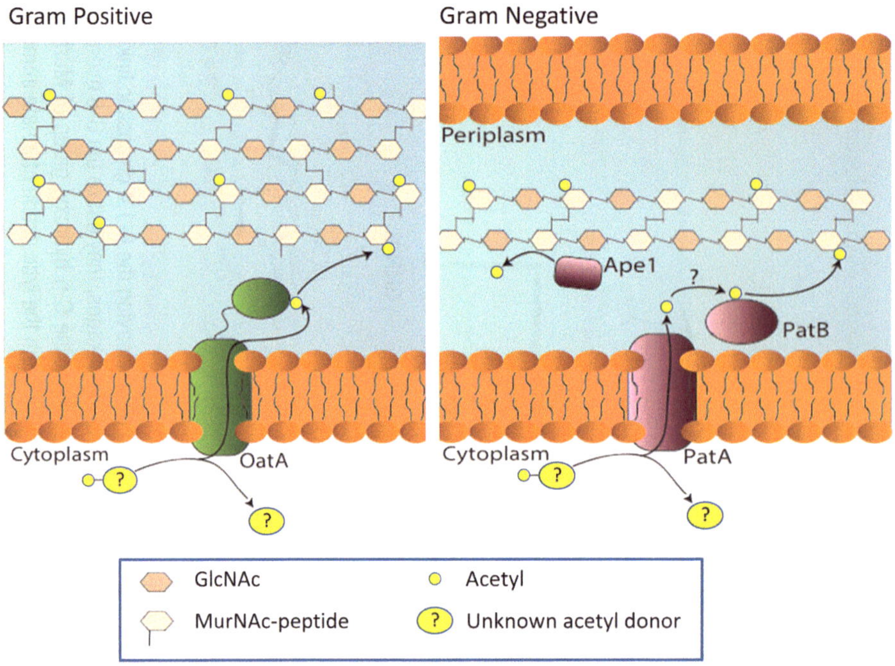

Fig. 2 Pathways for the O-acetylation of PG. In gram-positive bacteria, the N-terminal domain of OatA spans the cytoplasmic membrane and functions to translocate acetyl groups from a cytoplasmic source for their transfer to PG by the extracellular C-terminal domain. With gram-negative bacteria, these two activities are catalyzed by two distinct proteins, PatA and PatB. A third enzyme, Ape, is localized to the periplasm to remove blocking *O*-acetyl groups as required for continued PG metabolism. With both bacterial types, both the cytoplasmic source of acetyl groups and whether or not a membrane-associated membrane carrier is involved remain unknown

translocated from a source in the cytoplasm across the cytoplasmic membrane for its attachment to the existing PG sacculus by extracellular/periplasmic transferases (Fig. 2). Both of these activities are catalyzed by a single protein in gram-positive bacteria. Thus, the membrane-spanning N-terminal domain of *O*-acetyltransferase A (OatA) is thought to translocate acetyl groups from a yet to be identified cytoplasmic source across the cytoplasmic membrane for their subsequent transfer to muramoyl residues by the extracellular C-terminal domain [12, 25]. In gram-negative bacteria, these two functions are catalyzed by distinct proteins. Peptidoglycan *O*-acetyltransferase (Pat) A is a predicted integral membrane protein and a member of the MBOAT family of acyl-transferases. It is proposed to catalyze the translocation reaction, while periplasmic PatB receives the acetyl groups for their transfer to PG [8]. With these bacteria, an *O*-acetyl-PG esterase (Ape) also localized to the periplasm removes blocking acetyl groups as needed for continued PG biosynthesis or modification [7] (Fig. 2). Details of these enzymatic systems have only recently been discerned [2, 7–12], reflecting the complexity of investigating membrane-associated and -bound enzymes that act on their totally insoluble substrate. These advances were made possible with the development of appropriate in vitro assays that adopted soluble substrate mimics [9, 12]. Thus, the mechanism of action of both the transferases and esterases involves a catalytic triad of Ser-His-Asp which catalyze a ping-pong, bi–bi reaction that proceeds through a covalent acetyl-enzyme intermediate [11, 12] (Fig. 3). The following presents the assays used for these studies, as well as the more general protocols for PG analysis.

2 Materials

All solutions are prepared with ultrapure water (sensitivity of 18 MΩ-cm at 25 °C) and analytical grade reagents or better. Once prepared, store all reagents and enzymes at 4 °C and on ice, respectively (unless indicated otherwise).

2.1 Preparation of O-Acetylated Peptidoglycan (Sacculi)

1. Bacteria: Wash bacteria from 1 L of bacterial culture with 100 mL volumes of cold (4 °C) 10 mM sodium phosphate buffer, pH 6.0.
2. 25 mM sodium phosphate buffer, pH 6.0; 8% (*w/v*) SDS (prepared in the same phosphate buffer); 500 mL round-bottom flask fitted with a reflux condenser, ultracentrifuge.
3. 50 mM sodium phosphate buffer, pH 6.8 containing 20 mM $MgSO_4$ and 0.02% NaN_3 (*see* **Note 1**). DNase, RNase, and α-amylase.
4. Pronase.
5. 25 mM sodium phosphate buffer containing 2% (*w/v*) SDS. 48% (*v/v*) hydrogen fluoride (HF) (*see* **Note 2**). 1 mg/mL stock solution of Stains-all in 50% (*v/v*) isopropanol. 96-well microtiter plate, UV spectrophotometer, Bead beater homogenizer.

2.2 Soluble PG-Derived Muropeptides

1. Resuspension buffer: 50 mM sodium phosphate buffer, pH 6.2, containing 0.2% (*w/v*) NaN_3 and 5 mM $MgCl_2$.
2. Mutanolysin (Sigma-Aldrich).

2.3 Linear Homologous Oligomuropeptides

2.3.1 Muroglycan-5P

1. Lys-containing Lipid II: Suspend Lipid II (Antimicrobial Discovery Solutions Ltd., Coventry, UK) in 50 mM HEPES buffer, pH 7.5, containing 200 mM NaCl, 25 mM $MgCl_2$, 25% (*v/v*) dimethylsulfoxide (DMSO), and 0.02% Triton X-100.
2. 10 μM solution of recombinant *S. pneumoniae* PBP2a in 50 mM HEPES buffer, pH 7.5, containing 200 mM NaCl,

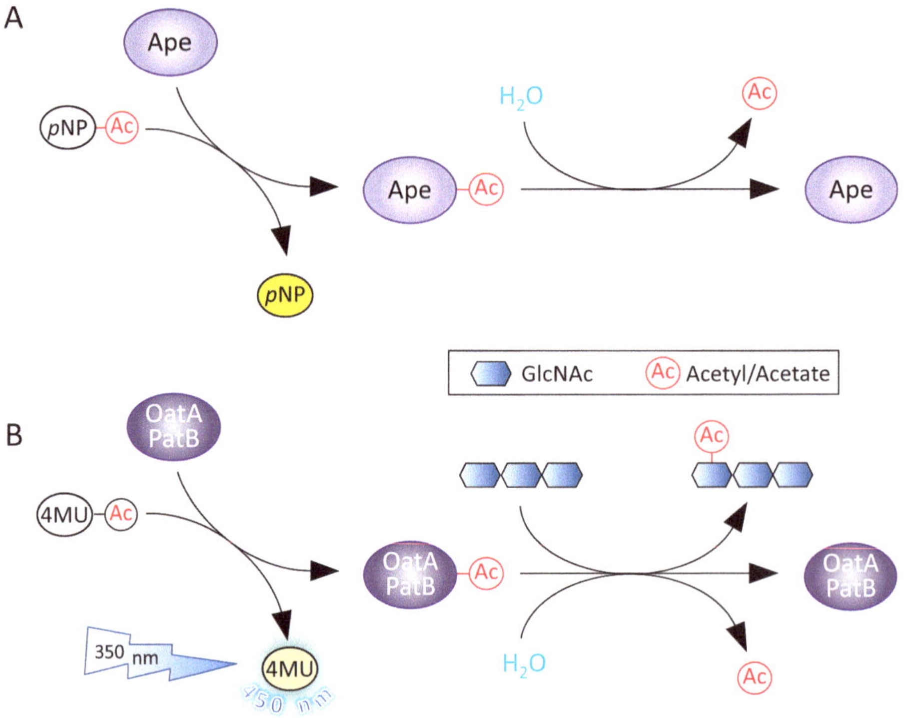

Fig. 3 Assays for *O*-acetyl-PG esterase (Ape) and PG *O*-acetyltransferases (OatA/PatB). Each of the three enzymes function like classic serine esterases and their mechanisms of action proceed through a covalent acetyl intermediate involving a ping-pong, bi–bi reaction pathway. The first step involves the enzymes acquiring the acetyl group from a donor (e.g., pNP-Ac or 4MU-Ac) and release the nonacyl group (e.g., pNP or 4MU, respectively) as the first product. In the second step the acetyl-enzyme intermediate binds an acceptor (e.g., water or a carbohydrate) to produce the second and final product, acetate or O-acetylated carbohydrate, respectively. (**a**) Reaction of Ape with pNP-Ac. The released pNP is chromogenic, absorbing maximally at 405–420 nm. (**b**) Reaction of OatA/PatB as either an esterase or a transferase. Like Ape, both OatA and PatB can use water as an acceptor to release acetate as the second product. However, oligosaccharides are the preferred acceptor substrate, leading to the production and release of the corresponding O-acetylated product. In both situations, the reporter group (e.g., 4MU as depicted) is released during the first half of the ping-pong, bi–bi reaction pathway (fluorescence detected at 450 nm)

25 mM $MgCl_2$ and 0.005–0.04% (*v/v*) Triton X-100, and store on ice until required.

3. Stock solutions (50 μM) of recombinant *S. pneumoniae* DacA and DacB in 50 mM sodium phosphate buffer pH 6.5.

2.3.2 Muroglycans-4P and 3P

1. Muroglycan-5P stock solution (Subheading 2.3.1) diluted to 15 μg/mL (final concentration) in 50 mM sodium phosphate buffer, pH 6.5.

2.4 Quantification of PG

1. Acid hydrolysis: 6 M HCl.
2. Instrumentation: High Performance Anion-Exchange Chromatography (HPAEC) with Pulsed-Amperometric Detection (PAD) [6].

2.5 Analysis of PG O-Acetylation

1. Release of *O*-acetyl ester: 0.5 M NaOH; 10 mM sodium phosphate buffer, pH 6.5 (for the negative control).
2. HPLC column: Aminex HPX-87H (Bio-Rad), or Rezex ROA-Organic Acid H+ (Phenomenex), equilibrated and in 5 mM H_2SO_4 at 60 °C.
3. Acetic acid analysis kit: K-ACETRM acetic acid analysis kit (Megazyme, Wicklow, Ireland).

2.6 Determination of Esterase Activity

1. Stock solution of 10–30 μM esterase in 50 mM sodium phosphate buffer at the desired pH; store on ice until required.
2. Stock pseudo-substrate solution of (1) 40 mM *p*-nitrophenylacetate (pNP-Ac) in ethanol, (2) 40 mM α-naphthyl acetate (αN-Ac) in ethanol, or (3) 10 mM 4-umbelliferyl-acetate (4MU-Ac) in DMSO. Using the respective stock substrate solution, prepare either 2 mM pNP-Ac, 2 mM αN-Ac, or 0.5 mM 4Mu-Ac in 50 mM sodium phosphate buffer, pH 7 with a final cosolvent (ethanol or DMSO) concentration of 5% (to retain the pseudosubstrate in solution).
3. Fluorescence spectrometer.

2.7 Determination of PG O-Acetyltransferase Activity

1. Stock solution of 10–30 μM enzyme in 50 mM sodium phosphate buffer, at the desired pH; store on ice until required.
2. Acetyl-donor stock solution at the desired concentration (e.g., 40 mM pNP-Ac in ethanol, or 10 mM 4MU-Ac in DMSO).
3. Stock solution of acceptor saccharide at the desired concentration (e.g., 30 mM chitooligosaccharide (Carbosynth, Berkshire, UK or Toronto Research Chemicals, Toronto, ON), or a concentrated pool of muropeptides in water as acceptor cosubstrate).
4. Quenching solution: cold perchloric acid.
5. HPLC: graphitized-carbon solid phase extraction column (Carbograph SPE) equilibrated in 0.1% (*v/v*) trifluoroacetic acid (TFA). Acetonitrile.
6. Elution solution: 2-propanol: acetonitrile in 0.1% TFA (or 0.1% formic acid), 50% tetrahydrofuran.
7. HPLC: Porous graphitized carbon (PGC) columns (Thermo Fisher Scientific).
8. Adsorption filtration using graphitized carbon: Supel-Tips™ Carbon micropipette tips (Sigma-Aldrich).
9. 0.1 M sodium phosphate buffer, pH 8.0, containing 1 mM EDTA.
10. 1 mM solution of DTNB in 50 mM sodium phosphate buffer (i.e., 0.4 mg/mL).

11. Mass spectrometry: Bruker Reflex III MALDI-TOF mass spectrometer (Bruker). SL ion-trap mass spectrometer (Amazon) for ESI-MS analysis. UHD 6530 Q-TOF mass spectrometer (Agilent). Bruker Compass tool (Bruker).

3 Methods

3.1 O-Acetylated Peptidoglycan (Sacculi)

The isolation and purification of insoluble PG sacculi is conducted by the boiling SDS procedure described by Hoyle and Beveridge [26] with modifications. Care needs to be taken to keep the pH of solutions below 7, especially at elevated temperatures, to preclude the ready hydrolysis of the ester-linked, *O*-acetyl groups. Also, it should be understood that PG O-acetylation levels are often growth-phase specific (e.g., [27]). The following protocol is described for a 1.0 L culture of bacteria which should yield 1–3 mg and approx. 10–15 mg of PG from gram-negative and gram-positive bacteria, respectively. It may be scaled appropriately for use with different culture volumes.

3.1.1 Isolation and Purification of Sacculi

1. Harvest bacteria cultured to the desired phase of growth by centrifugation (5000 × *g*, 15 min) at 4 °C. Wash the cell pellets twice by resuspending in 100 mL volumes of cold (4 °C) 10 mM sodium phosphate buffer, pH 6.0.
2. Resuspend the cell pellet in 150 mL 25 mM sodium phosphate buffer, pH 6.0 and add drop wise over 15 min into an equal volume of boiling 8% (*w/v*) SDS (prepared in the same phosphate buffer) in a 500 mL round-bottom flask (final SDS concentration, 4%). Fit with a reflux condenser and continue to boil for 2–3 h. Avoid using higher concentrations of phosphate buffer to preclude the formation of insoluble phosphate aggregates which cannot be separated from the insoluble PG product.
3. Allow suspension to cool to room temperature, and then transfer to ultracentrifuge tubes/bottles. Isolate insoluble PG sacculi by ultracentrifugation at ≥90,000 × *g* for 15 min at ambient temperature (*see* **Note 3**).
4. Wash pellet exhaustively with phosphate buffer by repeated ultracentrifugations at ambient temperature until traces of SDS are removed (usually a minimum of 3 washings are required—see below for SDS assay).
5. Using a Vortex, suspend pellet in 50 mM sodium phosphate buffer, pH 6.8 containing 20 mM $MgSO_4$ and 0.02% NaN_3 (*see* **Note 1**), and then incubate with (final concentrations of) 10 μg/mL DNase, 50 μg/mL RNase and 100 μg/mL

α-amylase for 3 h at 37 °C with nutation (to digest DNA, RNA and any glycogen trapped in the sacculi).

6. Add 100 μg/mL pronase (final concentration) and incubate with nutation at 37 °C overnight (for convenience; to digest added enzymes and any proteins covalently attached to PG).
7. Reisolate the PG sacculi by ultracentrifugation at ≥90,000 × *g*.
8. Resuspend the pellet in 5 mL phosphate buffer containing 2% SDS.
9. Incubate in a boiling water bath for 30 min.
10. Allow the suspension to cool to room temperature and reisolate the insoluble PG by ultracentrifugation at ≥90,000 × *g*.
11. Wash twice with water to remove traces of digestive enzymes and SDS, recovering PG each time by ultracentrifugation. This preparation of purified PG sacculi may be stored frozen, or in a lyophilized state (*see* **Notes 4–6**).

3.1.2 Removal of any O-Linked Glycans Associated with Gram-Positive Sacculi

1. Resuspend insoluble PG in a minimal volume of 48% (*v/v*) hydrogen fluoride (HF) precooled to 4 °C.
2. Incubate at 4 °C for 48 h with mixing.
3. Reisolate PG by ultracentrifugation (≥90,000 × *g*, 15 min, ambient temperature) and wash at least twice with water (or preferably with the buffer to be used in subsequent experiments) to remove excess HF, recovering the PG each time by centrifugation.

3.1.3 Assay for Residual SDS

1. Prepare a 1 mg/mL stock solution of Stains-all [28] in 50% isopropanol.
2. Prepare a working solution of Stains-all by first diluting a sample of the stock solution 1:1 in formamide, and then dilute 1:5 in water.
3. In a 96 well microtiter plate, mix 198 μL of sample with 2 μL of Stains-all working solution.
4. Measure absorbance at 438 nm for presence of SDS.

3.2 Soluble PG-Derived Muropeptides

Muropeptides are fragments, or combinations, of PG repeating units, GlcNAc-β-(1 → 4)-MurNAc with associated peptide, that may be attached through β-(1 → 4) glycosidic linkages and/or crosslinks between stem peptides. Pools of soluble muropeptides may be generated from isolated and purified PG by digestion with a muramidase. These pools will be highly heterogeneous given the varying crosslinking, amino acid substitutions and nonstoichiometric postsynthetic modifications that occur to the PG sacculus.

If the purpose is to produce O-acetylated muropeptides from an appropriate source PG (e.g., as a substrate for Ape), a muramidase that is not affected by the presence of O-acetylation should be used for these digestions. The commonly used enzyme in these situations is mutanolysin, a commercially available *N,O*-diacetylmuramidase from *Streptomyces globisporus.*

1. Resuspend purified PG in resuspension buffer.
2. Briefly sonicate (or vortex) the sample to provide an even suspension.
3. Add mutanolysin to a final concentration of 100 μg/mL and incubate at 37 °C overnight (for convenience).
4. Remove insoluble material (a limit digest will typically exist) by centrifugation at 20,000 × *g*, 4 °C and recover the supernatant containing the soluble muropeptides. Use immediately, or store at −20 °C until required (*see* **Note 7**).

3.3 Linear Homologous Oligomuropeptides

Pools of non-cross-linked strands of oligomuropeptides that differ only in their degree of polymerization are prepared by the enzymatic polymerization of the PG precursor Lipid II. Linear homopolymers are produced by truncated penicillin-binding protein (PBP) 2a (lacking its transmembrane anchor) from *Streptococcus pneumoniae* as its transpeptidase only catalyzes crosslinking after amidation of the Glu residue of the stem peptide [29, 30]; this amidation is catalyzed by a distinct amidase not included in the polymerization reaction. The degree of polymerization (DP) can be controlled by the Triton X-100 concentration: 0.04% and 0.005% concentrations will yield pools of muroglycans possessing associated stem pentapeptides (muroglycan-5P) with DP 2-10 and 8-22+, respectively [12]. The following presents a protocol that was recently described [12], and yields will be dependent on the amount of Lipid II starting material that can be obtained. Alternatively, it can be isolated in mg quantities from a bacterium of choice using a recently published protocol [31]. Also, recombinant soluble variants of *S. pneumoniae* PBP2a (covering residues 78-731) lacking its N-terminal transmembrane helix [29, 30], and both D,D-carboxypeptidase DacA (PBP3) and L,D-carboxypeptidase DacB [32] will need to be obtained, or generated [12].

3.3.1 Pools of Muroglycan-5P

1. Combine appropriate volumes of Lipid II suspension and PBP2a solution to give final concentrations of 15 μg/mL and 5 μM, respectively.
2. Incubate at 30 °C overnight (for convenience).
3. Heat inactivate the reaction mixture at 90 °C for 10 min.
4. Centrifuge at 10,000 × *g* (5 min) to pellet precipitated PBP2a.
5. Store supernatant containing the pool of muroglycan-5P at −20 °C until required (*see* **Note 8**).

3.3.2 Pools of Muroglycan-4P and Muroglycan-3P

1. To 100 μL of buffered stock muroglyan-5P add 10 μL buffered DacA alone (to generate muroglycan-4P) or 10 μL each of buffered DacA and DacB (to generate-3P) and incubate at 30 °C overnight (for convenience).
2. Heat inactivate the reaction mixture at 90 °C for 10 min.
3. Centrifuge at 10,000 × *g* (5 min) to pellet precipitated enzyme (s).
4. Store supernatant containing the respective muroglycans at −20 °C until required.

3.4 Quantification of PG

The most accurate determination of PG concentration is made by analysis for muramic acid (Mur) content as this amino sugar is almost exclusively found in PG. The amount (moles) of Mur present in a sample equates to an equivalent amount of PG repeating units (viz. GlcNAc-MurNAc-peptide). The method presented below is based on that first described in [6]. The acid hydrolysis procedure will convert all MurNAc and anhydro-MurNAc residues into muramic acid (Mur); GlcNAc will be hydrolyzed to glucosamine (GlcN).

3.4.1 Acid Hydrolysis

1. Transfer sample into a new a 10 mm × 75 mm test tube and dry in vacuo.
2. Add 200 μL of 6 M HCl.
3. With two hands, hold each end of the tube at a 45° angle (to hold the acid in) and slowly spin while heating its center using an oxy-natural gas torch.
4. When the center of the tube is soft and pliable, gently pull the two ends apart to constrict the tube, thereby creating a thin capillary space between the two halves (Fig. 4). Carefully place the constricted tube into a rack and set aside to cool.
5. When cooled sufficiently, lower the bottom half of the tube into a dry ice–acetone bath to freeze solid the acid.
6. While remaining frozen, attach a vacuum line to draw a vacuum, and then seal the tube by melting the glass in the middle of the constriction while pulling the two halves apart. The vacuum should draw the molten glass in to seal each end, but applying a quarter twist of the two ends relative to each other aids the sealing process.
7. Place the sealed tube in a heating block set at 100 °C for 1 h.
8. To open the tube, use a file (or electric engraver) to etch a short horizontal line toward the top of the tube. (The etch does not have to extend around the circumference.)

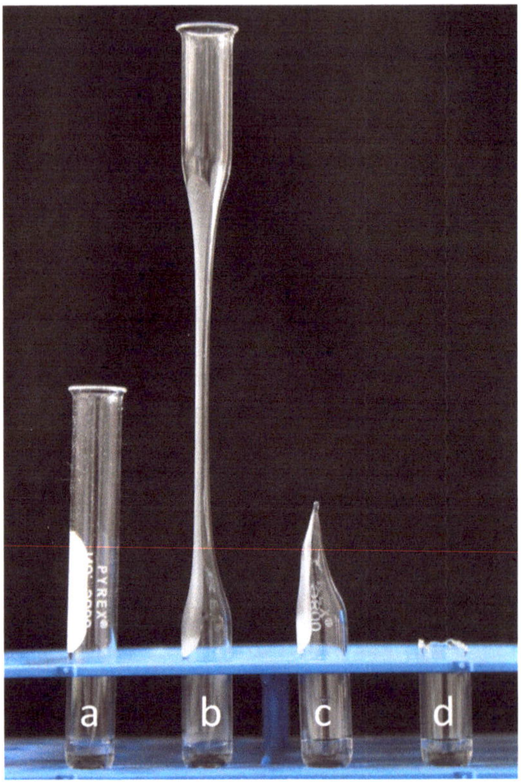

Fig. 4 Hydrolysis of PG in vacuo. Depicted are the various forms of the test tubes used for the in vacuo hydrolysis. (**a**) Sample and 6 M HCl in a total volume ≤200 μL are introduced to the bottom of a Pyrex® 10 mm × 75 mm tube. (**b**) A capillary is formed in the middle of the tube following heating with an oxy-natural gas flame. (**c**) Sealed tube. A seal is formed by melting the capillary in the flame while simultaneously drawing a vacuum in the tube. (**d**) Opened tube after incubation at 100 °C in a heating block

9. Add a drop of water to the etch and then apply a small drop of molten glass to the etch (e.g., a melted tip of a Pasteur pipette). A crack should form around the circumference of the tube, from one end of the etch to the other.
10. Snap off the top of the tube, and dry the sample in vacuo in a desiccator over NaOH pellets. Ensure that the vacuum pump is equipped with a trap so that the volatile acid does not reach the pump.
11. Cover the tube with Parafilm™ and store at −20 °C until required for analysis.

3.4.2 Muramic Acid Analysis

Quantification of Mur can be performed using an amino acid analyzer or by High Performance Anion-Exchange Chromatography (HPAEC) with Pulsed-Amperometric Detection (PAD) [6]. With HPAEC-PAD:

1. Equilibrate a 3 mm × 150 mm CarboPac™ PA20 column (Dionex; or equivalent) in 5 mM NaOH at 0.6 mL/min.
2. Inject sample (in water) and elute isocratically for 10 min.
3. At 10 min, apply a linear gradient to 150 mM sodium acetate in 100 mM NaOH over 40 min. For optimal detection of the amino sugars, the PAD (equipped with a gold electrode) settings should be: E1, 0.13 V; E2, 0.6 V; and E3, −0.6 V, with applied durations of 500, 10, and 5 ms, respectively. Program integration over 200 ms beginning at 300 ms and set the response time as1 s. Under the conditions described, GlcN elutes at approx. 14 min, while the Mur elutes as a sharp peak at 49 min.

3.5 Analysis of PG O-Acetylation

The level of PG O-acetylation in a sample is determined based on the amount of ester-linked (base-labile) acetate relative to its Mur content. The released acetate may be quantified by either HPLC-based organic acid analysis or using a commercially available coupled-enzyme acetic acid analysis kit following manufacturer's instructions.

3.5.1 Release of Ester-Linked Acetate

1. Transfer 100 μL of PG sample (ideally suspended in water) into a 500 μL microfuge tube.
2. Add 50 μL 0.5 M NaOH and incubate at 37 °C for 3 h with gentle rocking/nutation (substitute NaOH with 10 mM sodium phosphate buffer, pH 6.5 for the negative control).
3. Centrifuge samples at ≥20,000 × *g* to pellet insoluble PG.
4. Assay supernatant for acetate.

3.5.2 Acetate Analysis: HPLC-Based

1. Equilibrate an organic acid analysis HPLC column in 5 mM H_2SO_4 at 60 °C.
2. Filter the sample through a 0.5 mL centrifugal filter device (0.22 μm).
3. Inject the filtrate and elute isocratically at 0.6 mL/min.
4. Monitor absorbance at 205 nm or 210 nm and determine concentration from peak area using a response factor calculated from a standard curve generated using acetic acid.

3.6 Determination of O-Acetyl-PG Esterase Activity

For both routine and extensive kinetic analyses of esterase activity, monitoring the hydrolysis of chromogenic/fluorogenic pseudo-substrates can be employed (Fig. 3) (*see* **Note 9**). These substrates include *p*-nitrophenyl acetate (pNPAc), α-naphthyl acetate (αN-Ac), and 4-methylumbelliferyl acetate (4MU-Ac). All are chromogenic but 4MU-Ac has the advantage of also being fluorogenic, thereby increasing the sensitivity of the assay. The following presents the assay first developed by Weadge and Clarke [7] for use with microtiter plates. The assay was subsequently applied for the high-throughput screening of inhibitors of Ape from *N. gonorrhoeae* [33].

1. Transfer 200 μL aliquots of buffered substrate solution to individual wells of a microtiter plate.
2. Initiate (complete) reactions with the addition of 10–50 μL enzyme solution (as appropriate to provide a final concentration of 1–3 μM).
3. As a (negative) control for the spontaneous release of product, substitute the enzyme in some wells with the appropriate volume of buffer.
4. Using a microplate reader, incubate at the desired temperature and monitor with time (for at least 5 min) the release of: pNP by absorbance at 405–420 nm; αN-Ac by absorbance at 320 nm; or 4MU either by absorbance at 355 nm or fluorescence using excitation and emission wavelengths of 350 nm and 450 nm, respectively.
5. Calculate net rates of the esterase-catalyzed reaction by subtracting the rates of hydrolysis of the negative control from those of the complete reactions.
6. One unit (U) of esterase activity was typically defined as the amount of enzyme required to produce 1 μmol of product·min^{-1}·mg $protein^{-1}$.

3.7 Determination of PG *O*-Acetyltransferase Activity

Like Ape, the PG *O*-acetyltransferases from both gram-positive and gram-negative bacteria, OatA and PatB respectively, are able to utilize the chromogenic/fluorogenic acetyl pseudosubstrates as donor substrates. Moreover, in the absence of an acceptor substrate, OatA and PatB can function as weak esterases with the transfer of the acetyl group to water [9, 10, 12]. Hence, the esterase activity described in Subheading 3.3 involving the chromogenic/fluorogenic pseudosubstrates can be used to routinely assess the catalytic state of an enzyme preparation (*see* **Note 10**).

PG-based oligomers that could be used as acceptor substrates are not commercially available. Whereas it is theoretically possible to utilize insoluble PG as acetyl acceptor to assay transferase activity, the follow-up analysis of the insoluble *O*-acetyl-PG product is complex and precludes any meaningful kinetic analyses. Unfortunately, the use of soluble metabolites derived from a partial digestion of PG as acceptors is also impractical given the significant heterogeneity of derived pools (*see* **Note 11**). However, both OatA and PatB recognize the more (structurally) simple chitooligomers (oligomers of GlcNAc) as acceptors and these are available from several fine chemical suppliers. Using these, a direct HPLC-based analysis can be used to quantify the *O*-acetyl-chitooligomer products over time to assess kinetic parameters. More conveniently though, the ability of OatA and PatB to use the chromogenic/fluorogenic pseudosubstrates as acetyl donor can be exploited. Hence, kinetic data can be readily obtained using a chromogenic/fluorogenic assay with defined cosubstrates as acceptors that is based on monitoring the release of the reporter group (Fig. 3) (*see* **Note 12**).

3.7.1 General O-Acetyltransferase Assay

1. Using the prepared stock solutions of donors and acceptors, add appropriate volumes of the two cosubstrates to a 1.5 mL microfuge tube to give the desired final concentrations (e.g., 2 mM pNP-Ac or 0.5 mM 4MU-Ac; 2 mM chitooligosaccharide (if a cosolvent (ethanol or DMSO) is involved, ensure final concentrations will be 5% to retain the respective acetyl donor in solution).
2. Initiate reactions by adding the appropriate volume of the enzyme solution to provide the desired final concentration (typically 1–3 μM enzyme). In a separate reaction, substitute the enzyme solution with buffer alone to provide a negative control.
3. Incubate at the desired temperature (typically 37 °C) for an appropriate amount of time.
4. Quench the further reaction of aliquots removed over time, or the entire volume, by either flash freezing, or with 2% (final concentration) of cold perchloric acid (*see* **Note 13**).

3.7.2 Isolation of Saccharides from Reaction Mixtures

1. Wash 4 mL graphitized-carbon solid phase extraction columns with three column volumes of 80% acetonitrile in 0.1% trifluoroacetic acid (TFA).
2. Wash and equilibrate the columns with three volumes of water.
3. Apply samples to column and wash out unbound material (i.e., acetyl-based cosubstrates, buffer components) with six column volumes of water.
4. Elute nonacidic saccharides (native and *O*-acetylated) with 1 mL of 50% acetonitrile in 0.1% TFA (or 0.1% formic acid); elute PG-derived muropeptides with a 3:1 mixture of 2-propanol: acetonitrile in 0.1% TFA (or 0.1% formic acid), followed by 50% tetrahydrofuran to remove any material still bound.
5. Dry samples in vacuo (using a Speed Vac).

3.7.3 Direct Quantitative Analysis of O-Acetylated Oligosaccharides by Graphitized Carbon HPLC

1. At appropriate intervals of time, transfer 100 μL samples of an enzyme–substrate mixture to separate microfuge tubes containing 30 μL of cold 8% perchloric acid to quench further reaction (1 M HCl can substitute for the perchloric acid).

2. Inject 50 μL aliquots of the quenched reaction mixture onto a 4.6 mm × 100 mm Hypercarb PGC column (3 μm) equilibrated in water at 1 mL/min and 80 °C.
3. Apply a linear gradient of 0–50% acetonitrile over 50 min.
4. Monitor the elution of O-acetylated products spectrophotometrically at 210 nm.
5. Determine response factors for quantification, if desired, by collecting fractions and analyzing for O-acetylation (*see* Subheadings 3.4 and 3.5).

3.7.4 Chromogenic/ Fluorogenic (Indirect) Assay Monitoring Coproduct Formation

The assay described above monitors the turnover of the chromogenic/fluorogenic acetyl donor as a consequence of both esterase and transferase activities. It is thus important to establish that the difference calculated in **step 7** below truly reflects the activity of only acetyl transfer to oligosaccharide acceptor. This can be accomplished by assaying representative reaction mixtures directly for *O*-acetyl product and/or the quantity of acetate released to the bulk solvent. For the latter, the free acetate should be isolated from reaction materials and other products prior to its analysis. This is performed by adsorption filtration using graphitized carbon. For convenience, the following describes the protocol using Supel-Tips™ Carbon micropipette tips [34].

1. Using the prepared stock cosubstrate solutions, prepare either 2 mM pNP-Ac, 2 mM αN-Ac, or 0.5 mM 4MU-Ac in 50 mM sodium phosphate buffer, pH 7 (final cosolvent (ethanol or DMSO) concentrations will be 5% (to retain the acetyl donors in solution)).
2. Transfer 200 μL aliquots of buffered acetyl-donor substrates to wells of a microtiter plate.
3. Transfer 25 μL of acceptor substrate into one half of the (complete) wells (final concentration will be 1 mM). Reaction products will arise from both transferase and esterase activity.
4. Add 25 μL water into the other half of the (control) wells. These will serve as controls for just esterase activity (both enzyme catalyzed and spontaneous).
5. Initiate reactions by adding 25 μL of enzyme solution to each of the wells (final concentration will be 1–3 μM).
6. Using a microplate reader, continuously monitor spectrophotometrically at 405–420 nm or 320 nm for pNP or αN release, respectively, or fluorometrically for 4MU release (excitation and emission wavelengths, 350 nm and 450 nm, respectively) for at least 5 min.
7. Determine the rates of transferase activity from the difference of the rates for the complete and control reactions lacking chitooligosaccharide (Fig. 5).

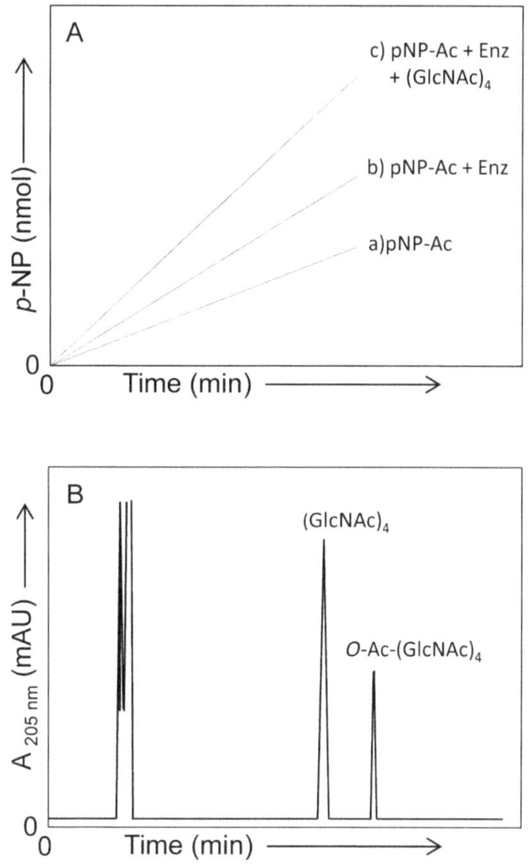

Fig. 5 Expected assay results. (**a**) Chromogenic assay of *O*-acetylesterase and *O*-acetyltransferase using pNP-Ac as substrate/co-substrate, respectively. (a) Spontaneous hydrolysis of pNP-Ac; control blank. (b) Increased release of pNP over control as a result of esterase activity catalyzed by an authentic esterase or an *O*-acetyltransferase. (c) pNP release as a result of both *O*-acetylesterase and *O*-acetyltransferase activities in the presence of an acceptor. The final (corrected) activity of an esterase and a transferase is determined from the difference between the rates of (b) and (a), and (c) and (b), respectively. (Note that pNP release in reaction (b) includes both spontaneous and enzyme-catalyzed hydrolysis which then serves as an appropriate control for the transferase activity.) (**b**) Separation of chitooligosaccharides by adsorption HLPC using a Hypercarb PGC column. Depicted are the reaction products following incubation of an *O*-acetyltransferase with an acetyl donor (e.g., 4MU-Ac) and chitotetraose. Under the conditions of the chromatography, the O-acetylated product elutes several minutes later

8. Attach a Supel-Tips Carbon tip to a 10 μL pipettor.
9. Draw and then dispense 10 μL of 100% acetonitrile, and then repeat.
10. In the same manner, wash three to five times with 10 μL water.

11. Submerge the tip into a $\leq$50 μL sample and without removing it, perform a series of ten aspirate and dispense cycles of 10 μL volumes (draw in and dispense out ten times). Oligosaccharides and acetyl-donor molecules will be retained on the carbon while acetate will remain in the dispensed solution.
12. Perform acetate analysis using a coupled-enzyme assay kit or by HPLC as described in Subheading 3.5.2.
13. If desired, elute oligosaccharides with 5 $\times$ 10 μL volumes of 50% acetonitrile in 0.1% TFA.

3.7.5 Assay for Acetyl-CoA Consumption (See **Note 14**)

1. Transfer 250 μL of phosphate buffer to the appropriate number of wells of a microtiter plate.
2. Add 15 μL of DTNB solution to each well.
3. Add $\leq$25 μL of sample to the wells prepared in **step 2**. For a blank(s), add an equal volume of phosphate buffer to separate (control) wells prepared in **step 2**.
4. Incubate at ambient temperature for 15 min.
5. Using a microplate reader, measure the absorbance at 412 nm of each sample/blank.
6. Calculate the concentration (M) of free sulfhydryl (CoASH) in each well using the molar extinction coefficient 14.150 M^{-1} cm^{-1}. Account for the dilution to determine the concentration of CoASH in the original sample (e.g., for a 25 μL sample tested, original concentration = $A_{412\ nm} \times 14.150\ M^{-1}\ cm^{-1} \times 290$ μL (total volume)/25 μL).

3.8 Mass-Spectrometric (MS) Analysis of Oligosaccharides

3.8.1 Matrix-Assisted Laser Desorption Ionization-Time of Flight (MALDI-TOF) MS

MALDI-TOF MS analysis of neutral sugars is achieved using 2,5-dihidroxybenzoic acid as a matrix, whereas 5-chloro-2-mercaptobenzothiazole should be used for amino sugars or PG derivatives. We have collected spectra in positive mode using a Bruker Reflex III MALDI-TOF mass spectrometer in reflectron mode equipped with a 337 nm nitrogen laser (set to 109–121 μJ output).

3.8.2 ESI-MS/MS

We conduct ESI-MS analysis using an Amazon SL ion-trap mass spectrometer. Samples are introduced by direct infusion at a flow rate of 5 μL/min with a spray voltage of 4.5 kV. The ion-trap is operated in positive ion mode with MS scans ranging from 200–2200 *m/z*. MS/MS scans are made on major ions with a fragmentation amplitude of 1.0. Mass spectra are analyzed using the Bruker Compass tool.

3.8.3 Electrospray Quadrupole-Time of Flight Mass Spectrometry (Q-TOF MS)

Q-TOF MS is performed on an Agilent UHD 6530 Q-TOF mass spectrometer. The MS electrospray capillary voltage is set at 4.0 kV and the flow rate was maintained at 0.2 mL/min. The drying gas temperature is 350 °C with a flow rate of 13 L/min, and the nebulizer pressure is 40 p.s.i. Mass/charge ratios are scanned across a range of 300–2000 *m/z* in 4 GHz in extended dynamic range positive ion mode. Precursor ions of O-acetylated muropeptides are selected for manual MS/MS with an isolation width of 1.5 *m/z* and a collision-induced dissociation reaction amplitude of 1 V.

4 Notes

1. NaN_3 is toxic and should be handled with care.
2. HF is extremely corrosive and should be handled with care.
3. Do not cool to 4 °C as SDS will precipitate.
4. With PG isolated from gram-positive bacteria, it may be necessary to remove cell wall-associated glycopolymers, such as wall teichoic acids or secondary cell-wall polysaccharides, as they may also be O-acetylated.
5. Residual SDS levels can be monitored using Stains-all dye as an indicator [28].
6. If PG fragments are desired rather than whole sacculi, a fragmentation protocol needs to be applied between **steps 4** and **5**. For gram-positive PG, resuspend the PG pellet in 5 mL of sodium phosphate buffer, mix 1:1 (volumetrically) with 0.1 mm silicon beads. Mechanically disrupt the sacculi using a bead beater homogenizer with 30 s bursts for a total of 3 min, cooling the sample on ice between bursts. With gram-negative PG, resuspend the PG pellet in 5 mL of sodium phosphate buffer. Sonicate PG using a microprobe for 2 min.
7. If the muropeptides are to be used as potential substrates for *O*-acetyltransferase assays, then the PG needs to be devoid of O-acetylation. This can be achieved by base hydrolysis of the purified insoluble PG prior to the muramidase digestion described above. Alternatively, PG sacculi could be isolated from a bacterium that does not O-acetylate its PG, such as *Escherichia coli*. The protocol described may be used to produce muropeptides from these PGs but with the option to replace mutanolysin with hen egg white lysozyme.
8. The oligomuropeptides obtained will be homogeneous in composition, each possessing stem pentapeptides (muroglycan-5P). Sequential treatment of muroglycan-5P with DacA and then DacB will provide uniformly tailored muroglycans with tetrapeptide (muroglycan-4P) and tripeptide (muroglycan-3P) stems, respectively.

9. Authentic *O*-acetyl-PG esterase (Ape) activity can be determined by monitoring the enzyme-catalyzed release of acetate from insoluble PG substrates (*see* Subheadings 3.4 and 3.5 for quantification of PG and acetate, respectively). However, the associated analyses of the PG and released acetate are somewhat laborious and hinder a full kinetic analysis of the enzyme reaction. Consequently, the use of insoluble PG as a substrate should be used to validate the specificity, and hence, identification of the enzyme as Ape.
10. Unlike the true esterase Ape, OatA and PatB are incapable of utilizing *O*-acetyl-PG as an esterase substrate.
11. These PG-based materials could, however, be used for qualitative analyses by for example mass spectrometry, to confirm the specificity, and hence identification, of the PG transferase. In these cases, it may be necessary to isolate the carbohydrate-based product(s) prior to analysis. This can be achieved by open-column adsorption chromatography on graphitized carbon. In other assays involving defined oligosaccharides as acceptors, reaction products can be separated and quantified directly by PGC-HPLC.
12. As this assay is indirect (in that it monitors only the release of the acetyl-donor product), a confirmatory assay should be conducted on a few samples to demonstrate directly the formation of an O-acetylated product.
13. Qualitative analysis of isolated reaction products can be conducted with the direct infusion of $\leq$10 μL of unquenched reaction mixtures into an electrospray ionization (ESI) mass spectrometer.
14. Acetyl-CoA does serve as an acetyl donor for *Neisseria gonorrhoeae* PatB [10] but it is not used by the C-terminal *O*-acetyltransferase domain of OatA from *Staphylococcus aureus* and *Streptococcus pneumoniae* [12] or secondary cell-wall PatB1 from *B. cereus* [13]. It and other acyl-CoAs, such as propionyl-, butyryl-, isobutyryl-, and hexanoyl-CoA, may be tested as an acetyl donor for other *O*-acetyltransferases. The routine assay for acyl-CoA consumption/turnover monitors the production of CoASH using the thiol reagent 5,5-′-dithiobis-(2-nitrobenzoic acid) (DTNB; Ellman's reagent) [35]. As sugar-based cosubstrates and products would not interfere with the thiol reagent, no prior sample preparation would be required.

Acknowledgments

Our research on *O*-acetylesterases and *O*-acetyltransferases has been supported by operating grants from the Canadian Institutes for Health Research, and more recently, the Canadian Glycomics Network, a National Centre of Excellence.

References

1. Gille S, Pauly M (2012) O-acetylation of plant cell wall polysaccharides. Front Plant Sci 3:12
2. Moynihan PJ, Sychantha D, Clarke AJ (2014) Chemical biology of peptidoglycan acetylation and deacetylation. Chemistry 54:44–50
3. Forsberg LS, Abshire TG, Friedlander A et al (2012) Localization and structural analysis of a conserved pyruvylated epitope in *Bacillus anthracis* secondary cell wall polysaccharides and characterization of the galactose-deficient wall polysaccharide from a virulent *B. anthracis* CDC 684. Glycobiology 22:1103–1117
4. Spiers AJ, Bohannon J, Gehrig SM et al (2003) Biofilm formation at the air-liquid interface by the *Pseudomonas fluorescens* SBW25 wrinkly spreader requires an acetylated form of cellulose. Mol Microbiol 50:15–27
5. Franklin MJ, Ohman DE (1993) Identification of algF in the alginate biosynthetic gene cluster of *Pseudomonas aeruginosa* which is required for alginate acetylation. J Bacteriol 175:5057–5065
6. Clarke AJ (1993) Compositional analysis of peptidoglycan by high-performance anion exchange chromatography. Anal Biochem 212:344–350
7. Weadge JT, Clarke AJ (2006) Identification and characterization of O-acetylpeptidoglycan esterase: a novel enzyme discovered in *Neisseria gonorrhoeae*. Biochemistry 45:839–851
8. Moynihan PJ, Clarke AJ (2010) O-acetylation of peptidoglycan in gram-negative bacteria: identification and characterization of peptidoglycan *O*-acetyltransferase in *Neisseria gonorrhoeae*. J Biol Chem 285:13264–13273
9. Moynihan PJ, Clarke AJ (2013) Assay for peptidoglycan *O*-acetyltransferase: a potential new antibacterial target. Anal Biochem 439:73–79
10. Moynihan PJ, Clarke AJ (2014) Substrate specificity and kinetic characterization of peptidoglycan *O*-acetyltransferase B from *Neisseria gonorrhoeae*. J Biol Chem 289:16748–16760
11. Moynihan PJ, Clarke AJ (2014) The mechanism of peptidoglycan *O*-acetyltransferase involves an Asp-His-Ser catalytic triad. Biochemistry 53:6243–6251
12. Sychantha D, Jones C, Little DJ et al (2017) Structure and molecular basis of catalysis of the peptidoglycan *O*-acetyltransferase A (OatA) catalytic domain. PLoS Pathog 13:e1006667
13. Sychantha D, Little DJ, Chapman RN et al (2017) Structural and mechanistic basis for the O-acetylation of SCWP, an essential activity for the proper assembly of cell walls in the *Bacillus cereus* group of pathogens. Nat Chem Biol 14:79–85
14. Baker P, Ricer T, Moynihan PJ et al (2014) *P. aeruginosa* SGNH hydrolase-like proteins AlgJ and AlgX have similar topology but separate and distinct roles in alginate acetylation. PLoS Path 10:e1004334
15. Maranha A, Moynihan PJ, Miranda V et al (2015) Octanoylation of early intermediates of mycobacterial methylglucose lipopolysaccharides. Sci Rep 5:13610
16. Seepersaud R, Sychantha D, Bensing BA et al (2017) O-acetylation of the serine rich repeat glycoprotein GspB is coordinated with accessory Sec transport. PLoS Path 13:e1006558
17. Urbanowicz BR, Peña MJ, Moniz HA et al (2014) Two *Arabidopsis* proteins synthesize acetylated xylan *in vitro*. Plant J 80:197–206
18. Baumann AM, Bakkers MJ, Buettner FF et al (2015) 9-O-acetylation of sialic acids is catalysed by CASD1 via a covalent acetyl-enzyme intermediate. Nat Commun 6:7673
19. Rogers HJ, Perkins HR (1959) Cell-wall mucopeptides of *Staphylococcus aureus* and *Micrococcus lysodeikticus*. Nature 184:520–524
20. Strominger JL, Park JT, Thompson RE (1959) Composition of the cell wall of *Staphylococcus aureus*: its relation to the mechanism of action of penicillin. J Biol Chem 234:3263–3268
21. Ghuysen J-M, Strominger JL (1963) Structure of the cell wall of *Staphylococcus aureus* strain Copenhagen. II. Separation and structure of disaccharides. Biochemistry 2:1119–11125
22. Bernard E, Rolain T, Courtin P et al (2011) Characterization of O-acetylation of *N*-acetylglucosamine: a novel structural variation of bacterial peptidoglycan. J Biol Chem 286:23950–23958

23. Scheurwater EM, Reid CW, Clarke AJ (2008) Lytic transglycosylases: bacterial space-making autolysins. Int J Biochem Cell Biol 40:586–591
24. Clarke AJ, Dupont C (1992) O-acetylated peptidoglycan: its occurrence, pathobiological significance, and biosynthesis. Can J Microbiol 38:85–91
25. Bera A, Herbert S, Jakob A et al (2005) Why are pathogenic staphylococci so lysozyme resistant? The peptidoglycan *O*-acetyltransferase OatA is the major determinant for lysozyme resistance of *Staphylococcus aureus*. Mol Microbiol 55:778–787
26. Hoyle BD, Beveridge TJ (1984) Metal binding by the peptidoglycan sacculus of *Escherichia coli* K-12. Can J Microbiol 30:204–211
27. Pfeffer JM, Strating H, Weadge JT et al (2006) Peptidoglycan O-acetylation and autolysin profile of *Enterococcus faecalis* in the viable but non-culturable state. J Bacteriol 188:902–908
28. Rusconi F, Valton E, Nguyen R et al (2001) Quantification of sodium dodecyl sulfate in microliter-volume biochemical samples by visible light spectroscopy. Anal Biochem 295:31–37
29. Helassa N, Vollmer W, Breukink E et al (2012) The membrane anchor of penicillin-binding protein PBP2a from *Streptococcus pneumoniae* influences peptidoglycan chain length. FEBS J 279:2071–2081
30. Di Guilmi AM, Dessen A, Dideberg O et al (2003) The glycosyltransferase domain of penicillin-binding protein 2a from *Streptococcus pneumoniae* catalyzes the polymerization of murein glycan chains. J Bacteriol 185:4418–4423
31. Qiao Y, Srisuknimit V, Rubino F et al (2017) Lipid II overproduction allows direct assay of transpeptidase inhibition by ß-lactams. Nat Chem Biol 13:793–798
32. Abdullah MR, Gutiérrez-Fernández J, Pribyl T et al (2014) Structure of the pneumococcal L,D-carboxypeptidase DacB and pathophysiological effects of disabled cell wall hydrolases DacA and DacB. Mol Microbiol 93:1183–1206
33. Pfeffer JM, Clarke AJ (2012) Identification of first-known inhibitors of *O*-acetylpeptidoglycan esterase: a potential new antibacterial target. Chembiochem 13:722–731
34. https://www.sigmaaldrich.com/content/dam/sigma-aldrich/docs/Supelco/Product_Information_Sheet/t707002.pdf
35. Ellman GL (1959) Tissue sulfhydryl groups. Arch Biochem Biophys 82:70–77

Chapter 11

Designing Glycosyltransferase Expression Constructs for Improved Purification, Protein Yield, and Crystallization

Taylor J. B. Forrester, Lana El Osta, Joseph S. Lam, and Matthew S. Kimber

Abstract

Glycosyltransferases in bacteria are built using only four known architectures, but this structural core is often supplemented by fusions with a wide variety of other domains, including those that help recruit them to the membrane. Structural and functional characterization of these proteins is often simplified by making a subconstruct that is better behaved in solution, and perhaps monofunctional. In this chapter we review bioinformatics tools and strategies that can be used for designing such constructs of glycosyltransferases.

Key words Construct design, Bioinformatics, Membrane, Domain boundaries, Glycosyltransferase

1 Introduction

Glycosyltransferases are a large family of enzymes that exhibit considerable mechanistic diversity despite only occurring as four distinct folds. In bacteria, these enzymes function primarily in the cytosol to assemble a diverse array of products. Many bacteria produce glycosylated secondary metabolites that influence other organisms in their environments, including molecules with antibiotic properties [1]. Bacteria also universally synthesize diverse oligosaccharides that are then exported outside of the cell, serving either as structural components of the cell wall (peptidoglycan, teichoic acids, secondary cell wall carbohydrates) surface decorating/barrier forming carbohydrate structures (lipopolysaccharide core oligosaccharide, O antigens, capsule or cell wall polysaccharides) [2], or as slime-like extracellular polysaccharides that contribute as major structural components of biofilms. Biosynthesis of these polysaccharide structures generally entails assembly on lipid carriers that act as recognition sites for initiating

Taylor Forrester and Lana El Osta contributed equally to the chapter and should be considered joint first authors.

Inka Brockhausen (ed.), *Bacterial Polysaccharides: Methods and Protocols*, Methods in Molecular Biology, vol. 1954, https://doi.org/10.1007/978-1-4939-9154-9_11, © Springer Science+Business Media, LLC, part of Springer Nature 2019

glycosyltransferases, as recognition handles for export machinery, and also as recyclable activated leaving groups that allow these carbohydrates to be either polymerized (as in the "Wzy-dependent" O antigen assembly model) or transferred onto a terminal acceptor [3]. For glycosyltransferases whose substrates are membrane embedded, membrane association confers a potential kinetic advantage. Such membrane association can be mediated directly (through the display of extended hydrophobic patches) or indirectly (through interacting with another protein that is directly targeted). In the literature, many bacterial membrane-associated glycosyltransferases exhibit interaction motifs within the Rossmann-fold domains, which form large hydrophobic patches that are often surrounded by basic amino acid residues. Interestingly, a trend has been noted in membrane-associating GT-B enzymes where the N-terminal domain, responsible for membrane interaction, not only has exposed patches of hydrophobic residues (most notably tryptophan) but also have a high pI (generally in the range of 8.5–10.5). This results in a large net-positive charge that preferentially maps to one face of the molecule, helping drive favorable interactions with the head groups of lipids [4]. Theoretical pI values can be readily predicted from the amino acid sequence, making this a potentially invaluable prediction tool. Well-studied examples of glycosyltransferases that interact with the membrane through extended hydrophobic patches include MurG, WaaA, and PimA [5–7]. Interested readers are referred to Albesa-Jove et al. for an in depth discussion of membrane association in GT-B proteins, including membrane association mechanisms, and a comprehensive list of examples [4].

Because most glycosyltransferases (other than GT-C members) have evolved from soluble enzymes, targeting to the membrane often occurs through peripheral elements. These can include the use of small numbers of (generally one) membrane-spanning helices, or amphipathic helices and/or patches of exposed nonpolar residues on the surface of the protein that allow the protein to interact with the periphery of the membrane [4]. In cases where membrane interaction occurs through extensive interactions involving an intrinsic nonpolar patch, solubilization of the protein using detergents is generally helpful for facile biochemical characterization. Where membrane interactions occur through transmembrane or amphipathic helices, these elements are very commonly localized at either the N- or the C-terminus. These can therefore be removed through appropriate construct design, in many cases resulting in a protein that is soluble, active, and can be expressed in high yields. It is worth noting that amphipathic helices can also be integrated into the protein's overall architecture and play important additional roles in the functional mechanism: for example, in the galactofuranosyltransferase GlfT2, an amphipathic helix integrated into the C-terminal domain positions the active site to face the membrane [8], while in GalNAc-transferase PglH, an internal amphipathic helix plays a direct role in regulating product chain length [9].

Glycosyltransferase domains are also sometimes found fused to additional domains to form a larger protein; isolating distinct functional units may allow one to begin to elucidate the structure and function of a particular protein. Multiple glycosyltransferase domains, for example, are commonly fused into a single polypeptide. In such cases, subcloning individual domains allow the assignment of specific biochemical functions to the corresponding regions (*see* **Note 1**). In other instances, additional domains were discovered that act as spacers or were important for oligomerization. An interesting example of this are coiled-coil domains, which, in at least some instances, are used to control the chain length of the product of glycosyltransferases [10, 11].

In this chapter, we explore several approaches to designing truncated constructs for the study of bacterial glycosyltransferases. We discuss some useful bioinformatics tools and decision making strategies for designing such constructs. We also review a series of useful case studies, focussing on those in which the authors were able to determine an x-ray structure, implying that they succeeded in producing significant quantities of soluble, active protein. Similar constructs would also likely support extensive enzymatic and biophysical characterization, or provide suitable material for demanding high throughput screening studies. Creating functional, well-behaved subconstructs of a larger protein is always a challenge. Judicious use of bioinformatics tools—both those that map known data onto a new sequence, and those that can highlight generalized patterns in protein sequences—can help improve the likelihood of success. Our recommendation is to start by exploring the target sequence with all pertinent tools, and then giving careful thought as to which tools give the most reliable insights in this specific case. A useful training exercise might be to analyze some sequences from pertinent examples (as given above) with the recommended tools, and to then try to understand how the resulting predictions might guide you toward the chosen (successful) strategies ultimately employed by the authors.

2 Materials

2.1 Access to the Internet and Bioinformatics Web Pages

No special computers are required, and for most purposes, all key programs (with the exception of a structure viewing program such as Pymol or ChimeraX which are required if three dimensional models are to be inspected in detail) can be run on online servers rather than needing local installation. Please see Table 1 for a list of useful programs and their server's addresses.

Table 1
Useful bioinformatics tools for construct design

	Program	URL	Purpose
1	NCBI-BLAST	blast.ncbi.nlm.nih.gov/Blast.cgi	Sequence searches; general sequence information portal
2	i-TASSER	zhanglab.ccmb.med.umich.edu/I-TASSER/	Homology modelling of protein structures
3	Rosetta	robetta.bakerlab.org	Homology modelling of protein structures
4	Pymol	pymol.org/2/	Visualizing protein structures
5	ChimeraX	www.cgl.ucsf.edu/chimerax/	Visualizing protein structures
6	JPred	www.compbio.dundee.ac.uk/jpred/	Secondary structure prediction
7	PsiPred	bioinf.cs.ucl.ac.uk/psipred/	Secondary structure prediction
8	CDD	www.ncbi.nlm.nih.gov/Structure/cdd/wrpsb.cgi	Compilation of analyses of conserved domains
9	PrDOS	prdos.hgc.jp/cgi-bin/top.cgi	Intrinsic disorder predictions
10	MEMSAT-SVM	bioinf.cs.ucl.ac.uk/psipred/?memsatsvm=1	Transmembrane helix prediction
11	Polyphobius	phobius.sbc.su.se/poly.html	Transmembrane helix prediction
12	HELIQUEST	heliquest.ipmc.cnrs.fr/cgi-bin/ComputParams.py	Analysis of peripheral and transmembrane helices
13	AmphiaSeek	npsa-prabi.ibcp.fr/cgi-bin/npsa_automat.pl?page=/NPSA/npsa_amphipaseek.html	Peripheral helix prediction
14	Coils	embnet.vital-it.ch/software/COILS_form.html	Coiled-coil predictions

3 Methods

3.1 Bioinformatics—Structural Homologs and Homology Modelling

The most straightforward prediction of domain boundaries utilizes comparisons of established structures to homologous proteins of interest. For glycosyltransferase families which now have representative structures that have been experimentally determined, these known structures can be used as a guide to the organization of other members of that protein family. NCBI-BLAST can be used to search the database of sequences derived from the Protein Data Bank (pdb) to find the closest structurally characterized homolog. In general, proteins with 40–50% sequence identity or better can be trusted to serve as reliable templates, at least for those regions that clearly correspond to the reference structure. Next, a homology model can be built using the algorithm from a server such as I-TASSER [12] or the ROSETTA-SERVER [13] (both highly ranked in recent blind Critical Assessment of Techniques for

Protein Structure Prediction competitions) to help visualize the likely organization of the overall structure. A potential pitfall in using this approach is that these tools will try to fill in regions where prior structures do not provide reliable guides. We would suggest paying close attention to the predicted reliability of key regions of the structure, and being cautious about building hypotheses that depend on the details of regions with low reliability. Comparing multiple predicted structures can help give some sense in the uncertainties that are inherent. These predicted structures can be used in multiple ways, including identifying potential functional residues for mutagenesis studies. For the purpose of designing new constructs, the predicted structures can help define likely domain boundaries, but can also help visualize the presence of extensive exposed surface nonpolar patches. The presence of these nonpolar patches suggest that the protein is likely intimately associated with the membrane, and that truncation strategies are unlikely to result in soluble protein. In these cases, a detergent-based purification strategy is generally required.

3.2 Secondary Structure Prediction

Secondary structures can be predicted with high accuracy by using tools such as Jpred or PSIpred [14, 15]. For glycosyltransferases, these tools can be useful for designating domain boundaries because most glycosyltransferases conform to one of two general patterns of structure, both of which are based on the Rossmann-fold—the GT-A and GT-B folds. Most glycosyltransferase structures are generally quite conserved, though variants with additional helices are fairly common among bacterial GTs, and some families have been found to truncate or extend these canonical structures in various ways. Usefully, however, extensions to GT domains rarely occur at the N-terminus. The N-terminal boundary for constructs designed to subclone the glycosyltransferase domain can generally safely be positioned 2-3 residues N-terminal to the first predicted ordered β-strand (the added residues allow for typical imprecision in the prediction of secondary structure boundaries). Otherwise, comparison of the predicted secondary structure to the most closely related structurally characterized homologs can help identify motifs outside of the core structure. These predictions are also useful as domain construct boundaries located in loop regions are less likely to disrupt the structure than those in the middle of predicted secondary structure elements (*see* **Note 2**).

3.3 Domain Prediction

Domains within a glycosyltransferase are defined in a variety of ways but, in general, they loosely correspond to independently folding regions of a larger protein, have independent evolutionary origins, and are responsible for one or more subprocesses within the protein's overall function. Glycosyltransferases are often found fused to a variety of other domains, including other glycosyltransferases. Because domains are often found to occur independently in a variety of different contexts, mapping the boundaries of sequence conservation can be a useful guide to predict how far they extend. A number of bioinformatics tools look to map such domain boundaries, including DOMPRED [16]. However, domain classification is a common bioinformatics problem, and domain analysis across all known sequences have been carried out using a variety of criteria and algorithms, resulting in the generation of databases including Pfam, COG and TIGRFAM, SMART, and PRK. These precomputed databases are collectively presented as the conserved domain database (CDD) [17], which is conveniently linked within the context of an NCBI-BLAST search through the domain icons presented. The CDD contains a detailed analysis, including multiple sequence alignments that can be reconfigured according to the user's priorities, and links to a wide variety of external resources [17] (*see* **Note 3**).

3.4 Intrinsically Disordered Protein Regions

It is now widely recognized that intrinsically disordered regions very commonly occur within proteins; within enzymes such as glycosyltransferases, these motifs show up in two contexts. Firstly, the first few and last few amino acids in otherwise structured proteins are very often disordered. Removing these regions can improve the solubility and the propensity of the construct to crystallize. Secondly, short disordered motifs can act as flexible linkers between structured domains of a multidomain protein. Siting the boundaries of constructs in these regions can help produce cleanly separated constructs with minimal solubility issues. A variety of programs are able to predict the presence of short range disorder, including PrDOS, which integrates information from homologous sequences, and allows the user to tune acceptability thresholds [18].

3.5 Transmembrane Helices

A common strategy used by organisms for localizing proteins to the membrane surface is to append a small subdomain which includes one or more transmembrane helices. These helices generally contain a central hydrophobic portion averaging 25 ± 7 amino acids long, with the periphery of such a region being enriched in aromatic amino acids [19]. Such helices can be removed by subcloning soluble domains, or, if the subdomain is located at the C-terminus, by introducing a stop codon by site directed mutagenesis. Identifying such transmembrane helices is a very common problem in biochemistry, and a large number of bioinformatics tools address this problem. Historically, these methods rely on variations identified in hydropathy plots, testing for long runs of nonpolar residues that can be used to span the membrane's width; TopPred2 is one widely used modern tool. More sophisticated prediction tools utilize hidden Markov models, neural networks, or support vector machines, and incorporate steps to discriminate between TM helices and extracellular localization signals. A recent survey found that many dedicated TM helix programs perform significantly worse when predicting prokaryotic TM helices [20]. Usefully, single-spanning membrane helices are, in general, more accurately predicted than multispanning membrane proteins since they tend to be more hydrophobic (due to the possibility in multispanning proteins of burying polar residues between helices not being available). MEMSAT-SVM and PolyPhobius reportedly perform well in predicting bacterial TM helices [21, 22]. We tend to prefer the latter program, as it runs faster and is less prone to overpredict possible "pore lining helices" in glycosyltransferases. HELIQUEST is a useful supplementary tool, calculating mean hydrophobicities as well as plotting helical wheels for proposed hydrophobic helices [23]. In general, single spanning transmembrane helices can be readily and unambiguously identified.

3.6 Peripheral Membrane Helices

As an alternative (or supplement) to transmembrane helices, amphipathic helices can also be used to localize proteins to the membrane. These helices are characterized by having one nonpolar, and one polar face; they are believed to sit at the periphery of the membrane, interacting with both the lipid tails and head groups; such motifs are also called in-plane membrane (IPM) anchors. Validated amphipathic helices are relatively rich in phenylalanine and tryptophan residues on the nonpolar face, and depleted in proline [24]. In addition, bacterial IPMs seem to generally have a net-positive charge, likely an adaptation to the net negative charge which characterizes bacterial cell membranes. Tools for predicting amphipathic helices are considerably less well developed than for predicting TM helices, in part because relatively few well-characterized examples are available from which to form a training set, and in part because their properties are less easily distinguishable from other protein motifs. In this regard, it is worth noting that many surface helices in soluble proteins have an amphipathic nature dictated by their need to interact with both the solvent and the protein's hydrophobic core; candidate membrane interacting amphipathic helices should be checked against secondary predictions and homology models to ensure that these motifs are not simply structural helices.

Amphipathic helices are classically predicted using hydrophobic moment, which quantifies the degree to which nonpolar residues are distributed preferentially to one face of the helix. The support vector classifier AmphipaSeeK [24] can be used to predict the presence of IPM motifs in a protein; however, this program is based on a 21 protein training set, with a bias toward sequences obtained from viral proteins. It is not clear how many prokaryotic sequences were included, a concern given that the lipid composition of bacteria differs markedly from eukaryotes. HELIQUEST is again a very useful tool in the context of identifying possible IPM motifs in a protein known to interact with the membrane [23]. This program maps all subsequences of a defined length in a protein to helical wheel projections, and then allows one to calculate a series of parameters—including hydrophobic moment and mean hydrophobicity—applicable to this wheel. The challenge then lies in interpreting these statistics. It has been suggested that a hydrophobic moment of at least 0.35 and a mean hydrophobicity between 0.28 and 0.6 are good indications of a possible IPM. Because IPMs do not span a membrane of defined thickness, they may be less constrained in their lengths. Sapay's survey of 21 examples suggests that these helices average 23 ± 10 amino acids. Longer candidate amphipathic helix segments provide more opportunities for interaction with the membrane, and may allow a slight relaxation of the criteria. In our experience, at least six nonpolar residues need to be mapped to a continuous face for these residues to represent a strong candidate for forming a membrane association motif (*see* **Note 4**).

4 Case Studies: WbbB

The Kdo-transferase WbbB from *Klebsiella pneumoniae* (GT99, GT102 and GT103; UNIPROT: Q6U8B0; pdb: 5FA0, 5FA1) has three glycosyltransferase domains within its 1106 residue length polypeptide. The full-length protein is poorly soluble; the authors therefore analyzed this protein predominantly as subconstructs containing one or more domains. The N-terminal Kdo--transferase domain (residues 2-401), with the domain boundary selected to exclude a predicted coiled-coil spacer domain (as predicted by COILS), has its truncation site located in a short predicted disordered region bordering this. This construct was enzymatically active and was successfully crystallized [25]. In a follow-up study, the C-terminal GT102 and GT103 domains were expressed separately as a construct spanning residues 540-1106 [11]. In addition, the GT102 domain was expressed in isolation as residues 576-871. Since no prior structures were known for these GT families, sequence conservation, coiled-coil and secondary structure predictions were used to define domain boundaries.

5 Case Studies: KpsC

For the dual-functional, *E. coli* group 2 capsule linker retaining Kdo transferase KpsC (GT107; Genbank KIE82100), the N-terminal domain of the full, dual GT99-like protein was isolated by subcloning residues 2-352 into an expression vector. On the other hand, attempts to subclone the homologous C-terminal domain using a similar strategy resulted in only insoluble, inactive protein constructs [26]. Structural homology was of little use in predicting domain boundaries as the closest homolog, WbbB, is only weakly similar, while the pattern of secondary structure at the N-terminus diverges from other GTs with few helices predicted. However, sequence similarity between the KpsC N-and C-terminal domains, as well as the existence of homologs where the two domains are found as separate proteins, was useful for defining domain boundaries. HELIQUEST predicts an extended, very basic amphipathic helix at the C-terminus of the protein (residues 656-671, with a hydrophobic moment of 0.50, a mean hydrophobicity of 0.519, and a net charge of +5), which may explain the observed tendency of constructs to partition to the membrane fractions.

6 Case Studies: α2,3/6-Sialyltransferase

Lin et al. [27] determined the structure of membrane associated α2,3/2,6 lipooligosaccharide sialyltransferase from *Neisseria meningitidis* serotype L1 (GT52; UNIPROT: U60662; pdb: 2YK4, 2YK5, 2YK6). This protein required deletion of the first 29 amino acids for successful crystallization. However, the truncated protein still partitioned to the membrane, and required the detergent, Triton-X100, for purification [27]. The structure showed that in addition to the deleted residues, amino acids 30-49 are also unstructured. For this protein, MEMSAT-SVM predicts a transmembrane helix in residues 15-30 with the C-terminus cytoplasmic and a score of 0.49. Jpred predicts residue 51 as the first residue of the first β-strand, one off from its position (residue 50) in the structure.

7 Case Studies: Chondroitin Synthase

Chondroitin synthase from *E. coli* (GT2; UNIPROT: Q8L0V4; pdb: 2Z86, 2Z87) was initially expressed as a full-length protein. It was found that 57 amino acids could be removed from the N-terminus by trypsin digestion to produce a protein which was fully catalytically active, though with altered oligomerization behavior. Subcloning the larger proteolysis fragment (residues 58-686) allowed successful crystallization [28].

The HELIQUEST server predicts an amphipathic helix spanning residues 34-51, with a hydrophobicity of 0.468 and a hydrophobic moment of 0.401. Jpred predicts four helices within the first 57 amino acid residues, with residue 152 as the first residue of the first β-strand. The structure confirms this prediction, and shows that residues 58-151 form an unusual extension that wraps around the C-terminal glycosyltransferase domain [28].

8 Case Studies: α2,6-Sialyltransferase

The α2,6-sialyltransferase from *Photobacterium damselae* (GT80; UNIPROT: O66375; pdb: 4R83, 4R84, 4R9V) is organized as a multidomain protein with an N-terminal Ig-like domain, and a C-terminal PhoU domain [29]. The authors studied this protein using several constructs, all of which lack the first 15 amino acid residues; the authors state that this region forms a membrane anchor, and HELIQUEST results are consistent with this motif forming an amphipathic helix (residues 1-18; hydrophobic moment of 0.356, mean hydrophobicity of 0.786, +2 net charge). The C-terminal most 178 residues (i.e., 498-675) form a predicted PhoU domain, and were also deleted. Constructs spanning residues 16-497 and 112-497 were made and the protein purified; both proved to be soluble and were successfully crystallized. Residues 24-111 form an Ig-like domain that dimerizes, while a GTB is formed by residues 112-497. Jpred predicts the first β-strand of the GT domain to begin at residue 114, while the crystal structures shows it to begin at 113.

9 Case Studies: α2,3-Sialyltransferase

The crystal structure of an α/β-galactoside α2,3-sialyltransferase from *Photobacterium phosphoreum* (GT80; UNIPROT: A5LHX0, pdb: 2ZWI) was determined with a truncation of the first 24 amino acid residues at the N-terminus [30]. The authors propose that residues 1-22 form a prelipoprotein signal peptide that is absent in homologous GT80 enzymes [31]. Polyphobius concurs with this analysis, while HELIQUEST highlights a highly hydrophobic stretch in residues 8-22. Jpred predicts residue 41 starts the first β-strand, while the crystal structure shows Lys39 as the first ordered residue. In addition, PrDOS predicts that residues 25-35 are disordered, in agreement with the location of the first ordered residue in the crystal structure. Possibly an even shorter construct (starting around residue 39) would also have worked well for this protein.

10 Case Studies: α2,3-Sialyltransferase CstII

The structure of the α2,3-sialyltransferase CstII from *Campylobacter jejuni* (GT42; UNIPROT Q9LAK3; pdb: 1Ro8)) was determined using a construct spanning residues 1-259, with 32 residues truncated from the C-terminus. The authors state that this region is rich in basic, hydrophobic, and aromatic residues, and therefore likely to drive interactions with the membrane [32]. The protein also contained an I52S point mutation designed to stabilize the enzyme. Jpred predicts a helix spanning residues 266-287 at the C-terminus, while HELIQUEST suggests that a clear amphipathic helix can be found (268-289) with a mean hydrophobicity of 0.480, and a hydrophobic moment of 0.613, and a net charge of +3.

11 Case Studies: α2,6-Sialyltransferase Cst-I

Chiu et al [33] determined the structure of the α2,3-sialyltransferase, Cst-I, from *Campylobacter jejuni* (GT42; UNIPROT: Q9RGF1; pdb: 2P2V). Full-length Cst-I (residues 1-430) contains a predicted N-terminal sialyltransferase domain, and a C-terminal domain that shows similarity only to other *Campylobacter* proteins. Since the N-terminal domain was 42% identical in sequence to the previously determined CST-II, identifying the domain boundary could rely upon structural information from its homolog. The authors subcloned the N-terminal domain, leaving a soluble and active 1-285 construct. PolyPhobis predicts an internal transmembrane helix from residues 334 to 352 in the C-terminal domain, with a mean hydrophobicity of 1.031; this is likely the source of the poor solubility observed for full-length constructs of this protein [33].

12 Case Studies: Dual Function Glycosyltransferase dGT1

Zhang et al. [34] structurally and biochemically characterized the dual domain of dual function glycosyltransferase dGT1 (582 residues) from *Streptococcus parasanguinis* (GT2, GT101; UNIPROT: I1ZPA1; pdb: 5HEC, 4PFX). The two domains were subcloned independently for structural studies. The N-terminal glucosyltransferase domain has no similarity to previously characterized glycosyltransferases, instead being classified as a member of the domain of unknown function family, DUF1792 which is conserved across several bacterial families, as well as archaea; the structure of this domain proved it possesses a novel GT-D fold [34]. The C-terminal galactosyltransferase domain has a GT-A fold and homology to CAZy family 2 and to the WcaA protein [35]. The domains were individually subcloned for crystallization: the DUF1792 domain was isolated as a construct spanning 1–272 and the C-terminal GT-A domain spanning 287–582. The authors do not state their rationale for the break point, but sequence conservation within each domain likely was a major driver (CDD annotates the DUF1792 domain as spanning residues 9–272). The omitted 14-amino-acid linker was possibly left out due to a tendency toward disorder: PrDOS predicts heightened disorder in the region of 274-287, though this only reaches the 90%

confidence level (0.444–0.463, threshold 0.5 for 95% confidence). DomPRED also predicts a domain boundary between these domains at position 287.

13 Case Studies: Mannosylglycerate Synthase

The structure of a mannosylglycerate synthase from *Rhodothermus marinus* (GT78; UNIPROT: D0MI02; pdb: 2Y4K, 2Bo7) has been studied [36, 37]. The N-terminal 174 residues show homology to GT-A type glycosyltransferases, with a unique extended C-terminal region that forms six helices which mediate oligomeric interactions. While the enzyme was originally crystallized as a full-length protein, electron density maps showed the last 15 residues to be disordered. The protein was subsequently recrystallized from a construct with these residues truncated (1-382) resulting in better diffraction [37]. In a follow up study, surface entropy reduction using SERp for guidance was used to identify one pair of residues (among four tried) that produced protein that crystallized with fewer molecules in the asymmetric unit and was therefore suited for high resolution substrate complex structures [36]. PrDOS predicts a highly disordered C-terminus, with the last 17 residues being disordered (381-397). Similarly, Jpred predicts the last 16 residues to not adopt a regular secondary structure.

14 Notes

1. In our laboratories, we generally make and screen several candidate constructs in parallel to maximize the chances of finding one that is well behaved. If the variants are to be generated at the C-terminus, then cloning with an N-terminal affinity tag allows variants to be rapidly generated simply by introducing new stop codons wherever needed.
2. In general, the features discussed here (domains boundaries, secondary structure, membrane anchors) will be conserved among homologs—or at least among the close ones. Checking predictions against multiple sequence alignments (generated by for example ClustalX) and/or running parallel analysis on several reasonably close homologs (say, 70–80% sequence identity) can help make predictions more robust.
3. Along the same lines, orthologous proteins from different organisms can vary markedly in terms of how well they express heterologously, and how amenable they are to construct engineering. If the initial target proves recalcitrant, it can be worth experimenting with a few homologous genes.

4. Be careful of relying too much on any single prediction method. In particular, domain boundaries can be quite ambiguous where sequence similarity becomes weak, homology models are very reliant on correct sequence alignments, and the criteria for predicting possible amphipathic helices or locally disordered regions are nonspecific enough to routinely give false positives. Having multiple lines of argument that converge on a single interpretation of the protein's configuration is preferable wherever possible.

Acknowledgments

This work was funded by grants from the National Science and Engineering Research Council of Canada (NSERC) and from the Canadian Glycomics Network, National Centers of Excellence (GlycoNet) to MSK and JSL.

References

1. Elshahawi SI, Shaaban KA, Kharel MK et al (2015) A comprehensive review of glycosylated bacterial natural products. Chem Soc Rev 44:7591–7697
2. Holst O, Moran AP, Brennan PJ (2010) Overview of the glycosylated components of the bacterial cell envelope. In: Microbial glycobiology. London, Elsevier, pp 1–13
3. Greenfield LK, Whitfield C (2012) Synthesis of lipopolysaccharide O-antigens by ABC transporter-dependent pathways. Carbohydr Res 356:12–24
4. Albesa-Jové D, Giganti D, Jackson M et al (2014) Structure-function relationships of membrane-associated GT-B glycosyltransferases. Glycobiology 24:108–124
5. Hu Y, Chen L, Ha S et al (2003) Crystal structure of the MurG:UDP-GlcNAc complex reveals common structural principles of a superfamily of glycosyltransferases. Proc Natl Acad Sci U S A 100:845–849
6. Schmidt H, Hansen G, Singh S et al (2012) Structural and mechanistic analysis of the membrane-embedded glycosyltransferase WaaA required for lipopolysaccharide synthesis. Proc Natl Acad Sci 109:6253–6258
7. Guerin ME, Kordulakova J, Schaeffer F et al (2007) Molecular recognition and interfacial catalysis by the essential phosphatidylinositol mannosyltransferase PimA from mycobacteria. J Biol Chem 282:20705–20714
8. Wheatley RW, Zheng RB, Richards MR et al (2012) Tetrameric structure of the GlfT2 galactofuranosyltransferase reveals a scaffold for the assembly of mycobacterial arabinogalactan. J Biol Chem 287:28132–28143
9. Ramírez AS, Boilevin J, Mehdipour AR et al (2018) Structural basis of the molecular ruler mechanism of a bacterial glycosyltransferase. Nat Commun 9:445
10. Hagelueken G, Clarke BR, Huang H et al (2015) A coiled-coil domain acts as a molecular ruler to regulate O-antigen chain length in lipopolysaccharide. Nat Struct Mol Biol 22:50–56
11. Williams DM, Ovchinnikova OG, Koizumi A et al (2017) Single polysaccharide assembly protein that integrates polymerization, termination, and chain-length quality control. Proc Natl Acad Sci U S A 114:E1215–E1223
12. Roy A, Kucukural A, Zhang Y (2010) I-TASSER: a unified platform for automated protein structure and function prediction. Nat Protoc 5:725–738
13. Moretti R, Lyskov S, Das R et al (2018) Web-accessible molecular modeling with Rosetta: the Rosetta online server that includes everyone (ROSIE). Protein Sci 27:259–268
14. Jones DT (1999) Protein secondary structure prediction based on position-specific scoring matrices. J Mol Biol 292:195–202
15. Drozdetskiy A, Cole C, Procter J et al (2015) JPred4: a protein secondary structure prediction server. Nucleic Acids Res 43:W389–W394
16. Yoo PD, Sikder AR, Taheri J et al (2008) DomNet: protein domain boundary prediction

using enhanced general regression network and new profiles. IEEE Trans Nanobioscience 7:172–181

17. Marchler-Bauer A, Bo Y, Han L et al (2017) CDD/SPARCLE: functional classification of proteins via subfamily domain architectures. Nucleic Acids Res 45:D200–D203
18. Ishida T, Kinoshita K (2007) PrDOS: prediction of disordered protein regions from amino acid sequence. Nucleic Acids Res 35: W460–W464
19. Saidijam M, Azizpour S, Patching SG (2018) Comprehensive analysis of the numbers, lengths and amino acid compositions of transmembrane helices in prokaryotic, eukaryotic and viral integral membrane proteins of high-resolution structure. J Biomol Struct Dyn 36:443–464
20. Reeb J, Kloppmann E, Bernhofer M et al (2015) Evaluation of transmembrane helix predictions in 2014. Proteins 83:473–484
21. Nugent T, Jones DT (2009) Transmembrane protein topology prediction using support vector machines. BMC Bioinformatics 10:159
22. Käll L, Krogh A, Sonnhammer ELL (2005) An HMM posterior decoder for sequence feature prediction that includes homology information. Bioinformatics 21(Suppl 1):i251–i257
23. Gautier R, Douguet D, Antonny B et al (2008) HELIQUEST: a web server to screen sequences with specific alpha-helical properties. Bioinformatics 24:2101–2102
24. Sapay N, Guermeur Y, Deléage G (2006) Prediction of amphipathic in-plane membrane anchors in monotopic proteins using a SVM classifier. BMC Bioinformatics 7:255
25. Ovchinnikova OG, Mallette E, Koizumi A et al (2016) Bacterial β-Kdo glycosyltransferases represent a new glycosyltransferase family (GT99). Proc Natl Acad Sci 113: E3120–E3129
26. Ovchinnikova OG, Doyle L, Huang B-S et al (2016) Biochemical characterization of Bifunctional 3-Deoxy-β-d-manno-oct-2-ulosonic acid (β-Kdo) Transferase KpsC from *Escherichia coli* involved in capsule biosynthesis. J Biol Chem 291:21519–21530
27. Lin LYC, Rakic B, Chiu CPC et al (2011) Structure and mechanism of the lipooligosaccharide sialyltransferase from *Neisseria meningitidis*. J Biol Chem 286:37237–37248
28. Osawa T, Sugiura N, Shimada H et al (2009) Crystal structure of chondroitin polymerase from *Escherichia coli* K4. Biochem Biophys Res Commun 378:10–14
29. Huynh N, Li Y, Yu H et al (2014) Crystal structures of sialyltransferase from *Photobacterium damselae*. FEBS Lett 588:4720–4729
30. Iwatani T, Okino N, Sakakura M et al (2009) Crystal structure of alpha/beta-galactoside alpha2,3-sialyltransferase from a luminous marine bacterium, *Photobacterium phosphoreum*. FEBS Lett 583:2083–2087
31. Tsukamoto H, Takakura Y, Yamamoto T (2007) Purification, cloning, and expression of an alpha/beta-galactoside alpha-2,3-sialyltransferase from a luminous marine bacterium, *Photobacterium phosphoreum*. J Biol Chem 282:29794–29802
32. Chiu CPC, Watts AG, Lairson LL et al (2004) Structural analysis of the sialyltransferase CstII from *Campylobacter jejuni* in complex with a substrate analog. Nat Struct Mol Biol 11:163–170
33. Chiu C, Lairson LL, Gilbert M et al (2007) Structural analysis of the α-2,3-sialyltransferase Cst-I from *Campylobacter jejuni* in apo and substrate-analogue bound forms. Biochemistry 46:7196–7204
34. Zhang H, Zhu F, Yang T et al (2014) The highly conserved domain of unknown function 1792 has a distinct glycosyltransferase fold. Nat Commun 5:4339
35. Zhang H, Zhou M, Yang T et al (2016) New helical binding domain mediates a glycosyltransferase activity of a bifunctional protein. J Biol Chem 291:22106–22117
36. Flint J, Taylor E, Yang M et al (2005) Structural dissection and high-throughput screening of mannosylglycerate synthase. Nat Struct Mol Biol 12:608–614
37. Nielsen MM, Suits MDL, Yang M et al (2011) Substrate and metal ion promiscuity in mannosylglycerate synthase. J Biol Chem 286:15155–15164

Chapter 12

Utilization of Fluorescently Tagged Synthetic Acceptor Molecules for In Vitro Characterization of a Dual-Domain Glycosyltransferase Enzyme, KpsC, from *Escherichia coli*

Liam Doyle, Olga G. Ovchinnikova, and Chris Whitfield

Abstract

The incorporation of fluorescent tags into synthetic acceptor molecules for in vitro biochemical assays allows quick and easy detection of enzyme activity. Reaction products can be separated via thin-layer chromatography and visualized under UV light for rapid detection of reaction progress. Subsequent structural analysis of these reaction products through the use of NMR spectroscopy and mass spectrometry allows for complete functional characterization of enzyme activity. Here we describe an application of this technique which was previously used to functionally characterize a dual-domain glycosyltransferase enzyme, KpsC, involved in capsular polysaccharide biosynthesis in *Escherichia coli*.

Key words Microbial glycobiology, Capsular polysaccharide, Glycosyltransferase, In vitro biochemical reaction, Fluorescently-tagged synthetic acceptor

1 Introduction

In vitro biochemical assays offer advantages over in vivo methodologies when it comes to elucidation of protein function. The main advantages of this in vitro method comes from working with purified proteins as conditions which may be unsuitable for cell survival can be used and activities of the purified proteins can be directly investigated. These assays are commonly employed in the fields of molecular biology and biochemistry to establish enzyme function and analyze enzyme kinetics.

In particular, the field of glycobiology has benefitted greatly from incorporation of natural or synthetic acceptor molecules into in vitro assays. Inclusion of synthetic acceptor molecules have allowed for a more in-depth analysis of a wider variety of biochemical reactions by eliminating difficulties associated with isolating endogenous acceptor molecules. Radiolabeled substrates have previously been utilized in order to monitor the transfer of sugar

Inka Brockhausen (ed.), *Bacterial Polysaccharides: Methods and Protocols*, Methods in Molecular Biology, vol. 1954, https://doi.org/10.1007/978-1-4939-9154-9_12,

residues, via glycosyltransferase (GT) enzymes, to the synthetic acceptor molecules [1]. Another technique for monitoring reaction progress utilizes chemical synthesis to attach fluorophores to components of the natural acceptor [2]. The use of these fluorophore-containing synthetic acceptor molecules has simplified detection of reaction progress. In particular, thin-layer chromatography allows quick and simple detection of reaction progress as the fluorescent tags can be easily visualized under UV light [2, 3]. Furthermore, more sophisticated analysis of reaction products, including mass spectrometry and nuclear magnetic resonance spectroscopy, allow enzyme functional characterization [2, 3]. Here we describe the use of these techniques in the in vitro functional characterization of an *Escherichia coli* GT enzyme, KpsC, using fluorophore-containing synthetic acceptor molecules.

Bacterial capsules are surface glycostructures important in host–pathogen interactions [4]. Extraintestinal pathogenic *Escherichia coli* (ExPEC) and other gram-negative pathogens including *Campylobacter jejuni*, *Haemophilus influenzae* and *Neisseria meningitidis*, assemble layers of protective capsular polysaccharides (CPS) via a conserved strategy (Fig. 1a) [4]. A defining feature of these CPS is their attachment to a terminal phospholipid by a chain of β-linked 3-deoxy-D-*manno*-oct-2-ulosonic acid (Kdo) sugar residues [5]. The β-Kdo linker is synthesized by two pathway-defining retaining β-Kdo GTs, KpsS, and KpsC [5]. KpsS adds the first Kdo residue to the phospholipid, and the dual-GT domain KpsC extends it with multiple residues [5]. Until recently, the individual activity of each KpsC domain as well as the precise structure of the β-Kdo linker was unknown.

To characterize the linkage-specificities catalyzed by each domain, three constructs were used in in vitro biochemical assays (Fig. 1b): the full-length enzyme, KpsC, an isolated N-terminal domain, KpsC-N and a mutant construct containing an active C-terminal and inactivated N-terminal catalytic site, KpsC-C [6]. The assay used here couples the production of the activated sugar donor, CMP-β-Kdo, with the subsequent addition of β-Kdo residues from this activated donor to a synthetic acceptor molecule (Fig. 1c), via KpsC [6]. CMP-β-Kdo is synthesized in situ from CTP, Kdo and a purified CMP-β-Kdo synthetase, KdsB, from *E. coli*. Through the use of the in vitro methods described here, KpsC was biochemically confirmed to possess β2,4- and β2,7-Kdo linkage specificities in the N-terminal and C-terminal GT domains, respectively [6]. Additionally, we confirmed that these activities account for the alternating β2,4- and β2,7-Kdo linkages found in the natural product [6].

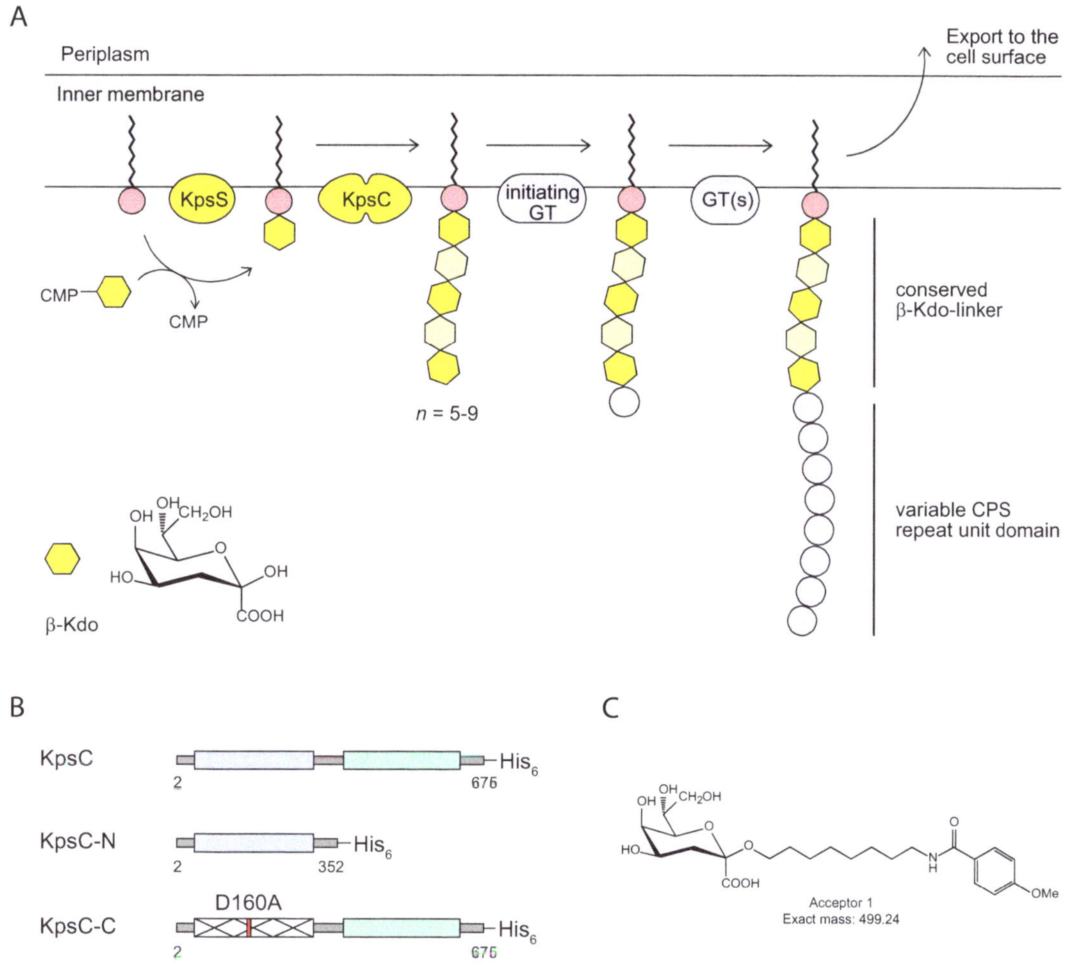

Fig. 1 Biochemical characterization of *E. coli* KpsC in synthesizing a conserved β-Kdo linker in group 2 capsular polysaccharides. (**a**) Group 2 capsule biosynthesis pathway. Biosynthesis begins with KpsS and KpsC contributing to synthesize the β-Kdo linker on a phospholipid acceptor molecule, as previously described. Following linker synthesis, serotype-specific glycosyltransferases synthesize the repeat unit domain of the capsular polysaccharides. (**b**) KpsC constructs used in the biochemical characterization of KpsC. (**c**) Structure and molecular mass of synthetic acceptor 1, used for analysis of the in vitro activity characterization of KpsC. Acceptor 1 consists of a single β-linked Kdo residue attached, via an eight carbon linker, to a *p*-methoxybenzamide tag. This research was originally published in the Journal of Biological Chemistry. [6]. © the American Society for Biochemistry and Molecular Biology

2 Materials

Prepare all solutions using ultrapure water and analytical grade reagents. All reagents should be prepared and stored at room temperature, unless otherwise indicated. Follow regulations for waste disposal when disposing of waste materials.

2.1 Protein Storage Buffers

1. KpsC storage buffer: 50 mM Tris–HCl, 250 mM NaCl, pH 7.4. Weigh out 6.06 g of Tris base, followed by 14.61 g of NaCl and transfer into a large beaker (>1 L). Add water to a volume of 900 mL. Mix thoroughly and adjust pH using concentrated HCl. Top up to 1 L with water and store buffer at room temperature.
2. KdsB storage buffer: 50 mM Tris–HCl, 250 mM NaCl, pH 8.0. Weigh out 6.06 g of Tris base, followed by 14.61 g of NaCl and transfer into a large beaker (>1 L). Add water to a volume of 900 mL. Mix thoroughly and adjust pH using concentrated HCl. Top up to 1 L with water and store buffer at room temperature (*see* **Notes 1–3**).

2.2 In Vitro Assay Components

1. 500 mM HEPES, pH 8.0: Weigh out 23.83 g HEPES and transfer into a beaker (>100 mL). Add water to a volume of 90 mL. Mix thoroughly and adjust pH using concentrated NaOH. Top up to 100 mL with water and store buffer at room temperature. Store 1 mL aliquot of this buffer at −20 °C for use in in vitro assay.
2. 100 mM $MgCl_2$: Weigh out 0.95 g $MgCl_2$ and transfer into a beaker (>100 mL). Add water to a volume of 100 mL. Store 1 mL aliquot of this buffer at −20 °C for use in in vitro assay.
3. 40 mM Kdo: Weigh out 5 mg Kdo (Sigma) and transfer into a microfuge tube. Add water to a volume of 524.79 μL. Store in aliquots at −20 °C.
4. 100 mM CTP: Weigh out 161.6 mg CTP (Sigma) and transfer into a 15 mL conical tube. Add water to a volume of 5 mL. Store in aliquots at −20 °C (*see* **Note 4**).
5. 10 mM synthetic acceptor 1. Store in aliquots at −20 °C (*see* **Notes 5–7**).

2.3 Thin-Layer Chromatography Analysis

1. TLC solvent: In a fume hood, measure out and mix thoroughly chloroform, methanol, water, and acetic acid in a ratio of 25:15:4:2, respectively. Store in a tightly sealed container in the fume hood at room temperature (*see* **Note 8**).
2. Silica Gel 60 F_{254} TLC plate (EDM Millipore).

2.4 Nuclear Magnetic Resonance Spectroscopy Sample Preparation

1. 99.96% D_2O (add supplier).
2. Speed-vac centrifuge.

3 Methods

3.1 KpsC-N In Vitro Assay with Acceptor 1, Small-Scale

1. On ice, combine, in a PCR tube, 50 mM HEPES, 10 mM $MgCl_2$, 2 mM Kdo, 5 mM CTP, 0.1 mM acceptor 1, and 10 μg of KpsC-N, to a final volume of 19 μL.
2. Finally, add 1 μg of KdsB to make up reaction volume to 20 μL and initiate CMP-Kdo synthesis. Immediately cap and incubate reaction at 37 °C for 30 min.

3.2 TLC Analysis

1. Based on the number of reactions, cut TLC plate (Fig. 2) to 6 cm length and appropriate width to accommodate all samples (leave 0.5 cm between samples and at each side of plate). Lightly draw line of origin 1.5 cm from the bottom of the plate where samples will be spotted.
2. Following 30 min incubation, move reaction mixtures to ice. Spot 1 μL of each reaction mixture on the TLC plate, along with a synthetic acceptor standard. Cover from light and allow 10–15 min for samples to dry.
3. In a fume hood, pour TLC solvent into chamber until solution covers the bottom of chamber. Place lid on chamber and allow equilibration time of 10–15 min.
4. Once samples have dried, carefully place TLC plate into chamber and allow solvent to run until approximately 1 cm from top of plate. Remove plate from chamber. In fume hood, cover from light and allow plate to dry.
5. Image TLC plate under UV light.

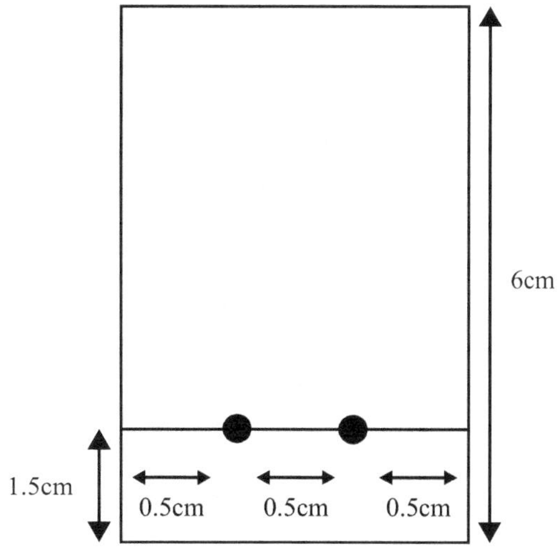

Fig. 2 Overview of TLC plate design. The dimensions of the plate and spacing between samples are outlined as mentioned in the text

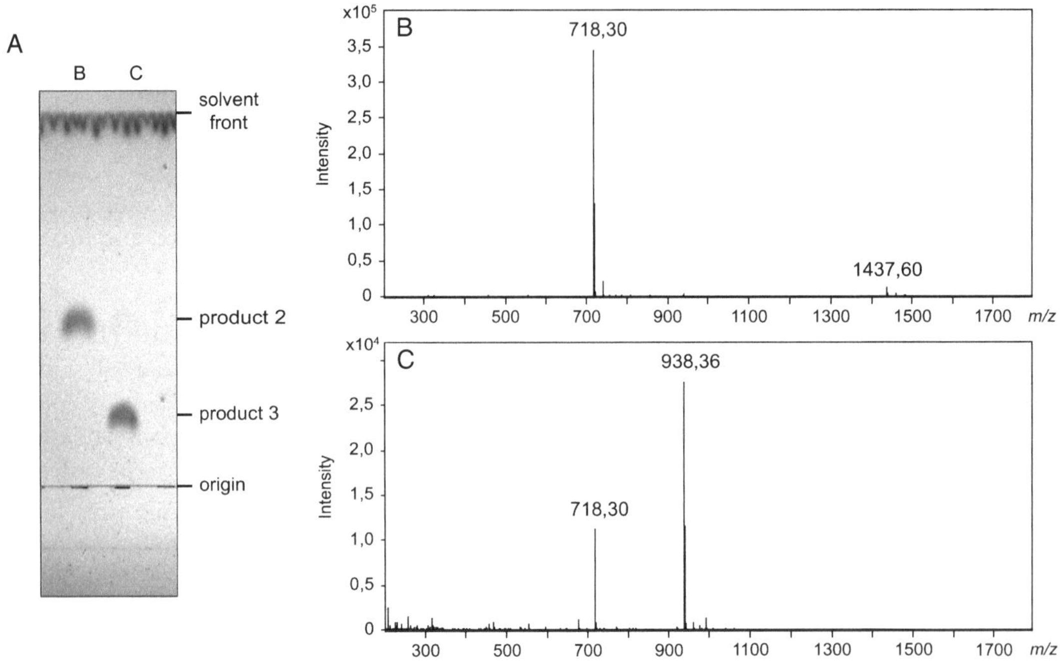

Fig. 3 Structural characterization of in vitro reaction products 2 and 3, synthesized using KpsC-N and KpsC-C, respectively. (**a**) TLC analysis of reaction products 2 and 3. Sample lanes are labeled according to their corresponding mass spectra discussed in parts B and C. (**b**) ESI mass spectrum in negative ion mode of reaction product 2, synthesized via KpsC-N and acceptor 1. Product 2 has a calculated molecular mass of 719.30 atomic mass units corresponding to the negative ion peaks at $[M-H]^-$ 718.30 and $[2M-H]^-$1437.60. (**c**) ESI mass spectrum in negative ion mode of reaction product 3, synthesized via KpsC-C and product 2. Product 3 has a calculated molecular mass of 939.36 atomic mass units corresponding to the negative ion peak at $[M-H]^-$ 938.36. This research was originally published in the Journal of Biological Chemistry [6]. © the American Society for Biochemistry and Molecular Biology

3.3 KpsC-N In Vitro Assay with Acceptor 1, Large-Scale

1. On ice, combine, in a microfuge tube, 50 mM HEPES, 10 mM $MgCl_2$, 2 mM Kdo, 5 mM CTP, 1 mM acceptor 1 and 5 mg of KpsC-N, to a final volume of 950 μL.
2. Finally, add 50 μg of KdsB to make up reaction volume to 1 mL and initiate CMP-Kdo synthesis. Immediately cap and incubate reaction mixture at 37 °C for 1 h.
3. Follow TLC analysis protocol listed above for inspection of reaction progress (Fig. 3a).

3.4 Mass Spectrometry Sample Preparation

1. Take a 5 μL aliquot of your purified in vitro reaction products and mix 1:1 with chloroform in a PCR tube.
2. Centrifuge mixture at 12,000 × *g* for 5 min. Remove the aqueous layer being careful not to agitate any precipitate and transfer mixture into a clean sample vial for mass spectrometry analysis (Fig. 3b, c) (*see* **Note 9**).

3.5 Purification of In Vitro Reaction Products

1. Combine completed in vitro reactions, if necessary, and dilute to 5 mL total volume in water.
2. Obtain a C_{18} Sep-Pak cartridge and equilibrate with 10 mL of 100% acetonitrile followed by 20 mL of water. Take care throughout this procedure to not let the cartridge run dry.
3. Apply your sample to the cartridge and collect the flow through.
4. Wash your sample by applying 20 mL water over the cartridge and collect wash fraction.
5. Elute your sample in 5 mL of a 1:1 acetonitrile–water mixture and collect eluent.
6. Dry sample using a SpeedVac centrifuge.

3.6 Nuclear Magnetic Resonance Spectroscopy Sample Preparation

1. Resuspend sample in 99.96% D_2O. Mix gently. Dry sample in SpeedVac centrifuge.
2. Repeat **step 1**.
3. Resuspend sample in 250 μL of 99.96% D_2O for NMR analysis (*see* **Note 10**).

3.7 KpsC-C In Vitro Assay with Acceptor 1, Large-Scale

1. On ice, combine, in a microfuge tube, 50 mM HEPES, 10 mM $MgCl_2$, 2 mM Kdo, 5 mM CTP, 1 mM purified product 2 (synthesized and purified as described above) and 150 μg of KpsC-C, to a final volume of 950 μL.
2. Finally, add 50 μg of KdsB to make up reaction volume to 1 mL and initiate CMP-Kdo synthesis. Immediately cap and incubate reaction at 37 °C for 1 h.
3. Follow protocols listed above, regarding product 3 purification as well as TLC, mass spectrometry and NMR spectroscopy analysis, for characterization of in vitro reaction (Fig. 3a, c).

3.8 KpsC In Vitro Assay with Product 2, Large-Scale

1. On ice, combine, in a microfuge tube, 50 mM HEPES, 10 mM $MgCl_2$, 2 mM Kdo, 5 mM CTP, 0.1 mM purified product 2 and 800 μg of KpsC, to a final volume of 3.8 mL.
2. Finally, add 200 μg of KdsB to make up reaction volume to 4 mL and initiate CMP-Kdo synthesis. Immediately cap and incubate reaction at 37 °C for 1 h to produce product 4.
3. Follow protocols listed above, regarding product 4 purification as well as TLC, mass spectrometry and NMR spectroscopy analysis, for characterization of in vitro reaction (Fig. 4a, b).

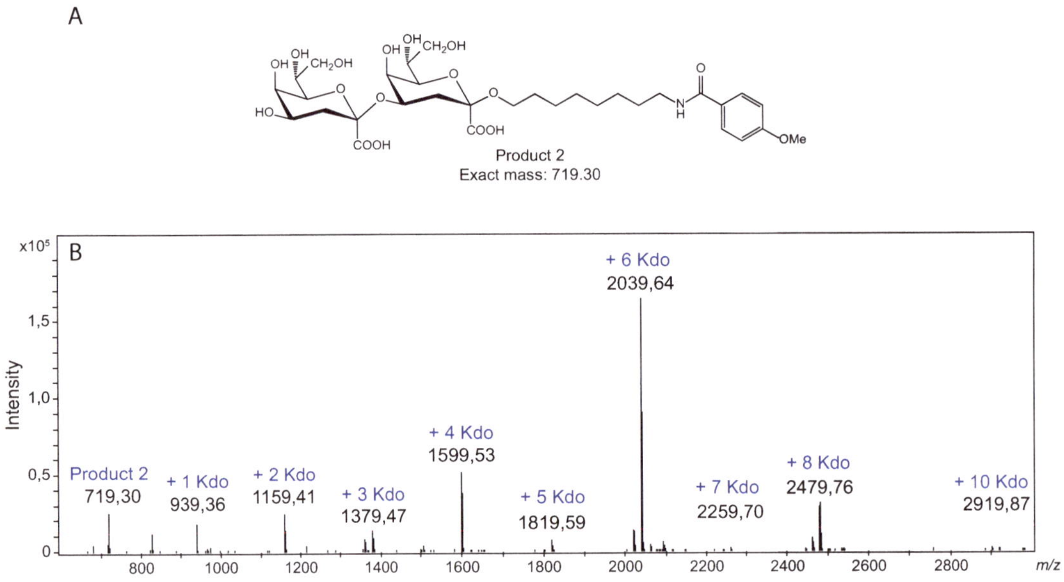

Fig. 4 Structural characterization of in vitro reaction product 4, synthesized using KpsC. (**a**) Structure and molecular mass of product 2 which was used as a substrate for reaction with the native KpsC enzyme. (**b**) Charge-deconvoluted liquid chromatography mass spectrum of reaction product 4. Product 2 was elongated with up to ten Kdo residues with predominant peaks corresponding to the addition of four, six, and eight sugars

4 Notes

1. The nucleotide sugar donor, CMP-β-Kdo, is inherently unstable and therefore must be synthesized in situ. This synthesis is accomplished via addition of CTP and Kdo along with a purified CMP-Kdo synthetase, KdsB, from *E. coli*.
2. To store KpsC and KdsB enzymes long-term (i.e., weeks to months), add glycerol to the purified enzymes to a final concentration of 10%. Store at −80 °C.
3. Long-term storage of enzymes may result in protein precipitation and/or loss of activity.
4. We recommend storing reaction components in small (~100 μL) aliquots at −20 °C to limit the number of freeze–thaw cycles.
5. Fluorescent-tagged synthetic acceptors are light-sensitive. Store aliquots away from light at −20 °C.
6. Structural differences in synthetic acceptor molecules may have an impact on KpsC activity. In this event, enzyme and synthetic acceptor concentrations may need to be optimized.

7. Synthetic acceptors [6] were synthesized by the research group of Todd L. Lowary at the University of Alberta, Department of Chemistry.
8. TLC solvent was stored in a fume hood for maximum of 1 week for future use. After this storage period fresh TLC solvent was prepared.
9. Mass spectra were analyzed using the Bruker Daltonik DataAnalysis program and the Agilent MassHunter Quantitative Analysis B.06.00 program.
10. NMR data was acquired and processed using the Bruker TopSpin 2.1 program.

Acknowledgments

We thank Bo-Shun Huang, Matthew S. Kimber, and Todd L. Lowary for their contributions to this project originally published as reference 6. This work was supported by operating funding from the Canadian Institutes of Health Research, the National Science and Engineering Research Council of Canada, and the Canadian Glycomics Network (GlycoNet, National Centres of Excellence Program).

References

1. Montoya-Peleaz PJ, Riley JG, Szarek WA et al (2005) Identification of a UDP-gal: GlcNAc-R galactosyltransferase activity in *Escherichia coli* VW187. Bioorg Med Chem Lett 15:1205–1211
2. Clarke BR, Richards MR, Greenfield LK et al (2011) In vitro reconstruction of the chain termination reaction in biosynthesis of the *Escherichia coli* O9a O-polysaccharide. J Biol Chem 286:41391–41401
3. Ovchinnikova OG, Mallette E, Koizumi A et al (2016) Bacterial β-Kdo glycosyltransferases represent a new glycosyltransferase family (GT99). Proc Natl Acad Sci 113:E3120–E3129
4. Whitfield C (2006) Biosynthesis and assembly of capsular polysaccharides in *Escherichia coli*. Annu Rev Biochem 75:39–68
5. Willis LM, Whitfield C (2013) KpsC and KpsS are retaining 3-deoxy-D-*manno*-oct-2-ulosonic acid (Kdo) transferases involved in synthesis of bacterial capsules. Proc Natl Acad Sci U S A 110:20753–20758
6. Ovchinnikova OG, Doyle L, Huang BS et al (2017) Biochemical characterization of bifunctional 3-deoxy-β-D-*manno*-oct-2-ulosonic acid (β-Kdo) transferase KpsC from *Escherichia coli* involved in capsule biosynthesis. J Biol Chem 291:21519–21530

Chapter 13

Synthesis of Phenoxyundecyl Diphosphate Disaccharides for Studies of the Biosynthesis of O Antigenic Polysaccharides in Enteric Bacteria

Vladimir Torgov, Leonid Danilov, Natalia Utkina, Vladimir Veselovsky, Alexander Kocev, and Inka Brockhausen

Abstract

The biosynthesis of O antigenic polysaccharides in enteric bacteria from nucleoside diphosphate sugars (donor substrates) is catalyzed by the corresponding glycosyltransferases and proceeds through the intermediate formation of undecaprenyl diphosphate sugars (acceptor substrates). To study this process, a chemical synthesis of the compounds having the natural structure or their modified analogs is necessary. The phosphoroimidazolidate method is a universal method for synthesis of lipid diphosphate disaccharides containing 2-acetamido-2-deoxyglycosyl residue at the reducing end of the disaccharide moiety and 11-phenoxyundecyl residue as lipid fragment of the molecule. We report here protocols to synthesize the disaccharides P^1-(11-phenoxyundecyl)-P^2-(2-acetamido-2-deoxy-3-O-α-D-rhamnopyranosyl-α-D-glucopyranosyl) diphosphate [D-Rha(α1-3)-D-GlcNAcα-PP-PhU, **Compound 1**] and P^1-(11-phenoxyundecyl)-P^2-(2-acetamido-2-deoxy-3-O-β-D-galactopyranosyl-α-D-galactopyranosyl) diphosphate [D-Gal(β1-3)-D-GalNAcα-PP-PhU, **Compound 6**]. We describe the procedures for identification and structure estimation of compounds by TLC, NMR, and MS. We also include the biochemical testing of **Compound 6** with α2,3-sialyltransferase WbwA from *Escherichia coli* O104.

Key words Chemical synthesis, Phosphoroimidazolidate method, D-Rha(α1-3)-D-GlcNAc-α-PP-PhU, D-Gal(β1-3)-D-GalNAcα-PP-PhU, TLC, NMR, MS, Sialyltransferase WbwA

1 Introduction

The biosynthesis of enterobacterial O antigenic polysaccharides occurs *via* intermediate formation of polyprenyl diphosphate sugars anchored into the cell membrane. The assembly of the repeating units of the polysaccharides is catalyzed by specific sequentially acting glycosyltransferases (GTs). Elucidating the mechanisms of this process opens a way to the development of a new generation of antibacterial drugs. The use of native polyprenyl diphosphate sugars for this purpose is practically impossible due to their extremely low concentrations and short lifetime in bacterial

Inka Brockhausen (ed.), *Bacterial Polysaccharides: Methods and Protocols*, Methods in Molecular Biology, vol. 1954, https://doi.org/10.1007/978-1-4939-9154-9_13,

cells. In addition, to study the specificity of GTs to the structure of acceptor substrates, it is necessary to have available a set of accordingly modified analogs of undecaprenyl diphosphate sugars. Such compounds cannot solely be obtained by biochemical methods. Therefore, the development of efficient methods for the chemical synthesis of polyprenyl diphosphate sugars and their analogs is essential for understanding the biosynthesis of O antigenic polysaccharides in enteric and other gram-negative bacteria.

In 1989 we have developed a simple and universal phosphoroimidazolidate method for the creation of a "pyrophosphate bridge" between glycosyl and hydrocarbyl residues of the compounds by condensation of glycosyl phosphates with polyprenyl phosphates [1]. This method allows the synthesis of a great number of polyprenyl diphosphate sugars and their modified analogs [2–5]. It has also been shown that the polyprenyl group can be replaced with a phenoxyundecyl radical to synthesize active GT acceptor substrates [6]. Introduction of chromophore groups in lipid diphosphate sugar molecules increases the sensitivity of detecting the biochemical reaction products and facilitates identification and purification of GT reaction products.

Previously, a number of phenoxyundecyl diphosphate monosaccharides Sug-PP-$(CH_2)_{11}$OPh (Sug = GlcNAc, GalNAc, Glc or Gal) were synthesized. It was demonstrated that these synthetic analogs can function as acceptor substrates for a number of bacterial GTs that catalyze the second step of O antigenic polysaccharide repeating unit biosynthesis [7–12].

To identify and to study GTs that catalyze the third step of biosynthesis of O antigenic polysaccharide repeating units, two new compounds containing a 11-phenoxyundecyl residue instead of undecaprenyl as lipid fragment and disaccharide residues as the sugar moiety were synthesized: P^1-(11-phenoxyundecyl)-P^2-(2-acetamido-2-deoxy-3-O-α-D-rhamnopyranosyl-α -D-glucopyranosyl) diphosphate [D-Rha(α1-3)-D-GlcNAcα-PP-PhU, **Compound 1**] and P^1-(11-phenoxyundecyl)-P^2-(2-acetamido-2-deoxy-3-O-β-D-galactopyranosyl-α-D-galactopyranosyl) diphosphate [D-Gal(β1-3)-D-GalNAcα-PP-PhU, **Compound 6**] [13]. These compounds were prepared to investigate D-rhamnosyltransferases WbpX and WbpY from *Pseudomonas aeruginosa* and sialyltransferase WbwA from *Escherichia coli* O104, respectively. Structural analyses of the O antigen repeating units (ECODAB data base) suggest that these compounds are suitable as acceptor substrates for several other bacterial GTs. The compounds also provide standards to identify biochemical reaction products and isolated bacterial oligosaccharides or polysaccharides.

In this protocol we describe the synthesis of a lipid diphosphate disaccharide (**Compound 1**) by the phosphoroimidazolidate method (Fig. 1) that can be used to study the extension of the Common Polysaccharide Antigen in *Pseudomonas aeruginosa*. In

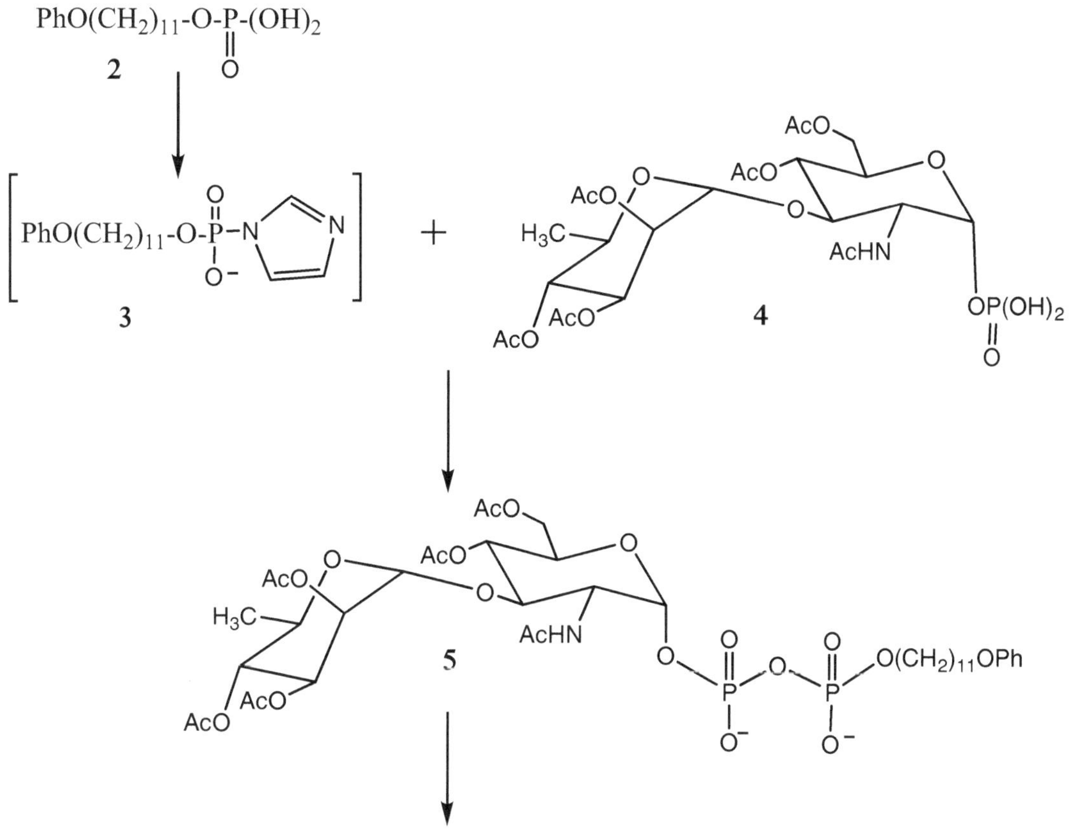

D-Rhap(α1–3)–D-GlcpNAcα–OP(O)$_3$-P(O)$_3$-(CH$_2$)$_{11}$-OPh

1

Fig. 1 Synthesis of P[1]-(11-phenoxyundecyl)-P[2]-(2-acetamido-2-deoxy-3-O-α-D-rhamnopyranosyl-α-D-glucopyranosyl) diphosphate (**Compound 1**)

addition, we give recommendations on the synthesis of an acceptor substrate (**Compound 6**) for sialyltransferase WbwA from *E.coli* O104.

2 Materials

2.1 Activating 11-Phenoxyundecyl Dihydrogen Phosphate

1. 11-Phenoxyundecyl dihydrogen phosphate (**2**, Fig. 1) was obtained according to Druzhinina et al. [7] (*see* **Note 1**).
2. Diisopropylamine [$(CH_3)_2HC–NH–CH(CH_3)_2$] (Aldrich), benzene (C_6H_6) (Acros Organics), 1.1′-carbonyldiimidazole (CDI) (Aldrich), methanol (MeOH) (Acros Organics), tetrahydrofuran (THF) (Acros Organics) (*see* **Note 2**).

3. Thin-layer chromatography plates 20 × 50 mM (Silica Gel 60 F_{254}) (Merck). System: Chloroform ($CHCl_3$)–MeOH–H_2O (6:4:0.5).
4. Lyophilizer.
5. Rotary evaporator.

2.2 Preparation of P[1]-(11-Phenoxyundecyl)-P[2]-(2-Acetamido-2-Deoxy-3-O-α-D-Rhamnopyranosyl-α-D-Glucopyranosyl) Diphosphate

1. 2-Acetamido-4,6-di-O-acetyl-2-deoxy-3-O-(2,3,4,6-tetra-O-acetyl-α-D-rhamnopyranosyl)-α-D-glucopyranosyl dihydrogen phosphate (**4**, Fig. 1) was obtained according to Torgov et al. [13] (*see* **Note 3**).
2. Crude 11-phenoxyundecyl phosphoroimidazolidate (**3,** Fig. 1) was obtained according to Torgov et al. [13]
3. 1 M solution of sodium methoxide (MeONa) in dry MeOH (*see* **Note 4).**
4. Glacial Acetic acid (AcOH), ethanol (EtOH) (Acros Organics).
5. C18 Sep-Pak cartridges (5 mL, Waters).
6. Thin-layer chromatography system: $CHCl_3$, MeOH, H_2O (10:10:3).
7. Optical rotations: JASCO P-2000 automatic digital polarimeter for solutions in $CHCl_3$ with concentration $c = 1$.
8. Mass spectrometer: Bruker micrOTOF II (electrospray ionization, HRESIMS). Electrospray Ionization source run in negative mode.
9. NMR spectrometer: Bruker AVANCE 600 (for ^{1}H, 600.13, for ^{13}C, 150.90, and for ^{31}P, 242.9 MHz). CD_3OD.
10. Vacuum pump.

2.3 Preparation of P[1]-(11-Phenoxyundecyl)-P[2]-(2-Acetamido-2-Deoxy-3-O-β-D-Galactopyranosyl-α-D-Galactopyranosyl) Diphosphate

1. 2-Acetamido-4,6-di-O-acetyl-2-deoxy-3-O-(2,3,4,6-tetra-O-acetyl-β-D-galactopyranosyl)-α-D-galactopyranosyl dihydrogen phosphate was obtained according to Torgov et al. [13] (*see* **Note 5**).
2. Reagents and equipment for the synthesis and detection of **Compound 6** are described in Subheading 2.2, **items 2–10**.

2.4 Sialyltransferase Purification

1. Sialyltransferase WbwA: *E. coli* BL21 bacteria expressing the *wbwA* gene in pRSETA plasmid are obtained as described in Czuchry et al. [14].
2. Affinity column: HisPure Co-resin (Thermo).
3. Lysis buffer: 50 mM HEPES, pH 7.5, 150 mM NaCl, 1 mM dithiothreitol (DTT), 1 protease inhibitor cap, Complete Mini, EDTA-free (Roche) per 100 mL buffer.

4. Wash buffer: 50 mM HEPES, pH 7.5, 150 mM NaCl, 1 mM DTT, 10 mM imidazole.
5. Elution buffer: 50 mM HEPES, pH 7.5, 150 mM NaCl, 1 mM DTT, 250 mM imidazole.
6. Storage buffer: 10% glycerol in lysis buffer.
7. Filtration: Centricon 10,000 cutoff (Merck).
8. Bradford protein assay (Bio-Rad).
9. Bovine serum albumin.

2.5 Sialyltransferase Activity Assay

1. Purified WbwA in storage buffer, stored at −20 °C.
2. Donor substrate: CMP-[6-^{3}H]sialic acid (American Radiolabeled Chemicals); CMP-sialic acid (EMD Millipore), mixed to yield 10 mM solution, 2 nCi/nmol.
3. Acceptor substrate: 5 mM **Compound 6.**
4. Reaction buffer: 4 μL 0.5 M Tris–HCl, pH 7, 2 μL 0.1 M $MnCl_2$, 2 μL 10 mM CMP-[6-^{3}H]sialic acid (2 nCi/nmol).
5. C18 Sep-Pak cartridges, short (Waters).
6. Methanol.
7. Scintillation counting: scintillation fluid, 7 mL scintillation vials, scintillation counter.
8. Control acceptor: 5 mM Galβ1-3GalNAcα-Bn (Sigma).

3 Methods

3.1 Activating 11-Phenoxyundecyl Dihydrogen Phosphate

1. Dissolve 76 mg (0.22 mmol) 11-phenoxyundecyl dihydrogen phosphate (**2,** Fig. 1) in 4 mL dry benzene.
2. Add 0.3 mL dry diisopropylamine to solution stirred under argon atmosphere for 5 min, freeze, and then lyophilize in vacuum of oil pump with LN_2 trap (*see* **Note 6**).
3. Dissolve the resulting white powder in 6 mL of dry THF.
4. Add 270 mg (1.66 mmol) CDI, stir the reaction mixture under argon atmosphere at rt.
5. Analyze the reaction mixture by TLC (*see* **Note 7**). After 2 h TLC shows a total conversion of phosphate **2** (Fig. 1) (R_f 0.50) into intermediate phosphoroimidazolidate **3** (R_f 0.85).
6. Add 1 mL dry MeOH, stir the reaction mixture for 30 min at rt.
7. Remove solvent with vacuum pump.
8. Dissolve the residue in 6 mL dry benzene, freeze and lyophilize with vacuum pump.

3.2 Synthesis of P[1]-(11-Phenoxyundecyl)-P[2]-(2-Acetamido-2-deoxy-3-O-α-D-Rhamnopyranosyl-α-D-Glucopyranosyl) Diphosphate (1)

1. Add 6 mL dry benzene and 0.3 mL dry diisopropylamine to 92 mg (0.14 mmol) of freshly prepared 2-acetamido-4,6-di-O-acetyl-2-deoxy-3-O-(2,3,4,6-tetra-O-acetyl-α-D-rhamnopyranosyl)-α-D-glucopyranosyl dihydrogen phosphate (**4,** Fig. 1) (*see* **Note 8**), stir the mixture under argon atmosphere for 5 min, freeze, then lyophilize with vacuum pump.
2. Dissolve crude 11-phenoxyundecyl phosphoroimidazolidate (**3**) in 3 mL dry THF and add to the lyophilizate with stirring under argon atmosphere. The reaction mixture is left at 37 ° C for 48 h (*see* **Note 9**).
3. Remove the solvent by rotary evaporation. For deacetylation, dissolve the oily residue in 3 mL MeOH, add 0.2 mL 1 M solution of MeONa in dry MeOH, and stir for 1 h at rt. Monitor by TLC.
4. For neutralization add 0.6 mL of glacial AcOH, stir the solution for 15 min and remove the solvents by rotary evaporation.

3.3 Purification of Compound 1

1. The reaction product is isolated using a C18 Sep-Pak cartridge (5 mL). The solid residue obtained after deacetylation and neutralization is dissolved in 20 mL MeOH and divided into two portions of 10 mL which are evaporated to dryness. Each portion is dissolved in 40 mL of distilled water and applied to the preactivated cartridge (*see* **Note 10**). Components of the reaction mixture which are not absorbed on C18 cartridge are eluted with 40 mL water.
2. Then 5, 10, 15, 20, and 30% v/v solutions of EtOH in water (20 mL of each) are passed sequentially through the cartridge. Fractions of 20 mL are collected and analyzed by TLC (*see* **Note 11**). The desired product, **Compound 1** is eluted with 10% solution of EtOH in water (*see* **Note 12**).
3. Fractions containing **Compound 1** are combined, EtOH is removed by rotary evaporation followed by lyophilization to remove H_2O.
4. Desired **Compound 1** is obtained as a white powder (68 mg, 60%).

3.4 Analysis of Compound 1

1. Optical rotation: 10 mg of **Compound 1** is dissolved in 1 mL $CHCl_3$, the solution ($c = 1$) is placed in 100 mm cell of an automatic polarimeter. $[\alpha]_D^{20} + 62.0$.
2. Mass spectrometry: Mass spectrometry allows to determine the molecular weight and to estimate the purity of the product. Dissolve 1 mg of **Compound 1** in 1 mL of MeOH. High-resolution mass spectra are recorded in the negative mode by MS (Fig. 2).

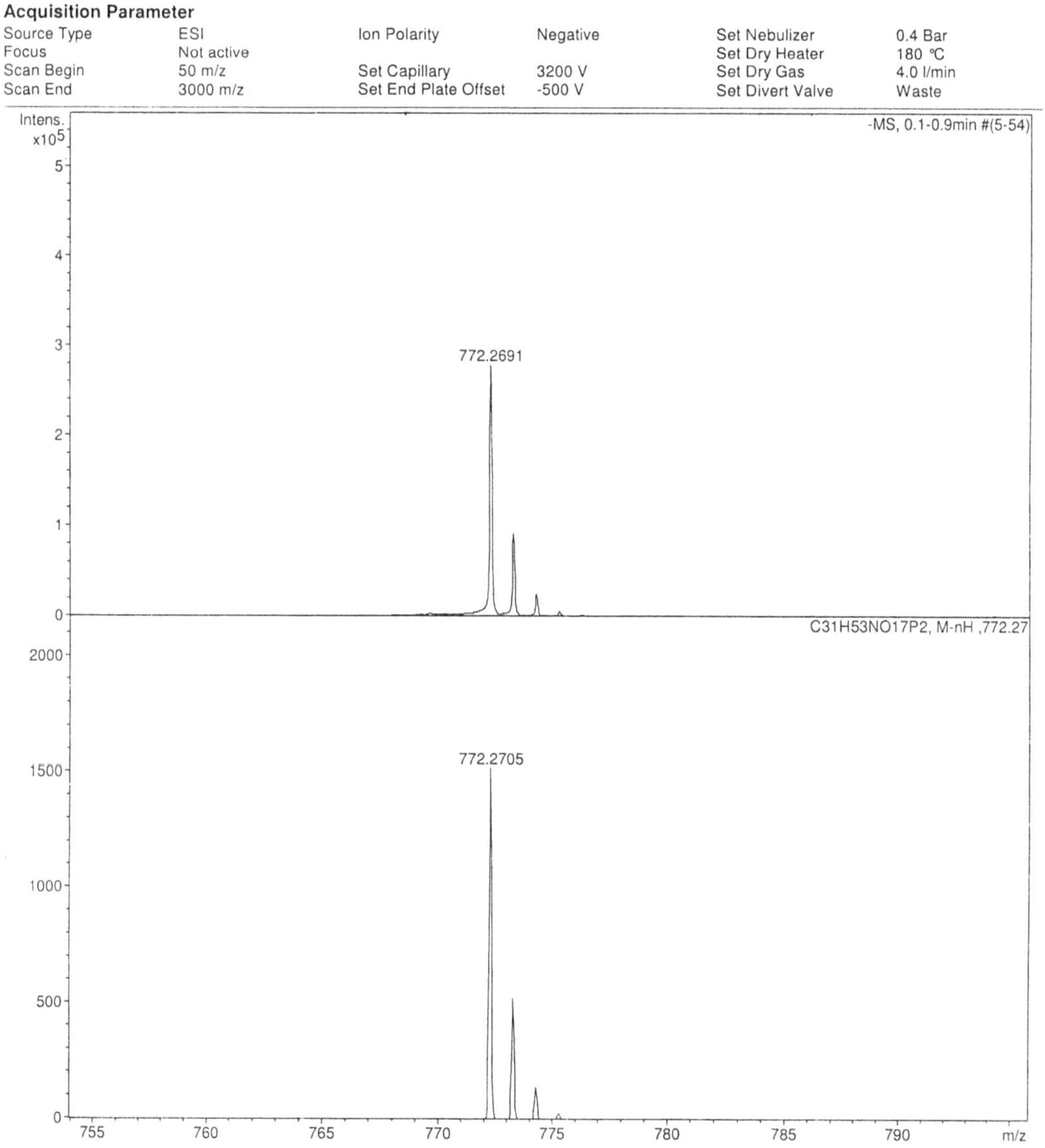

Fig. 2 Mass spectrum of D-Rha(α1-3)-D-GlcNAcα-O–PO_3–PO_3–$(CH_2)_{11}$-O-Ph by electrospray ionization (ESI) in the negative ion mode (**Compound 1**)

3. NMR: 15 mg **Compound 1** is dissolved in 0.6 mL of CD_3OD and placed in an NMR tube. All ^1H and ^{13}C spectra are recorded by 1D and 2D NMR methods. The Heteronuclear Single Quantum Coherence (HSQC) spectrum is shown in Fig. 3. The HSQC experiment is required for the further determination of carbohydrate structures. It provides sensitive assignments of ^{13}C chemical shifts with proton correlation.

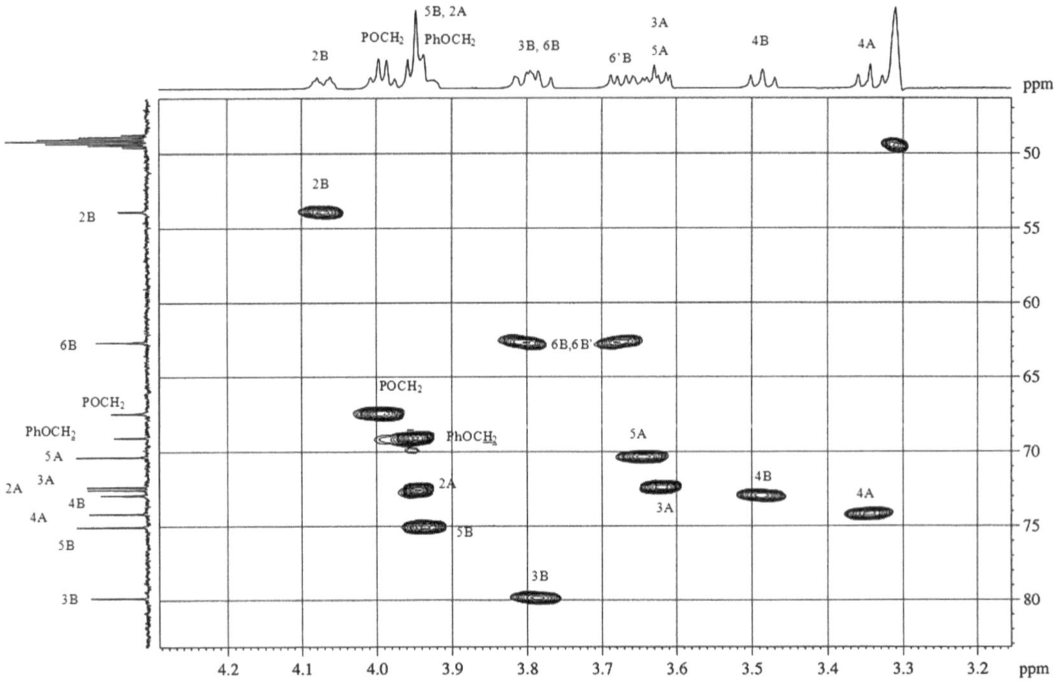

Fig. 3 Heteronuclear single quantum coherence (EdHSQC) spectrum of D-Rha(α1-3)-D-GlcNAcα--O–PO_3–PO_3–$(CH_2)_{11}$-O-Ph, D-Rha = A, D-GlcNAc = B

3.5 Synthesis and Analysis of Compound 6, P[1]-(11-Phenoxyundecyl)-P[2]-(2-Acetamido-2-Deoxy-3-O-β-D-Galactopyranosyl-α-D-Galactopyranosyl) Diphosphate

The first step of the synthesis of **Compound 6**, the activation of 11-phenoxyundecyl dihydrogen phosphate has been described in Subheading 3.1. The subsequent steps are similar to those for **Compound 1**, except that the sugar acceptor is benzyl 2-acetamido-4,6-O-cyclohexylidene-2-deoxy-α-D-galactopyranoside which is condensed with 2,3,4,6-tetra-O-acetyl-α-D-galactopyranosyl bromide [13]. **Compound 6** is purified (Subheading 3.3) and analyzed (Subheading 3.4) using the same methods as for **Compound 1**.

3.6 Sialyltransferase Purification

1. Lysis of bacteria: lyse pellet of bacteria expressing His-tagged WbwA (from 250 mL culture) by sonication (5 times 15 s) in Lysis buffer.
2. Affinity chromatography: centrifuge lysate for 10 min at 4 °C at 8000 × *g*. Apply the supernatant to 2 mL Co-resin and wash the resin six times with 1 mL Wash buffer. Enzyme is eluted four times with 1 mL Elution buffer.
3. Remove imidazole by filtration/centrifugation through a Centricon filter and replacement of the elution buffer with lysis buffer containing 10% glycerol.

Table 1
^{1}H and ^{13}C NMR parameters of D-Rha(α1-3)-D-GlcNAcα-O–PO_3–PO_3–$(CH_2)_{11}$-O-Ph

Residue	^{1}H (ppm)	^{13}C (ppm)
Rha-1	5.13	103.2
Rha-2	3.90–4.00	72.4
Rha-3	3.60–3.70	72.6
Rha-4	3.34 $J_{3.4} = 10.2$ $J_{4.5} = 10.2$	74.2
Rha-5	3.60–3.70	70.3
Rha-6	1.27 3H $J_{5.6} = 6.0$	18.3
GlcNAc-1	5.51 $J_{1.2}$3.6 $J_{1.P} = 6.6$	96.6 $J_{1.P} = 6.3$
GlcNAc-2	4.10–4.05	53.9 $J_{2.P} = 8.5$
GlcNAc-3	3.76–3.83	79.8
GlcNAc-4	3.46 $J_{3.4} = 10.2$ $J_{4.5} = 10.2$	72.9
GlcNAc-5	3.90–4.00	75.0
GlcNAc-6	3.76–3.83	62.6
GlcNAc-6’	3.60–3.70	
PhO	7.22–7.25 2H 6.85–6.90 3H	115.7, 120.6, 121.6
PhO-$\underline{CH_2}$	4.00–3.90	69.1
$\underline{CH_2}$-OP	3.98–4.02	67.4 $J_{P.C} = 6.0$
$\underline{CH_2}$-CH_2-OP		32.0
NHCO$\underline{CH_3}$	2.05	23.4
CO		174.1
CH_2	1.80–1.30	27.04, 27.31, 30.58, 30.64, 30.39, 30.83, 30.85

4. Analysis of protein: SDS-PAGE and Bradford protein assays, using bovine serum albumin as a standard [10].
5. Store aliquots of the purified enzyme at −20 °C.

3.7 Sialyltransferase Activity Assay

1. Donor substrate: carefully dry 10 μCuries of radioactive CMP-[6-^{3}H]sialic acid under a nitrogen stream; add 0.5 mL 10 mM (5 μmol) CMP-sialic acid and mix. Test radioactivity (cpm) of 2 μL by scintillation counting. Dilute 2 μL with 2 mL water and determine the absorbance at 271 nm in a quartz cuvette. A 10 mM solution, diluted 1:1000, has an absorbance of 0.091 (Molar extinction coefficient of CMP is 9100). From the cpm and the concentration, calculate the specific

radioactivity which should be 2 μCi/μmol or about 2000 cpm/nmol (*see* **Note 13**). Store at −20 °C.

2. Acceptor substrate: Prepare 5 mM acceptor substrate (**Compound 6**). Store at −20 °C.
3. Assay mixtures: all assays should be done in duplicates. Prepare six Eppendorf or small glass tubes and keep all tubes on ice. Tubes 1 and 2 are negative controls that have no acceptor. Add 8 μL reaction buffer to each tube and fill volume to 30 μL with water. Then add 10 μL purified WbwA (1.5 μg protein) and gently mix 2 s on a Vortex mixer. Tubes 3 and 4: add 4 μL 5 mM **Compound 6,** add 8 μL reaction buffer and fill volume to 30 μL with water, then thaw one aliquot of purified WbwA and add 10 μL. Tubes 5 and 6 are identical to Tubes 3 and 4, except that a control acceptor (Galβ1-3GalNAcα-Bn) is added instead of **Compound 6**.
4. Incubation: incubate Tubes 1 to 6 for 30 min at 37 °C. Stop reaction by adding 700 μL cold water.
5. Hydrophobic chromatography: Apply the assay mixture very slowly (10 drop per min) to C18 Sep-Pak cartridge using a 3 mL syringe (*see* **Note 14**). Both, acceptor **Compound 6** and reaction product have a hydrophobic group that binds to the C18 column. Wash cartridge with 1 mL water, then 2 mL water, collect into vial A. Wash again with 1 mL water, collect into vial B. Elute hydrophobic reaction product with 1.5 mL MeOH, collect into vial C. Repeat elution with 1.5 mL MeOH, collect into vial D. The acceptor **Compound 6** and reaction product will be present in vials C and D.
6. Quantification of reaction product: add 4.5 mL scintillation fluid to tubes B, C and D. Invert vials carefully 4 times to mix. Count in [^{3}H] channel. Take the average cpm from duplicate tubes. Calculate the cpm per assay for **Compound 6**: cpm (tubes 3,4) minus cpm (tubes 1,2) × 0.5 h^{-1} × mg protein^{-1}. For the control substrate (not active) calculate cpm (tubes 5,6) minus cpm (tubes 1,2) × 0.5 h^{-1} × mg protein^{-1}. To calculate the sialyltransferase activity in nmol/h/mg, divide cpm per assay by the specific radioactivity (cpm/nmol) (*see* **Note 15**).

4 Notes

1. 11-Phenoxyundecyl dihydrogen phosphate (**2,** Fig. 1) was obtained in a total yield of 58% according to Druzhinina et al. [7] by gently reducing 11-phenoxyundecanoic acid with diborane generated in situ from $BF_3 \cdot Et_2O$ and $NaBH_4$ to give 11-phenoxyundecan-1-ol, which was phosphorylated with $POCl_3$ in the presence of triethylamine in dry THF followed

by hydrolysis with aqueous NaOH, acidification, and crystallization.

2. Diisopropylamine and benzene are distilled over with calcium hydride, methanol is distilled over Mg chips, tetrahydrofuran is distilled over $LiAlH_4$ just before use.

3. 2-Acetamido-4,6-di-O-acetyl-2-deoxy-3-O-(2,3,4,6-tetra-O-acetyl-α-D-rhamnopyranosyl)-α-D-glucopyranosyl dihydrogen phosphate (**4**, Fig. 1) was obtained according to Torgov et al. [13]. 2,3,4-Tri-O-acetyl-α-D-rhamnopyranosyl bromide was obtained by reaction of 1,2,3,4-tetra-O-acetyl-α-D-rhamnopyranose with 40% solution of HBr in glacial AcOH and without purification was condensed with benzyl 2-acetamido-4,6-O-cyclohexylidene-2-deoxy-α-D-glucopyranoside in dry acetonitrile in the presence of $Hg(CN)_2$ affording benzyl 2-acetamido-4,6-O-cyclohexylidene-2-deoxy-3-O-(2,3,4-tri-O-acetyl-α-D-rhamnopyranosyl)-α-D-glucopyranoside in a yield of 40%. After removal of the cyclohexylidene protecting group by 90% CF_3COOH, 4,6-diol was obtained in a yield of 95% and was further converted to benzyl 2-acetamido-4,6-di-O-acetyl-2-deoxy-3-O-(2,3,4-tri-O-acetyl-α-D-rhamnopyranosyl)-α-D-glucopyranoside in a yield of 98%. Removing the benzyl group by hydrogenolysis afforded 2-acetamido-4,6-di-O-acetyl-2-deoxy-3-O-(2,3,4-tri-O-acetyl-α-D-rhamnopyranosyl)-α-D-glucopyranose in a yield of 90%. The derivative was treated with bis(benzyloxy)(diisopropylamino) phosphine in the presence of tetrazole followed by oxidation with m-chloroperbenzoic acid (mCPBA) affording 2-acetamido-4,6-di-O-acetyl-2-deoxy-3-O-(2,3,4-tri-O-acetyl-α-D-rhamnopyranosyl)-α-D-glucopyranosyl dibenzyl phosphate in a yield of 85%. Hydrogenolysis of the latter compound in the presence of 10% $Pd(OH)_2/C$ in glacial acetic acid resulted in acetylated α-glycosyl dihydrogen phosphate (**4**) in a yield of 90%. The product was used at the next step of the synthesis immediately without additional purification.

4. 2.3 g of metallic Na is dissolved in 100 mL of dry MeOH for preparing a 1 M solution of MeONa.

5. 2-Acetamido-4,6-di-O-acetyl-2-deoxy-3-O-(2,3,4,6-tetra-O-acetyl-β-D-galactopyranosyl)-α-D-galactopyranosyl phosphate was obtained according to Torgov et al. [13]. Benzyl 2-acetamido-4,6-O-cyclohexylidene-2-deoxy-α-D-galactopyranoside was glycosylated with 2,3,4,6-tetra-O-acetyl-α-D-galactopyranosyl bromide as indicated in **Note 3** affording benzyl 2-acetamido-4,6-O-cyclohexylidene-2-deoxy-3-O-(2,3,4,6-tetra-O-acetyl-β-D-galactopyranosyl)-α-D-galactopyranoside in a yield of 42%. After removal of the cyclohexylidene protecting group by 90% CF_3COOH benzyl 2-acetamido 2 deoxy 3 O (2,3,4,6-tetra-O-acetyl-β-D-galactopyranosyl)-α-D-

galactopyranoside was obtained in a yield of 94% and was further converted to benzyl 2-acetamido-4,6-di-O-acetyl-2-deoxy-3-O-(2,3,4,6-tetra-O-acetyl-β-D-galactopyranosyl)-α-D-galactopyranoside in a yield of 98%. Benzyl group was removed by hydrogenolysis affording 2-acetamido-4,6-di-O-acetyl-2-deoxy-3-O-(2,3,4,6-tetra-O-acetyl-β-D-galactopyranosyl)-α-D-galactopyranose in a yield of 90%. Subsequent treatment with bis(benzyloxy)(diisopropylamino) phosphine, oxidation, and hydrogenolysis according to the procedure described in **Note 3** resulted in 2-acetamido-4,6-di-O-acetyl-2-deoxy-3-O-(2,3,4,6-tetra-O-acetyl-β-D-galactopyranosyl)-α-D-galactopyranosyl dihydrogen phosphate in a yield of 90%, which was used further immediately without additional purification.

6. The flask containing the benzene solution of the substance 6 with diisopropylamine is frozen in liquid nitrogen and lyophilized in vacuum of an oil pump equipped with a trap cooled with liquid nitrogen.
7. For TLC Silica Gel 60 F_{254} plates 20 × 50 mM were used, solvent: $CHCl_3$–MeOH–H_2O (6:4:0.5, v/v/v). 11-Phenoxyundecyl-containing derivatives were detected by UV illumination and further spraying of 5% sulfuric acid in MeOH followed by heating.
8. Dihydrogen phosphate (**4**) must be freshly prepared and converted immediately into diisopropylammonium salt because of its extreme instability.
9. Reaction mixture is analyzed by TLC in $CHCl_3$–MeOH–H_2O (6:4:0.5). Acetylated sugar phosphate (**4**) have R_f 0.30, phosphoroimidazolidate (**3**) R_f 0.85, target product (a protected diphosphate **5**) have R_f 0.60, a by-product [the symmetrical bic-(11-phenoxyundecyl) diphosphate] R_f 0.70. After 48 h at 37 ° C the main component of the reaction mixture was diphosphate (**5**).
10. For purification use C18 Sep-Pak cartridge (5 mL) activated by washing with 20 mL EtOH, letting stand 0.5 h in EtOH, followed by washing with 50 mL of distilled water. The reaction mixture is divided into portions containing about 20 mg of the product in 40 mL of distilled water. Each portion applies to a cartridge. After use, the C18 Sep-Pak columns are regenerated by washing with 20 mL EtOH, letting stand 0.5 h in EtOH, followed by washing with 50 mL of distilled water.
11. Since concentrations of compounds in fractions are very small, all fractions are concentrated by rotary evaporation to dryness, then dissolved in 0.3 mL MeOH and analyzed by TLC in $CHCl_3$–MeOH–H_2O (10:10:3); product (**1**) has R_f 0.65.

12. Eluate from 15% solution of EtOH in water contains the desired product (**1**) (R_f 0.65) as well as impurities. These fractions from two purification steps are combined, evaporated to dryness, dissolved in 40 mL of distilled water and reapplied to the cartridge.
13. The final cpm of the radioactive CMP-sialic acid solution depend on the efficiency of counting that varies between scintillation counters and programs. For the assays, always use the same scintillation counter and scintillation fluid used to determine cpm/nmol. CMP-[^{14}C]-labeled sialic acid has a much higher efficiency of counting but is more expensive than CMP-[^{3}H]sialic acid.

    The volume ratio (assay mixture–scintillation fluid) and the solvents present could also affect the cpm. Keep these variables as constant as possible.
14. The application of the assay mixture to the C18 Sep-Pak cartridge has to be done very slowly in order to ensure that the reaction product binds. Its phenoxyundecyl group will bind to C18 although it has negative charges of two phosphate groups and sialic acid. The subsequent washes also have to be done slowly and carefully. Most of the CMP-sialic acid and free sialic acid will be eluted with the water fractions. Another type of assay could be carried out by C18 HPLC in 7% acetonitrile–93% water mixtures as the liquid phase. The elution times of CMP-sialic acid, **Compound 2** and WbwA reaction product need to be established with standard compounds. Fractions are collected and counted for radioactivity after transfer to scintillation vials and adding scintillation fluid.
15. Controls in assays are essential. Negative controls are those lacking the acceptor, or enzyme, or adding a structurally related but inactive substrate. Galβ1-3GalNAcα-Bn is an excellent acceptor for human sialyltransferase ST3Gal1 but does not work for WbwA. The presence of the reaction product can be established by MS (ESI negative mode) from assays using nonradioactive CMP-sialic acid. Bacterial homogenates can be used initially to establish activity. Often, the enzymes are highly active in homogenates, and it may be difficult to lyse the enzymes that are often tightly associated with the bacterial membranes. Purified enzymes may also lose activity upon storage. Once the activity has been established, conditions can be altered to characterize the enzyme.

Acknowledgments

This work was supported by the Russian Foundation for Basic Research, Project 16-04-00372 and the Canadian Institutes of Health Research (GlycoNet) (to IB).

References

1. Danilov LL, Druzhinina TN, Kalinchuk NA et al (1989) Polyprenyl phosphates: synthesis and structure-activity relationship for a biosynthetic system of *Salmonella anatum* O-specific polysaccharide. Chem Phys Lipids 51:191–203
2. Danilov LL, Shibaev VN (1991) Phosphopolyprenols and their glycosyl esters: chemical synthesis and biochemical application. In: Studies in natural products chemistry (Attaur-Rahman ed.). V 8. Stereoselective synthesis (Part E). Elsevier, pp 68–114
3. Shibaev VN, Danilov LL (1992) New developments in the synthesis of phosphopolyprenols and their glycosyl esters. Biochem Cell Biol 70:429–437
4. Shibaev VN, Danilov LL (1997) Synthesis of intermediates in the dolichol pathway of protein glycosylation. In: Large DC, Warren CD (eds) Glycopeptides and related compounds. Synthesis, analysis and applications. Marcel Dekker, New York, pp 427–504
5. Danilov LL, Druzhinina TN (2007) Chemical synthesis of dolichyl phosphates, their analogues and derivatives and application of these compounds in biochemical assays. Acta Biochim Pol 54:696–701
6. Montoya-Peleaz P, Riley JG, Szarek WA et al (2005) Identification of a UDP-Gal: GlcNAc-R galactosyltransferase activity in *Escherichia coli* VW187. Bioorg Med Chem Lett 15:1205–1211
7. Druzhinina TN, Danilov LL, Torgov VI et al (2010) 11-Phenoxyundecyl phosphate as a 2-acetamido-2-deoxy-α-D-glucopyranosyl phosphate acceptor in O-antigen repeating unit assembly of *Salmonella arizonae* O:59. Carbohydr Res 345:2636–2640
8. Utkina NS, Danilov LL, Druzhinina TN et al (2010) Simple synthesis of P^1-(11-phenoxyundecyl)-P^2-(2-acetamido-2-deoxy-α-*D*-galactopyranosyl) diphosphate. Russ J Bioorg Chem 36:783–785
9. Wang S, Hao Y, Lam JZ et al (2015) Biosynthesis of the common polysaccharide antigen of *Pseudomonas aeruginosa* PAO1: characterization and role of WbpZ—α GDP-D-rhamnose: GlcNAc/GalNAc-diphosphate-lipid α1,3-D-rhamnosyltransferase WbpZ. J Bacteriol 197:2012–2019
10. Wang S, Czuchry D, Liu B et al (2014) Characterization of two UDP-Gal: GalNAc-diphosphate-lipid β1,3-galactosyltransferases WbwC from *Escherichia coli* serotypes O104 and O5. J Bacteriol 196:3122–3133
11. Utkina NS, Danilov LL, Veselovsky VV et al (2012) Synthesis of P^1-(11-phenoxyundecyl)-P^2-(α-D-galactopyranosyl) diphosphate and P^1-(11-phenoxyundecyl)-P^2-(α-D-glucopyranosyl) diphosphate and investigation of their acceptor properties in the reaction of mannosyl residue transfer catalyzed by mannosyltransferase from *Salmonella newport*. Russ J Bioorg Chem 4:412–416
12. Gao Y, Liu B, Strum S et al (2012) Biochemical characterization of WbdN, a β 1,3-glucosyltransferase involved in O-antigen synthesis in enterohemorrhagic *Escherichia coli* O157. Glycobiology 22:1092–1102
13. Torgov V, Danilov L, Utkina N et al (2017) Synthesis of P^1-(11-phenoxyundecyl)-P^2-(2-acetamido-2-deoxy-3-O-α-D-rhamnopyranosyl-α-D-glucopyranosyl) diphosphate and P^1-(11-phenoxyundecyl)-P^2-(2-acetamido-2-deoxy-3-O-β-D-galactopyranosyl-α-D-galactopyranosyl) diphosphate for the investigation of biosynthesis of O-antigenic polysaccharides *Pseudomonas aeruginosa* and *Escherichia coli* O104. Carbohydr Res 453-454:19–25
14. Czuchry D, Desormeaux P, Stuart M et al (2015) Identification and biochemical characterization of the novel α2,3-sialyltransferase WbwA from pathogenic *Escherichia coli* serotype O104. J Bacteriol 197:3760–3768

Chapter 14

Characterization of Arabinosyl Transfer Reactions in the Biosynthesis of Mycobacterial Cell Envelope (Lipo) Polysaccharides

Shiva Kumar Angala and Mary Jackson

Abstract

D-Arabinofuranose is a major glycosyl constituent of mycobacteria found in two essential cell envelope heteropolysaccharides, arabinogalactan and lipoarabinomannan. Seven different arabinosyltransferases at least are required to synthesize the arabinan domain of these two major glycans. Because of their interest from the perspective of drug development, these enzymes have been the object of intense investigations. In this chapter, we describe the protocols used to perform nonradioactive arabinosyltransferase assays with synthetic acceptor and donor substrates and characterize the enzymatic products of the reactions by mass spectrometry.

Key words Arabinosyltransferase, D-Arabinose, Mycobacteria, Synthetic arabinoside acceptor, Lipid donor

1 Introduction

Mycobacterium tuberculosis causes tuberculosis (TB) in humans and kills about 1.7 million people annually. The emergence of multi-drug-resistant *M. tuberculosis* strains threatens TB control [1]. There is an unmet need for novel drugs to treat *M. tuberculosis* infections. Mycobacteria are known to produce a complex cell envelope which plays various important roles during infection and further forms an impermeable barrier to anti-TB drug treatment [2, 3]. The unique cell envelope of mycobacteria is also the site of action of many anti-TB drugs and drug candidates [4]. Two major D-arabinofuranose-containing heteropolysaccharides, arabinogalactan (AG) and lipoarabinomannan (LAM), found in the cell envelope of all mycobacteria are essential to the structural integrity of the bacilli [5]. The structural organization of the arabinan domains of AG and LAM is highly conserved, consisting of three different glycosidic bonds *viz.* α-(1, 5), α-(1, 3) and β-(1, 2)

Inka Brockhausen (ed.), *Bacterial Polysaccharides: Methods and Protocols*, Methods in Molecular Biology, vol. 1954,
https://doi.org/10.1007/978-1-4939-9154-9_14,

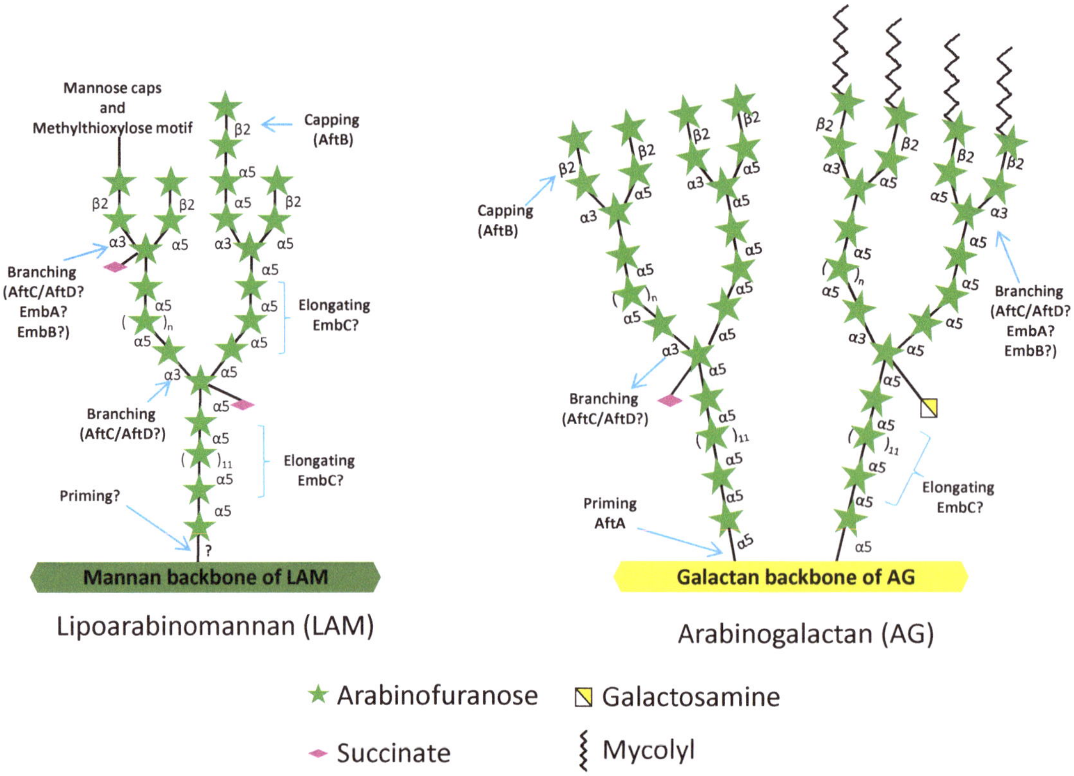

Fig. 1 Arabinosyltransferases involved in the biosynthesis of the arabinan domains of lipoarabinomannan (left panel) and arabinogalactan (right panel) in *M. tuberculosis*

[5] whose introduction in the course of the elongation of the two glycans is coordinated by a series of linkage-specific arabinosyltransferases (Fig. 1). The arabinosyltransferases AftA, AftB, AftC, AftD, EmbA, and EmbB are involved in AG biosynthesis [6–11], whereas AftB, AftC, AftD, and EmbC participate in the biosynthesis of the arabinan domain of LAM [10–13] (Fig. 1). Although EmbA, EmbB, and EmbC are known targets of the frontline anti-TB drug, ethambutol, the enzymatic activities of these proteins have not yet been directly demonstrated in vitro. All mycobacterial arabinosyltransferases identified to date utilize decaprenylphosphoryl-β-D-arabinose (DPA) (Fig. 2a) as sole arabinose donor [14]. The catalytic steps leading to DPA synthesis have been elucidated and shown to be targeted by the TB drug candidates benzothiazinones and dinitrobenzamides [15, 16]. Thus, both the DPA biosynthetic enzymes and arabinosyltransferases involved in AG and LAM biosynthesis are validated targets for the development of new anti-TB drugs. Up until now, the in vitro characterization of mycobacterial arabinosyltransferases has primarily relied on the use of rather cumbersome radioactive assays using mycobacterial lysates as enzyme source to monitor the transfer of ^{14}C-labeled arabinose from DPA onto synthetic neoglycolipid acceptors

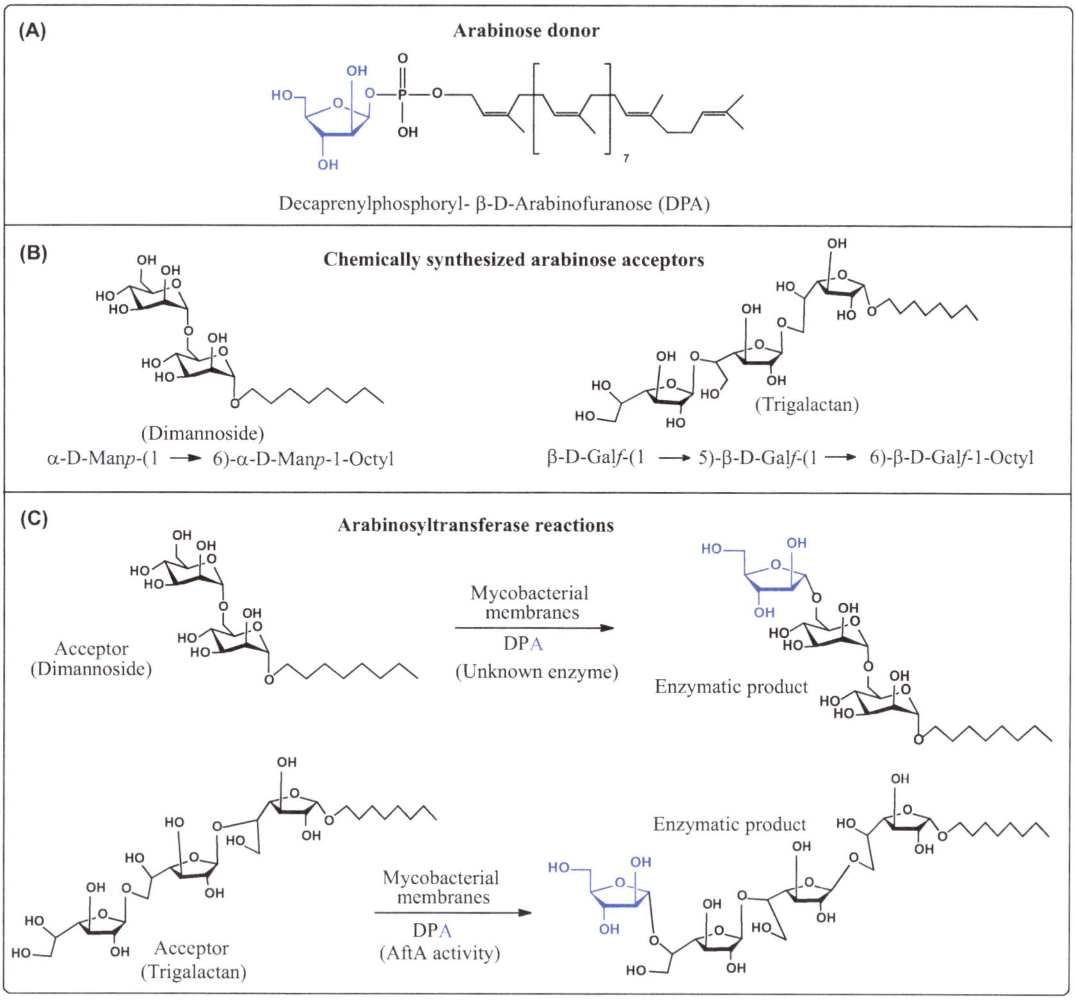

Fig. 2 Structures of the synthetic acceptor and donor substrates used in the arabinosyltransferase reactions described in this chapter. (**a**) Chemical structure of decaprenylphosphoryl-β-D-arabinose (DPA), the only known arabinose donor in mycobacteria. (**b**) Structures of synthetic dimannoside (α-D-Man*p*-(1→6)-α-D-Man*p*-$O(CH_2)_7CH_3$) and trigalactan (β-D-Galf-(1→5)-β-D-Gal*f*-(1→6)-β-D-Gal*f*-$O(CH_2)_7CH_3$) acceptor substrates. (**c**) Schematic representation of arabinosyltransferase reactions using *M. smegmatis* membranes as enzyme source, DPA as the arabinose donor and the synthetic dimannoside and trigalactan acceptors shown in (**b**)

[17]. ^{14}C-labeled DPA is generated in situ from 5-phosphoribosyl pyrophosphate p[^{14}C]Rpp [18] and product formation is monitored by thin-layer chromatography and autoradiography upon partial purification [17]. The synthetic neoglycolipid acceptors mimic the native structural organization of the arabinan domains of AG and LAM with their aglycon chain decreasing their polarity to allow for their partitioning in organic solvents. While relatively straightforward, the radioactive nature of the enzymatic products

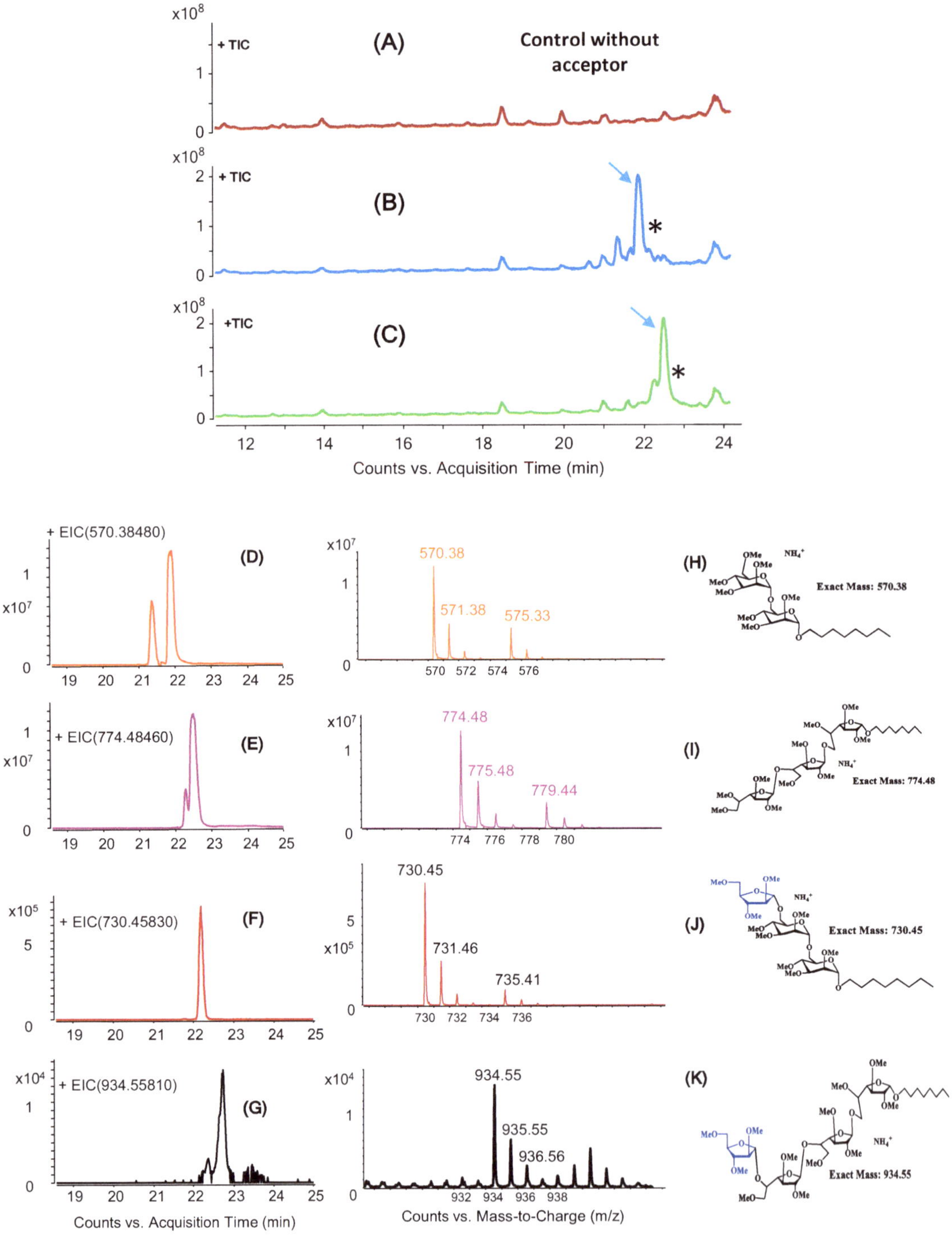

Fig. 3 LC-MS analysis of permethylated reaction products from in vitro cell-free arabinosyltransferase reactions. The arabinose acceptors shown in Fig. 2b and their enzymatic products are mostly detected as their ammonium adducts $[M + NH_4]^+$ in the positive ion mode. (**a**–**c**) Total ion chromatograms. The two major peaks indicated with an arrow correspond to the dimannoside (**b**) and trigalactan (**c**) acceptors. These peaks are lacking in the control reaction devoid of acceptor substrate (**a**). The asterisks indicate their corresponding

generated by these assays limits their further structural characterization.

We describe here a nonradioactive method to measure arabinosyltransferase activity that makes use of the same synthetic arabinoside acceptors and nonradiolabeled arabinose donor substrate (s) (Fig. 2). The enzymatic reaction products formed are extracted with ethanol and separated from interfering compounds using a strong anion exchange column. The partially purified enzymatic products are then permethylated and analyzed using liquid chromatography (LC)–mass spectrometry (MS)-based methods (Fig. 3). Compared to radioactive assays, the nonradioactive LC-MS-based assays described herein not only offer the advantage of not generating any radioactive waste but also allow for the separation of the synthetic arabinose acceptors from their enzymatic products. The purified enzymatic products may be eluted at a preparative scale and their structure subsequently characterized by mass fragmentation (LC-MS/MS). Alternatively, the permethylated eluted products may be derivatized into permethylated alditol acetates to confirm the linkage of the enzymatically transferred arabinosyl residue [19]. Importantly, multiple glycosyltransferase activities on the same acceptor substrate may be analyzed in a single LC-MS run.

2 Materials

2.1 Bacterial Growth Conditions

1. Frozen stock of *M. smegmatis* mc^{2}155 (American type culture collection) (*see* **Note 1**).
2. Luria Bertani (LB) agar medium.
3. Luria–Bertani (LB) broth medium.
4. Phosphate buffer saline pH 7.0.
5. 37 °C incubator with shaking.
6. 250 mL harvesting bottles.
7. Benchtop centrifuge.

Fig. 3 (continued) enzymatic products. Elution profiles (**d–g**) and extracted ion chromatograms (**h–k**). The elution profiles of the dimannoside acceptor (*m/z* 570.38) at 21.9 min (**d**) and trigalactan acceptor (*m/z* 770.48) at 22.4 min (**e**) are shown. Two new peaks with retention times 22.2 min (**f**) and 22.7 min (**g**) correspond to the addition of an arabinosyl residue to the dimannoside and trigalactan acceptors, respectively. The presence of ions with *m/z* 730.45 (**j**) and 934.55 (**k**) further confirms the enzymatic transfer of a single arabinosyl residue onto the synthetic dimannoside and trigalactan acceptors, respectively. The extracted ion chromatograms of the two acceptor substrates are shown in panels H and I. The chemical structures of the acceptors and their possible enzymatic products are shown alongside their corresponding extracted ion chromatograms

2.2 Preparation of Enzymatically Active Membrane Fractions

1. Endotoxin-free water.
2. Buffer A: 50 mM MOPS (pH 8.0 with KOH), 5 mM 2-mercaptoethanol, and 10 mM $MgCl_2$ (Buffer A should be kept at 4 °C at all times).
3. Probe tip sonicator (Soniprep 150; MSE Ltd., Crawley Sussex, UK; 1-cm probe).
4. Oak Ridge round-bottom FEP tubes (30 mL).
5. Bench top centrifuge.
6. Ultra-centrifuge.
7. 1 mL and 7 mL Dounce glass homogenizers.
8. Bicinchoninic acid (BCA) protein estimation kit.
9. Spectrophotometer.

2.3 Arabinosyltransferase Assay

1. Buffer B: 50 mM MOPS (adjust pH 8.0 with KOH), 5 mM 2-mercaptoethanol, 10 mM $MgCl_2$, and 1 mM ATP.
2. Chemically synthesized α-D-Man*p*-(1→6)-α-D-Man*p*-1-Octyl (dimannoside) as arabinose acceptor [20] (*see* **Note 2**).
3. Chemically synthesized β-D-Gal*f*-(1→5)-β-D-Gal*f*-(1→6)-β-D-Gal*f*-1-Octyl (trigalactan) as arabinose acceptor [7] (*see* **Note 2**).
4. Chemically synthesized decaprenylphosphoryl-β-D-arabinose (DPA) as arabinose donor [21] (*see* **Note 3**).
5. Anhydrous ethyl alcohol, absolute.
6. 37 °C water bath.
7. Benchtop centrifuge.
8. HyperSep strong anion exchange (SAX) columns (Thermo Scientific).
9. Vacuum concentrator, speed—vac (Savant).
10. Teflon-lined screw capped 13 × 100 mM glass tubes (Hot air oven dried).

2.4 Permethylation of Enzymatic Products

1. Vacuum concentrator, speed—vac (Savant).
2. Dried sample.
3. Anhydrous dimethylsulfoxide (DMSO).
4. NaOH pellets.
5. Iodomethane.
6. Water bath sonicator.
7. Endotoxin-free water.
8. Mortar and Pestle (Hot air oven dried).
9. Glass syringe with needle.

10. Ice-cold deionized water.
11. Teflon-lined screw capped 100 × 13 mM glass tubes (Hot air oven dried).

2.5 LC-MS Analysis

1. Methanol.
2. Mass spectrometer: Agilent 6220A time-of-flight (TOF) mass spectrometer equipped with an electrospray ionization/atmospheric pressure chemical ionization (ESI/APCI) multimode source.
3. Liquid chromatography: Agilent 1200 with binary pump and autosampler (Agilent technologies; Palo Alto, CA).
4. HPLC Column: 2.1 mM inner diameter (ID) × 150 mM, 3.5 μm X-Bridge reverse phase C18 column.
5. Solvent A: LCMS-grade water with 0.1% formic acid.
6. Solvent B: LC-MS-grade methanol with 0.1% formic acid.
7. 2 mL autosampler vials.
8. 250 μL glass inserts with springs.

3 Methods

3.1 Bacterial Growth Conditions

1. Grow a *M. smegmatis* preculture from frozen stock by spreading on LB agar plate and incubating at 37 °C until a lawn has formed.
2. Aseptically transfer the entire plate of *M. smegmatis* into 2 L of LB broth and incubate at 37 °C with shaking for 48–72 h to obtain optimal biomass (*see* **Note 4**).
3. Harvest cells by centrifugation at 3500 rpm (2600 × *g*) for 15 min at 4 °C in a 250 mL conical tube.
4. Resuspend the harvested cell pellet with phosphate buffered saline (PBS) pH 7.0 and repeat centrifugation at 3500 rpm (2600 × *g*) for 15 min at 4 °C.
5. Store PBS-washed cells at −80 °C.

3.2 Preparation of Enzymatically Active Membrane Fractions

All of the following steps are to be carried out on ice.

1. Thaw frozen *M. smegmatis* cells (10 g wet weight) on ice for 10 min and add prechilled 30 mL of buffer A to the thawed cell pellet. Vortex every 30 s to create a homogeneous solution.
2. Transfer 30 mL of suspension to the sonication tube and break the cells using a 1-cm probe tip sonicator at 50% output for 10 cycles consisting of 60 s pulses with 90 s cooling intervals between pulses.

3. Transfer cell lysates into prechilled Oakridge round-bottom tubes on ice and remove the cell wall material and unbroken cells by centrifugation at 27,000 × *g* for 60 min at 4 °C.
4. Carefully transfer the supernatant containing the membrane proteins to ultracentrifuge tubes and spin at 100,000 × *g* for 2 h at 4 °C.
5. Discard the supernatant containing cytosolic proteins and resuspend the pellet containing the enzymatically active membrane fraction in 0.5–1.0 mL of Buffer A with gentle homogenization using a Dounce homogenizer. Prepare aliquots of 50–100 μL of homogenized membranes and store until further use at −80 °C (*see* **Note 5**).
6. Measure membrane proteins concentration using the Pierce BCA protein assay kit as described by the manufacturer (*see* **Note 6**).

3.3 Arabinosyltransferase Assay

1. The typical arabinosyltransferase reaction mixture contains the following in a total volume of 200 μL: buffer B, 200 μM acceptor substrate, 1 mM donor substrate, and 0.5 mg of *M. smegmatis* membrane proteins (*see* **Note 7**).
2. As a negative control, perform an enzymatic reaction lacking the acceptor substrate.
3. Incubate the enzymatic reaction mixtures at 37 °C in a water bath for 2 h (*see* **Note 8**).
4. After 2 h incubation, stop the enzymatic reaction by adding 200 μL of 100% ethanol and vortex briefly.
5. Centrifuge the tube at 14,000 rpm (20,000 × *g*) for 10 min to clarify the mixture and transfer the supernatant containing the enzymatic product(s) into a clean Eppendorf tube.
6. Equilibrate a strong anion exchange (SAX) column with 5 mL of water, according to the manufacturer's instructions.
7. Load the sample into the SAX column and elute with 5 mL 50% ethanol.
8. Collect the flow-through and the eluate in the same tube.
9. Transfer the eluate containing the enzymatic product(s) into clean 13 × 100 mM glass tube and dry in a savant.

3.4 Permethylation of the Enzymatic Reaction Products for LC-MS Analysis

The following permethylation procedure is to be performed inside a chemical hood.

1. Completely dry the extracted reaction products into a Teflon-lined screw capped glass tubes on the savant.
2. Add 100 μL of methanol and dry under nitrogen.

3. Add 100 μL DMSO, sonicate and allow to stand at room temperature for 10 min.
4. Mark a Pasteur pipette with 0.5 mL volume.
5. Place a small magnetic stir bar into each sample tubes.
6. NaOH/DMSO slurry: Place five NaOH pellets in a dry mortar and transfer 3 mL anhydrous DMSO into the mortar using a glass syringe. With a pestle, grind the NaOH pellets until a slurry is formed (*see* **Note 9**).
7. Quickly add 0.5 mL of the NaOH/DMSO slurry directly on the sample without touching the side walls of the glass tube using a Pasteur pipette (0.5 mL marked), Cap the tubes and sonicate in water bath for 5 min.
8. Add 0.5 mL methyl iodide. Cap and allow to stir on a magnetic stir plate for 30 min at room temperature. The sample should look as a milky white slurry.
9. Add 200 μL of methyl iodide to the same tube and allow it to stir for another 30 min.
10. Quench the methylation reactions (exothermic) by dropwise addition of 1 mL ice cold water.
11. Add 2 mL of chloroform, vortex well and allow the mixture to settle into two layers.
12. Extract the permethylated products by spinning at 3500 rpm (2600 × *g*) for 5 min at room temperature.
13. Carefully remove the upper water layer and wash the bottom chloroform phase twice with 1 mL water (*see* **Note 10**).
14. Carefully transfer the lower chloroform layer containing the enzymatic product(s) into clean 13 × 100 mM glass tube and evaporate under a stream of nitrogen.

3.5 LC-MS Analysis

1. Resuspend the dried sample containing the enzymatic product (s) in 100 μL of methanol; briefly vortex and centrifuge at 3500 rpm (2600 × *g*) for 5 min at room temperature.
2. Using a Pasteur pipette, transfer the 100 μL methanol suspension into an autosampler vial with small volume insert.
3. Analyze the permethylated samples by LC-MS using an Agilent 6220A time-of-flight (TOF) mass spectrometer. Instrument parameters for the ion source: Gas temperature of 310 °C, vaporizer drying gas flow of 10 L/min; nebulizer pressure of 45 psig, capillary voltage of 2000 V. Acquire the scanning signals in positive ion mode. Select a mass range between *m/z* 115–2500 with a scanning rate of 1.32 spectra/s.
4. Conditions for column chromatography: Set the column flow rate at 0.320 mL/min and column temperature to 45 °C. Set initial first min at 90% solvent A and 10% solvent B. Then, ramp

solvent B from 10 to 100% over 20 min and hold at 100% solvent B for 4 min. After 24 min, bring back to 10% solvent B in 0.10 min.

5. Set the post run to 10 min using 90% solvent A.
6. After all samples have been run, rinse the column prior to storage with 100% acetonitrile for 30 min at a flow rate of 0.5 mL/min.
7. Prior to running analytical samples, calibrate mass spectrometer in positive mode using the Agilent ESI-L tune mix.
8. Set the autosampler to inject 1 μL of permethylated reaction products into the LC-MS.
9. For LC-MS data analysis use the MassHunter Workstation Software Qualitative Analysis version B. 07.00 (Fig. 3).

4 Notes

1. *Mycobacterium smegmatis* is a nonpathogenic (BSL-2) fast-growing mycobacterial species. The structural organization and biosynthesis of the arabinan domains of AG and LAM is highly conserved across the *Mycobacterium* genus. For these reasons, *M. smegmatis* is often used as a surrogate model of *M. tuberculosis* to study arabinosyltransferase activities.
2. For the purpose of this chapter, we chose to present two examples of arabinosyl transfer reactions present in mycobacteria. One uses synthetic α-D-Man*p*-(1→6)-α-D-Man*p*-1-Octyl (dimannoside) [20] as the acceptor substrate and the other β-D-Gal*f*-(1→5)-β-D-Gal*f*-(1→6)-β-D-Gal*f*-1-Octyl (trigalactan) [7] (Fig. 2).
3. Other functional arabinose donors harboring shorter acyl chains have also been reported [22].
4. Avoid adding Tween-80 to the LB broth as this detergent may interfere with LC-MS analysis.
5. Membranes may be stored at −80 °C. However, for best enzyme activity, it is recommended to use fresh membranes. Indeed, the freeze-thawing of membranes may lead to loss of enzymatic activity.
6. Usually 10 g of wet weight *M. smegmatis* cells yield nearly 15–20 mg membrane proteins.
7. DPA (stored in chloroform–methanol (2:1, by vol.)) is quickly evaporated to dryness under a stream of nitrogen in an Eppendorf tube. The dried DPA is then resuspended in 2% DMSO (final concentration) by gentle pipetting followed by a brief

water bath sonication. Then add the remaining components of the arabinosyltransferase reaction mixture.

8. To increase product yield, it is recommended to perform overnight incubations by replenishing the reaction mixture with fresh membranes every 4 to 6 h.
9. Avoid weighing the NaOH pellets. They are very hygroscopic and presence of water negatively impacts the methylation reaction.
10. The water layer contains unreacted methyl iodide. It is a toxic chemical and should be discarded as hazardous waste.

Acknowledgments

Studies on mycobacterial arabinosyltransferases in the authors' laboratory are supported by the National Institute of Allergy and Infectious Diseases (NIAID)/National Institutes of Health (NIH) grant AI064798. The content is solely the responsibility of the authors and does not necessarily represent the official views of the NIH.

References

1. World Health Organization (2016) Global tuberculosis report 2016. WHO Press
2. Queiroz A, Riley LW (2017) Bacterial immunostat: mycobacterium tuberculosis lipids and their role in the host immune response. Rev Soc Bras Med Trop 50:9–18
3. Brennan PJ, Nikaido H (1995) The envelope of mycobacteria. Annu Rev Biochem 64:29–63
4. Lohrasbi V, Talebi M, Bialvaei AZ et al (2017) Trends in the discovery of new drugs for Mycobacterium tuberculosis therapy with a glance at resistance. Tuberculosis 109:17–27
5. Angala SK, Belardinelli JM, Huc-Claustre E et al (2014) The cell envelope glycoconjugates of mycobacterium tuberculosis. Crit Rev Biochem Mol Biol 49:361–399
6. Alderwick LJ, Seidel M, Sahm H et al (2006) Identification of a novel arabinofuranosyltransferase (AftA) involved in cell wall arabinan biosynthesis in mycobacterium tuberculosis. J Biol Chem 281:15653–15661
7. Shi L, Zhou R, Liu Z et al (2008) Transfer of the first arabinofuranose residue to galactan is essential for mycobacterium smegmatis viability. J Bact 190:5248–5255
8. Seidel M, Alderwick LJ, Birch HL et al (2007) Identification of a novel arabinofuranosyltransferase AftB involved in a terminal step of cell wall arabinan biosynthesis in corynebacterianeae, such as corynebacterium glutamicum and mycobacterium tuberculosis. J Biol Chem 282:14729–14740
9. Birch HL, Alderwick LJ, Bhatt A et al (2008) Biosynthesis of mycobacterial arabinogalactan: identification of a novel α (1→3) arabinofuranosyltransferase. Mol Microbiol 69:1191–1206
10. Škovierová H, Larrouy-Maumus G, Zhang J et al (2009) AftD, a novel essential arabinofuranosyltransferase from mycobacteria. Glycobiology 19:1235–1247
11. Escuyer VE, Lety M-A, Torrelles JB et al (2001) The role of the embA and embB gene products in the biosynthesis of the terminal hexaarabinofuranosyl motif of mycobacterium smegmatis arabinogalactan. J Biol Chem 276:48854–48862
12. Jankute M, Alderwick LJ, Noack S et al (2016) Disruption of mycobacterial AftB results in complete loss of terminal β (1→2)

arabinofuranose residues of lipoarabinomannan. ACS Chem Biol 12:183–190
13. Alderwick LJ, Lloyd GS, Ghadbane H et al (2011) The C-terminal domain of the arabinosyltransferase mycobacterium tuberculosis EmbC is a lectin-like carbohydrate binding module. PLoS Pathog 7:e1001299
14. Wolucka BA, McNeil MR, de Hoffmann E et al (1994) Recognition of the lipid intermediate for arabinogalactan/arabinomannan biosynthesis and its relation to the mode of action of ethambutol on mycobacteria. J Biol Chem 269:23328–23335
15. Makarov V, Manina G, Mikusova K et al (2009) Benzothiazinones kill Mycobacterium tuberculosis by blocking arabinan synthesis. Science 324:801–804
16. Christophe T, Jackson M, Jeon HK et al (2009) High content screening identifies decaprenyl-phosphoribose 2′ epimerase as a target for intracellular antimycobacterial inhibitors. PLoS Pathog 5:e1000645
17. Lee R, Mikusova K, Brennan PJ et al (1995) Synthesis of the arabinose donor. beta.-d-arabinofuranosyl-1-monophosphoryldecaprenol, development of a basic arabinosyl-transferase assay, and identification of ethambutol as an arabinosyl transferase inhibitor. J Am Chem Soc 117:11829–11832
18. Scherman MS, Kalbe-Bournonville L, Bush D et al (1996) Polyprenylphosphate-pentoses in in mycobacteria are synthesized from 5-phosphoribose pyrophosphate. J Biol Chem 271:29652–29658
19. Angala SK, McNeil MR, Zou L et al (2016) Identification of a novel mycobacterial activity which adds an arabinosyl residue to alpha-D-mannosyl residues. ACS Chem Biol 11:1518–1524
20. Tam PH, Besra GS, Lowary TL (2008) Exploring the substrate specificity of a mycobacterial polyprenol monophosphomannose-dependent α-(1→6)-mannosyltransferase. Chem Bio Chem 9:267–278
21. Liav A, Huang H, Ciepichal E et al (2006) Stereoselective synthesis of decaprenylphosphoryl β-D-arabinofuranose. Tetrahedron Lett 47:545–547
22. Zhang J, Angala SK, Pramanik PK et al (2011) Reconstitution of functional mycobacterial arabinosyltransferase AftC proteoliposome and assessment of decaprenylphosphorylarabinose analogues as arabinofuranosyl donors. ACS Chem Biol 6:819–828

Chapter 15

Enzymatic Synthesis of Repeating Unit Oligosaccharides of *Escherichia coli* O104

Inka Brockhausen and Diana Czuchry

Abstract

Escherichia coli serotype O104:H4 (ECO104) is a potent intestinal pathogen that causes severe bloody diarrhea and hemolytic-uremic syndrome. The O antigenic polysaccharides of ECO104 consist of repeating units with the structure [4Galα1-4Neu5,7,9Ac3α2-3Galβ1-3GalNAcβ1-]$_n$. These repeating units are assembled sequentially by specific glycosyltransferases on a diphosphate-undecaprenol intermediate. Internal structures include mimics of the human T and sialyl-T antigen. This protocol describes the in vitro synthesis of the repeating unit by β1,3-Gal-transferase WbwC, α2,3-sialyltransferase WbwA, and α1,4-Gal-transferase WbwB. All of these enzymes require acceptor substrates based on GalNAc-diphosphate-lipid. These methods are applicable for the assembly of bacterial polysaccharides of gram-negative bacteria that require sugar-diphosphate intermediates and are a basis for vaccine synthesis.

Key words O antigen, *Escherichia coli* O104, Repeating unit, Glycosyltransferases, Assays, In vitro synthesis

1 Introduction

The human pathogen *Escherichia coli* (*E. coli*) O104:H4 (ECO104) is a Shiga toxin-producing *E. coli* (STEC) or enterohemorrhagic *E. coli* (EHEC) that can cause bloody diarrhea and lead to hemolytic uremic syndrome and kidney failure. A massive outbreak of foodborne illness caused by *E. coli* O104:H4 in Germany in 2011 resulted in more than 3000 people infected and more than 50 deaths [1]. The pathogenic, potentially fatal, and multiantibiotic-resistant nature of *E. coli* O104:H4 and other STECs requires specific prevention and treatment methods. Vaccines are not yet available for this pathogen but could be based on the structure of the serotype-specific O104 antigen.

The lipopolysaccharides (LPS) of gram-negative bacteria are essential for growth and virulence. The external antigenic polysaccharides (O antigens) of LPS are involved in interactions of bacteria with the environment and are considered to be virulence factors

Inka Brockhausen (ed.), *Bacterial Polysaccharides: Methods and Protocols*, Methods in Molecular Biology, vol. 1954, https://doi.org/10.1007/978-1-4939-9154-9_15,

[2]. The ECO104 antigen has the repeating unit [4Galα1-4Neu5,7,9Ac_3α2-3Galβ1-3GalNAcβ1-$]_n$ [3] where sialic acid is di-O-acetylated. The repeating unit structure is unusual because it includes mimics of the human T and sialyl-T antigens and has a Gal residue linked to sialic acid which has only been found in the capsule of *Neisseria meningitidis* [4].

The repeating units of O antigens are thought to be assembled on an undecaprenol-phosphate (P-Und) lipid carrier by sequentially acting specific glycosyltransferases (GTs) [5]. The ECO104 antigen biosynthesis gene cluster [6] contains the corresponding GT genes *wbwC*, *wbwA* and *wbwB*. In addition, the *wckD* gene may be responsible for the 7-, 9-O-acetylation of sialic acid. The presence of a translocase gene *wzx* and polymerase gene *wxy* in the ECO104 biosynthesis gene cluster suggests that the synthesis of the ECO104 antigen follows the polymerase-dependent pathway [5]. In this pathway (Fig. 1), the O antigen repeating unit is synthesized by the sequential action of GTs on the inner leaflet of the inner membrane, utilizing the membrane-bound lipid carrier P-Und and nucleotide donor substrates that are synthesized in the cytosolic compartment [5, 7]. The first reaction of the ECO104 repeating unit synthesis appears to be the reversible transfer of GlcNAc-phosphate from UDP-GlcNAc to P-Und by an analog of WecA [8] to form GlcNAc-diphosphate undecaprenol (GlcNAc-PP-Und). In ECO104 and several related strains, it is likely that GlcNAc-PP-Und is converted to GalNAc-PP-Und by a 4-epimerase that is identical to the enzyme identified in *E. coli* O157 [9]. We have synthesized a natural GalNAc-PP-Und analog, GalNAc–PO_3–PO_3–$(CH_2)_{11}$-O-phenyl (GalNAc-PP-PhU), that has been successful as an acceptor substrate for a number of GTs from gram-negative bacteria that catalyze the second step of repeating unit synthesis [10–12]. Other enzymes that catalyze the second step of repeating unit assembly require the GlcNAc-PP-lipid [13–16]. However, in some bacteria, the addition of sugar residues does not require the PP-lipid, and oligosaccharides can also be appropriate acceptors [17].

Using the GalNAc-PP-PhU acceptor, we have shown that the repeating unit of ECO104 is assembled by enzymes that all have a strict requirement for the sugar–PP–lipid structure in the acceptor substrate. The second reaction of the ECO104 repeating unit oligosaccharide assembly pathway is catalyzed by β1,3-Gal-transferase WbwC [12] that forms the mammalian cancer-associated T antigen, Galβ1-3GalNAcα- [18]. Subsequently, sialic acid (Neu5Ac) is transferred by α2,3-sialyltransferase WbwA that synthesizes the equivalent of the mammalian sialyl-T antigen, Neu5Acα2-3Galβ1-3GalNAcα-R [19]. The final GT in the ECO104 pathway, α1,4-Gal-transferase WbwB synthesizes the unusual linkage of Galα1-4Neu5Ac-R [20].

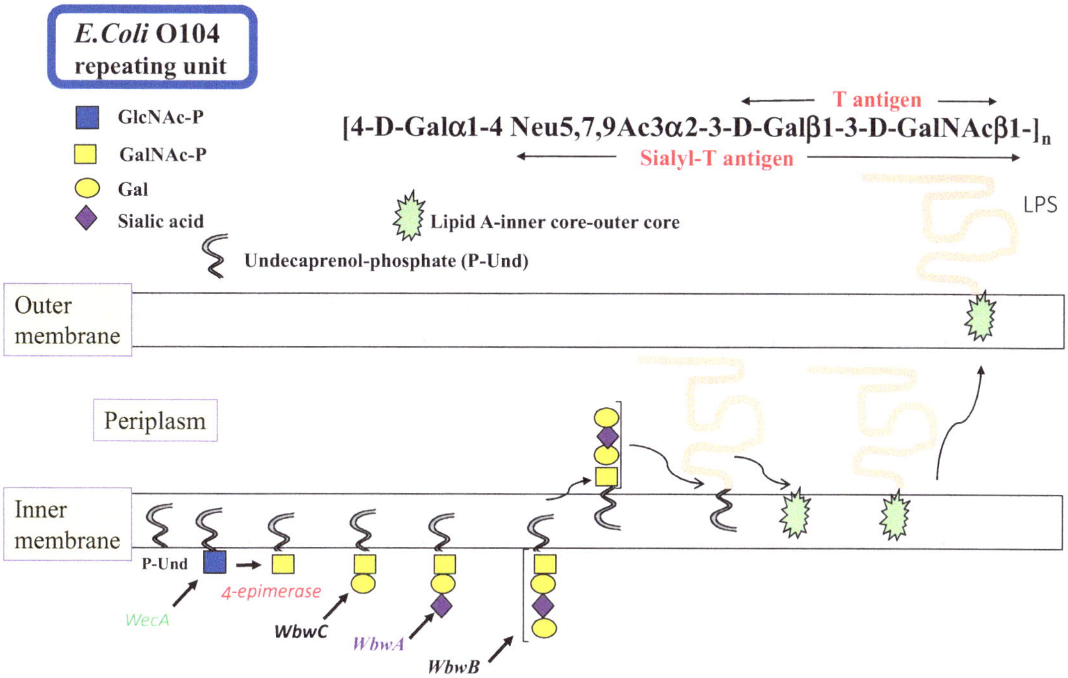

Fig. 1 Proposed pathway in the biosynthesis of the *E. coli* 0104 antigen repeating unit. The repeating unit of *E. coli* 0104 (EC0104) is thought to be assembled at the inner face of the cytoplasmic membrane on an undecaprenol-phosphate (UndP) lipid intermediate [5]. The first reaction in the pathway of EC0104 synthesis is the transfer of GlcNAc-phosphate from UDP-GlcNAc to P-Und by GlcNAc-1-P transferase WecA. A 4-epimerase may then convert GlcNAc-PP-Und to GalNAc-PP-Und which acts as an acceptor substrate for β1,3-Gal-transferase WbwC that synthesizes a structure identical to the human O-glycan core 1 or T antigen [12]. The α2,3-sialyltransferase WbwA then transfers sialic acid from CMP-sialic acid to Gal to form the sialyl-T antigen [19]. The α1,4-Gal-transferase WbwB adds galactose to sialic acid [20]. It is not known yet where the two O-acetyl groups are added to the oligosaccharide. The flippase Wzx translocates the repeating unit to the periplasm where polymerase Wzy polymerizes the repeating units to form the O antigen. The polysaccharide is then transferred by a ligase to the outer core of lipid A to form LPS which is then extruded to the outer membrane where the O antigen is exposed to the external environment

In the polymerase-dependent pathway, the completed repeating unit oligosaccharide linked to PP-Und is translocated from the cytoplasmic side to the periplasm by Wzx (Fig. 1) [5]. Polymerase Wzy then inserts a new repeating unit at the reducing end of the growing O antigen chain. This process is repeated multiple times and results in an O antigen polysaccharide chain containing variable numbers of repeating units. Finally, the O antigen polymer is transferred to the core oligosaccharide linked to lipid A by ligase WaaL while the membrane-bound PP-lipid is recycled. The completed LPS is then transported to the outer surface of the cell, exposing the O antigen to the environment.

In this chapter we describe the efficient synthesis and structural identification of the ECO104 repeating unit using recombinant His_6-tagged WbwC, WbwA and WbwB. The methods can be used to enzymatically and sequentially synthesize the ECO104 repeating unit, as well as the intermediate structures, based on a chemically synthesized acceptor substrate. We describe the enzymatic synthesis of disaccharides, trisaccharides, and tetrasaccharides using the individual purified enzymes that yield intermediate structures of the repeating unit which can be used as acceptors for the next step in the pathway. It is important to have these acceptors to verify the activities of individual enzymes before using them in a large-scale synthesis. We found that homogenates have high activities and can also be used to synthesize the intermediate structures. The combination of WbwC and WbwA, or all three enzymes, yields an unexpectedly high yield of enzyme products suggesting that enzymes either form an active complex or that the presence of the next enzyme in the pathway makes reactions very efficient. Thus, we describe here a large-scale synthesis of the repeating unit oligosaccharide in a one-pot reaction containing bacterial homogenates of all three enzymes, the necessary sugar-nucleotides, cofactors, and GalNAc-PP-PhU as the initial acceptor. The oligosaccharide products can be cleaved off the PP-lipid moiety to yield a reducing oligosaccharide lacking the diphosphate-lipid moiety for further studies.

Several repeating units, containing up to six sugar residues, have been chemically synthesized [21–25]. However, the expression and purification of bacterial GTs is economical and large amounts of proteins can be produced. Bacterial enzymes are relatively easy to use and usually are highly active in vitro. In addition, they have reliable specificities and usually catalyze only one linkage between sugars. The protocol can be used to synthesize both bacterial and mammalian antigens, and is a basis for vaccine development directed against cancer antigens or to protect from bacterial pathogens.

2 Materials

2.1 WbwC, Genes, Expression and Purification

1. pET28a plasmids having a kanamycin resistance gene and the *wbwC* gene (Thermo Fisher). Expressed WbwC protein has a C-terminal His_6 tag [12] (*see* **Note 1**).
2. BL21 expression strain of *E. coli* (*see* **Note 2**).
3. Isopropyl-1-thio-β-galactoside (IPTG), 100 mM (BioShop Canada Inc.).
4. Lysogeny (LB) broth (Bioshop Canada).
5. Kanamycin (BioShop Canada Inc.): 50 mg/mL stock solution.
6. Bacterial pellet containing WbwC.
7. Phosphate-buffered saline (PBS).

8. Bacteria buffer: PBS, 20% glycerol.
9. Ni-nitrilotriacetic acid (NTA)-Sepharose (Thermo Fisher).
10. Elution buffer: 200 mM imidazole in PBS.
11. Protein concentration: protein assay kit (Bio-Rad); bovine serum albumin standard solution (Sigma-Aldrich).

2.2 WbwA, Genes, Expression and Purification

1. pRSETA plasmid having an ampicillin resistance gene and the *wbwA* gene (Thermo Fisher). Expressed WbwA protein has an N-terminal His_6 tag [19] (*see* **Note 1**).
2. BL21 expression strain of *E. coli* (*see* **Note 2**).
3. 100 mM IPTG.
4. Ampicillin: 100 mg/mL.
5. Bacterial pellet containing WbwA.
6. Bacteria buffer: PBS, 20% glycerol.
7. HisPur Co-resin (Thermo Fisher) (*see* **Note 3**).
8. Lysis buffer: 50 mM 4-(2-hydroxyethyl)-1-piperazineethanesulfonic acid (HEPES), pH 7.5, 150 mM NaCl, 1 mM dithiothreitol (DTT), 1 protease inhibitor cap, Complete Mini, EDTA-free (Roche) per 100 mL buffer.
9. Wash buffer: 50 mM HEPES, pH 7.5, 150 mM NaCl, 1 mM DTT, 10 mM imidazole.
10. Elution buffer: 50 mM HEPES, pH 7.5, 150 mM NaCl, 1 mM DTT, 250 mM imidazole.
11. Filtration: Centricon 10,000 cutoff (Merck).

2.3 WbwB, Genes, Expression and Purification

1. pRSETA plasmids containing an ampicillin resistance gene and the *wbwB* gene (Thermo Fisher). Expressed WbwB protein has an N-terminal His_6 tag [20].
2. BL21 expression strain of *E. coli* (*see* **Note 2**).
3. PBS.
4. 100 mM IPTG.
5. Ampicillin: 100 mg/mL.
6. Bacterial pellet containing WbwB.
7. Bacteria buffer: PBS, 20% glycerol.
8. HisPur Ni-nitrilotriacetic acid (Ni-NTA) resin (Thermo Fisher).
9. Lysis buffer: 50 mM Na-phosphate, pH 7.4, 5% glycerol, 300 mM NaCl.
10. Wash buffer for Ni-NTA: 50 mM Na-phosphate, pH 7.4, 300 mM NaCl.

11. Elution buffer: 150 mM imidazole, 50 mM Na-phosphate, pH 7.4, 300 mM NaCl.
12. Dialysis buffer: 50 mM Na-phosphate, 1 mM dithiothreitol (DTT).

2.4 Gal-Transferase Assay for WbwC: Production of Galβ1-3GalNAc-PP-PhU

1. Donor substrate: UDP-Gal (Sigma-Aldrich) and UDP-[^{3}H]Gal (American Radiolabeled Chemicals), mixed to achieve a solution of 10 mM UDP-Gal and 2000 cpm/nmol (*see* Chapter 20).
2. Acceptor substrate: 1 mM synthetic GalNAc-PP-PhU dissolved in water [12].
3. Assay buffer: 0.5 M 2-(*N*-morpholino)ethanesulfonic acid (MES) pH 7,
4. 100 mM $MnCl_2$.
5. Bacterial homogenate in PBS, 20% glycerol.
6. Purified WbwC.
7. 50 mM sucrose (*see* **Note 4**).
8. Sonicator.
9. Purified WbwC in 200 mM imidazole, 50 mM Na-phosphate, pH 7.4, 300 mM NaCl.
10. C18 Sep-Pak cartridges, short (Waters).
11. Methanol.
12. Measurement of radioactivity: scintillation fluid, 7 mL scintillation vials, scintillation counter.

2.5 Sialyltransferase Assay for WbwA: Production of Sialylα2-3Galβ1-3GalNAc-PP-PhU

1. Donor substrate: CMP-sialic acid (EMD Millipore) and CMP-[^{3}H]sialic acid (American Radiolabeled Chemicals), mixed to achieve a solution of 10 mM CMP-sialic acid and 2000 cpm/nmol (*see* Chapter 20).
2. Acceptor substrate: 1 mM Galβ1-3GalNAc-PP-PhU dissolved in water [12]. This can be synthesized chemically (*see* Chapter 13) or enzymatically with WbwC (*see* Subheading 2.4)
3. Reaction buffer: 0.5 M Tris–HCl buffer pH 7.
4. 100 mM $MnCl_2$.
5. Bacterial homogenate.
6. Purified WbwA.
7. 50 mM sucrose (*see* **Note 3**).
8. Sonicator.
9. C18 Sep-Pak cartridges, short (Waters).
10. Methanol.

2.6 Gal-Transferase Assays for WbwB: Production of Galα1-4sialylα2-3Galβ1-3GalNAc-PP-PhU

1. Donor substrate: UDP-Gal (Sigma-Aldrich) and UDP-[^{3}H]Gal (American Radiolabeled Chemicals), mixed to achieve a solution of 10 mM UDP-Gal and 2000 cpm/nmol (*see* Chapter 20).
2. Acceptor substrate: 0.1 mM sialylα2-3Galβ1-3GalNAc-PP-PhU dissolved in water [19, 20]. This compound is synthesized with WbwC and WbwA (Subheadings 2.4 and 2.5) (*see* **Note 5**).
3. Assay buffer: 0.5 M Tris–HCl, pH 7.
4. 100 mM $MnCl_2$.
5. Bacterial homogenate.
6. Purified WbwB.
7. 50 mM sucrose (*see* **Note 4**).
8. Sonciator.
9. C18 Sep-Pak cartridges (Waters).
10. Methanol.

2.7 Large-Scale Production of Galα1-4sialylα2-3Galβ1-3GalNAc-PP-PhU

1. Bacterial homogenates containing WbwC in 50 mM sucrose.
2. Bacterial homogenates containing WbwA in 50 mM sucrose.
3. Bacterial homogenates containing WbwB in 50 mM sucrose.
4. Donor substrates: 100 mM UDP-Gal, 100 mM CMP-sialic acid.
5. Acceptor substrate: 10 mM GalNAc-PP-PhU [12].
6. 1 M MES pH 7.
7. 1 M $MnCl_2$.
8. CMP-[^{3}H]sialic acid.
9. C18 Sep-Pak cartridges (Waters).
10. Methanol.

2.8 Purification of ECO104 Repeating Unit

1. HPLC apparatus: C18 column, acetonitrile/degassed water mobile phase, UV absorbance monitor, fraction collector.
2. HPLC standards: 5 mM GlcNAc, 5 mM Galβ1-3GalNAcα-Bn (Toronto Research Chemicals), 1 mM Galβ1-3GalNAc-PP-PhU, 0.1 mM Siaα2-3Galβ1-3GalNAc-PP-PhU, 0.1 mM Galα1-4sialylα2-3Galβ1-3GalNAc-PP-Ph, 1 mM GalNAc-PP-PhU (*see* **Note 6**).
3. C18 column (4.5 mm × 25 cm).
4. Radioactivity measurement: scintillation fluid, scintillation vials, scintillation counter.

2.9 Analysis of Enzyme Products

1. Mass spectrometry (MS): Electrospray ionization (ESI) in negative ion mode.
2. NMR: 600 MHz Avance-600 (Bruker).
3. HPLC.

3 Methods

Although purified enzymes have less contaminants and side reactions that might degrade substrates and reaction products, the use of sonicated bacteria (homogenates) is simple, fast, and yields high activities. Both, purified enzymes and enzymes in homogenates can be used to produce the repeating unit intermediates. Assays are carried out with radioactive sugar-nucleotides for sensitive detection of reaction product, and all assays should be done in duplicate. Radioactive reaction products serve as standard compounds for HPLC identification.

Nonradioactive reaction products (from assays using nonradioactive nucleotide sugars) can be used as the acceptor substrate for the next step in the pathway and for analysis of oligosaccharide products by MS and NMR, or to generate enzyme substrate for further modification of the oligosaccharide (e.g., for O-acetylation), for testing as an antigen or for polymerization of the repeating unit to form the O antigen.

3.1 WbwC: Genes, Expression and Purification

1. Bacterial growth: The pET28a plasmids containing a kanamycin resistance gene and the *wbwC* gene are transformed into *E. coli* BL21 for protein expression. Bacteria are grown at 37 °C overnight on an agar plate containing 50 μg kanamycin/mL [12].
2. A single colony is then selected and grown at 37 °C overnight in 5 mL LB broth containing 50 μg kanamycin/mL. Bacteria from 1 mL overnight culture are added to 250 mL LB broth containing 50 μg kanamycin/mL and are grown at 37 °C until the absorbance at 600 nm is 0.6.
3. To induce protein expression, IPTG is added to a final concentration of 1 mM and cells are grown another 4 h. After centrifugation of bacteria at 3000 × *g* for 10 min, pellets are washed with PBS, and 5 mL Bacteria buffer are added. Aliquots of bacterial suspension (500 μL) are stored at −20 °C for enzyme assays.
4. For enzyme purification, bacteria are harvested by centrifugation, washed with PBS, and then lysed in PBS by sonication [12].

5. The His_6-tagged WbwC in lysates is purified using a Ni-NTA column. The resin is washed with PBS, and bound proteins are eluted with Elution buffer and used for enzyme assays.
6. The presence of WbwC is monitored by SDS-PAGE and Western blot as described [12], and by measuring Gal-transferase activity (*see* **Note** 7).

3.2 WbwA: Genes, Expression and Purification

1. Bacterial growth: pRSETA plasmids containing an ampicillin resistance gene and the *wbwA* gene are transformed into *E. coli* BL21 for protein expression. Bacteria are grown overnight on an LB Agar plate containing 100 µg ampicillin/mL [19].
2. A single colony is then selected and grown at 37 °C overnight in 5 mL LB broth containing 100 µg ampicillin/mL [19]. One mL of the overnight culture is then added to 250 mL LB broth containing 100 µg ampicillin/mL, and bacteria are grown at 37 °C until the absorbance at 600 nm is 0.6.
3. To induce protein expression, IPTG is added to a final concentration of 1 mM and cells are grown another 4 h. After centrifugation of bacteria at 3000 × *g* for 10 min, pellets are washed with PBS, and 5 mL Bacteria buffer are added. Aliquots of bacterial suspension (500 µL) are stored at −20 °C for enzyme assays.
4. For enzyme purification, bacteria are harvested, washed, lysed by sonication in Lysis buffer, and purified by Co-resin as described in Subheading 3.1 and in Chapter 13, Subheading 3.7. After elution in Elution buffer, enzyme solution is centrifuged through a Centricon filter and stored in aliquots in Lysis buffer at −80 °C.
5. The presence of WbwA is monitored by SDS-PAGE and Western blot as described [12], and by measuring sialyltransferase activity.

3.3 WbwB: Genes, Expression and Purification

1. Bacterial growth: pRSETA plasmids containing an ampicillin resistance gene and the *wbwB* gene are transformed into *E. coli* BL21 for protein expression. Bacteria are grown at 37 °C overnight on an LB Agar plate containing 100 µg ampicillin/mL.
2. A single colony is then selected and grown at 37 °C overnight in 5 mL LB broth containing 100 µg ampicillin/mL. One mL of overnight culture is then added to 250 mL LB broth containing 100 µg ampicillin/mL and bacteria are grown at 37 °C [20] until the absorbance at 600 nm is 0.6.
3. To induce protein expression, IPTG is added to a final concentration of 1 mM and cells are grown another 12 h at 25 °C. After centrifugation of bacteria at 3000 × *g* for 10 min, pellets are washed with PBS, and 5 mL Bacteria buffer is added.

Aliquots of bacterial suspension (500 μL) are stored at −20 °C for enzyme assays.

4. For enzyme purification, bacteria are harvested, washed, lysed by sonication in Lysis buffer, and purified by Ni-NTA as described in Subheading 3.1. The resin (5 mL) is washed with 50 mL Wash buffer and bound protein is eluted with 50 mL Elution buffer.
5. The presence of WbwB is monitored by SDS-PAGE and Western blot as described [12], and by measuring Gal-transferase activity.

3.4 Gal-Transferase Assay for WbwC: Production of [^{3}H]Galβ1-3GalNAc-PP-PhU and Galβ1-3GalNAc-PP-PhU

3.4.1 Radioactive Assays for WbwC

Radioactive assays require background controls that do not contain the acceptor substrate.

1. Bacterial homogenate: Thaw a 500 μL aliquot of bacteria expressing WbwC and centrifuge at 8000 × *g* for 10 min. Add 500 μL 50 mM sucrose and sonicate suspension three times for 15 s with 2 min on ice in between. Determine the protein content and use as the enzyme source in assays (*see* **Note 8**).
2. Prepare the assay mixture in a total volume of 40 μL: 4 μL 10 mM UDP-[^{3}H]Gal (2000 cpm/nmol) donor substrate, 10 μL 0.5 M MES buffer, pH 7, 4 μL 100 mM $MnCl_2$, 10 μL enzyme homogenate (1–5 mg protein) or purified enzyme (5–10 μg protein), 4 μL 1 mM GalNAc-PP-PhU acceptor substrate (or water in control assays), 8 μL water.
3. Incubate for 20 min at 37 °C, quench reaction by adding 700 μL water.
4. Recycle C18 Sep-Pak cartridges by passing 3 mL methanol through it followed by 6 mL water.
5. Add assay mixture to a 3 mL syringe and apply onto a recycled C18 Sep-Pak cartridge, push through very slowly and collect into a vial, then wash cartridge slowly with 3 × 1 mL water. This will elute the water soluble components (e.g., UDP-Gal) in fraction A.
6. Wash cartridge slowly with another 1 mL water and collect fraction B into a scintillation vial (background). Elute reaction product with 2 × 1.5 mL methanol and collect into scintillation vials (fractions C and D).
7. Add 4 mL scintillation fluid to vials of fractions B, C, D, carefully mix by inverting the vial three times, count cpm of [^{3}H] in scintillation counter.
8. Calculate the Gal-transferase activity (nmol × h^{-1} incubation time x mg^{-1} protein). Cpm of fractions C plus D from assays containing acceptor substrate minus cpm of fractions C plus D

from control assays = cpm of product. Divide by cpm/nmol of UDP-Gal to obtain nmol of product.

9. In order to produce sufficient nonradioactive WbwC product as an acceptor substrate for WbwA, scale up the assay volume 50 times and incubate for 1 h. This should produce at least 100 nmol of Galβ1-3GalNAc-PP-PhU. This disaccharide is purified by HPLC and its concentration estimated by comparison of its absorbance at 195 nm with that of standard compounds (e.g., GalNAc-PP-PhU). The fractions containing Galβ1-3GalNAc-PP-PhU are dried and dissolved in water to make a 1 mM solution.

3.4.2 Nonradioactive Assays for WbwC

Nonradioactive assays are carried out like radioactive assays, except that the radioactive sugar-nucleotide is replaced by the same concentration of nonradioactive sugar-nucleotide. Control assays (without acceptor) are usually not necessary. If the purpose of these assays is to obtain product for MS, a quarter of the assay volume is sufficient to obtain a good signal. The product Galβ1-3GalNAc-PP-PhU is analyzed and purified by HPLC (*see* Subheading 3.8) and used as an acceptor for WbwA.

3.5 Sialyltransferase Assay for WbwA: Production of [^{3}H] sialylα2-3Galβ1-3GalNAc-PP-PhU and sialylα2-3Galβ1-3GalNAc-PP-PhU

Radioactive assays require background controls that do not contain the acceptor substrate. Assays using a chemically synthesized acceptor (Galβ1-3GalNAc-PP-PhU) are described in Chapter 13, Subheadings 2.5 and 3.7. In the assays below the product of WbwC is used as the acceptor substrate for WbwA.

3.5.1 Assays for WbwA

1. Prepare enzyme homogenate with WbwA as described in Subheading 3.4.1, **step 1**.
2. Prepare the assay mixture in a total volume of 40 μL: 4 μL 10 mM CMP-[^{3}H]sialic acid (2000 cpm/nmol) donor substrate, 10 μL 0.5 M Tris–HCl buffer, pH 7, 4 μL 100 mM $MnCl_2$, 10 μL enzyme homogenate (4 mg protein) or purified enzyme (2 μg protein), 4 μL 1 mM Galβ1-3GalNAc-PP-PhU acceptor substrate (or water in control assays), 8 μL water.
3. Incubate for 30 min at 37 °C, quench reaction by adding 700 μL water.
4. Separate mixture using C18 Sep-Pak cartridges as described in Subheading 3.4.1, **steps 4–6**.
5. Count cpm of fractions B, C, D and calculate sialyltransferase activity (nmol/h/mg).
6. Nonradioactive assays are carried out as described in Subheading 3.3, **step 2**.
7. The WbwA product, sialylα2-3Galβ1-3GalNAc-PP-PhU, is purified by HPLC and used as an acceptor for WbwB (*see* Subheading 3.4.2).

3.6 Gal-Transferase Assays for WbwB: Production of [^{3}H]Galα1-4sialylα2-3Galβ1-3GalNAc-PP-PhU and Galα1-4sialylα2-3Galβ1-3GalNAc-PP-PhU

Radioactive assays require background controls that do not have the acceptor substrate. Assays using a chemically synthesized acceptor (Galβ1-3GalNAc-PP-PhU) are described in Chapter 13, Subheadings 2.5 and 3.7. In the assays below the product of WbwA is used as the acceptor substrate for WbwB.

1. Prepare enzyme homogenate with WbwB as described in Subheading 3.4.1, **step 1**.
2. Prepare the assay mixture in a total volume of 40 μL: 4 μL 10 mM UDP-[^{3}H]Gal (2000 cpm/nmol) donor substrate, 10 μL 0.5 M Tris–HCl buffer, pH 7, 2 μL 100 mM $MnCl_2$, 10 μL enzyme homogenate (2 mg protein) or purified enzyme (2 μg protein), 4 μL 0.1 mM sialylα2-3Galβ1-3GalNAc-PP-PhU acceptor substrate (or water in control assays), 10 μL water.
3. Incubate for 20 min at 37 °C, stop reaction by adding 700 μL water.
4. Separate mixture using C18 Sep-Pak cartridges as described in Subheading 3.4.1, **steps 4–6**.
5. Count cpm of fractions B, C, D and calculate Gal-transferase activity (nmol/h/mg).
6. Nonradioactive assays are carried out as described in Subheading 3.3, **step 2**. Nonradioactive WbwB product, Galα1-3sialylα2-3Galβ1-3GalNAc-PP-PhU, is analyzed by MS to confirm its structure. For further studies and to confirm its purity, it can be purified by HPLC.

3.7 Large-Scale Production of Galα1-4sialylα2-3Galβ1-3GalNAc-PP-PhU

Before carrying out a large-scale synthesis, the individual enzymes in homogenates need to be tested to verify their activities, as described in Subheadings 3.4–3.6.

1. Prepare bacterial homogenates with WbwC, WbwA and WbwB in 50 mM sucrose.
2. Dry 7 μmol GalNAc-PP-PhU (700 μL 10 mM) acceptor substrate.
3. Assay: to the dried acceptor add 7 mL WbwC homogenate, 7 mL WbwA homogenate, 7 mL WbwB homogenate, 0.2 mL 100 mM CMP-sialic acid, 0.28 mL 100 mM UDP-Gal, 0.14 mL 1 M $MnCl_2$, 3.5 mL 1 M MES, pH 7. Add 2.88 mL water to a total volume of 28 mL.
4. Incubate for 1 h at 37 °C, add 30 mL water.
5. Centrifuge mixture for 15 min at 12,000 × *g*, remove supernatant containing enzyme product.
6. Prepare 60 C18 Sep-Pak columns. Apply 1/60 of the total mixture to each column as described for the assays in Subheading 3.4.1. Elute product with methanol into the same round

bottom flask. Rotor-evaporate the eluate to remove methanol, then lyophilize. Take up the oligosaccharide for analysis in a minimal volume (1 mL) of water. Centrifuge again to remove insoluble precipitate.

7. Perform small-scale radioactive assays (using radioactive CMP-[^{3}H]sialic acid, *see* **Note 9**) in parallel to estimate the amount of product synthesized in the large-scale assays. Radioactive product can also be used to test susceptibility to hydrolases and for HPLC standard.

3.8 Purification of Oligosaccharides

1. Set up the HPLC apparatus: C18 column, acetonitrile/ degassed water = 28/72 run at 1 mL/min, UV absorbance monitor set to 195 nm, fraction collector to collect 2 min fractions to collect radioactive compounds.
2. Inject HPLC standards: 5 μL 5 mM GlcNAc, 5 μL 5 mM Galβ1-3GalNAcα-Bn, 5 μL 1 mM GalNAc-PP-PhU. Oligosaccharides will elute before GalNAc-PP-PhU which elutes at about 30 min. The acetonitrile concentration has to be reduced accordingly to achieve a good separation of enzyme products and acceptors and elution in 20–40 min.
3. Inject an aliquot of radioactive reaction product (2000 cpm) as a standard and collect 2 min fractions. Count cpm of fractions to determine the top of the peak which is used as the elution time (*see* **Note 10**). Inject an aliquot (containing at least 10 nmol) of nonradioactive product, monitor by absorbance at 195 nm and collect fractions. Keep all fractions until the structure is verified by MS. Then pool fractions containing product and concentrate by flash-evaporation and lyophilization.

3.9 Analysis of Enzyme Products

1. HPLC: Compare elution times of oligosaccharides measured by absorbance at 195 nm.
2. Mass spectrometry: Electrospray ionization (ESI) in the negative ion mode is a convenient method to analyze diphosphates. Only picomoles of compounds are required (*see* **Note 11**).
3. 600 MHz NMR: 1D proton-NMR will show the anomeric linkages of sugars. By 2D methods ^{13}C carbon and ^{1}H proton resonances are assigned. A comparison to NMR parameters of known compounds in the literature helps in the assignments. Nuclear Overhauser Enhancement experiments allow the determination of the linkages and the total structure of the oligosaccharide (Fig. 2) (*see* **Note 12**).
4. Aliquots of radioactive sugars can be analyzed by digestion with linkage-specific glycosidases (*see* **Note 13**).

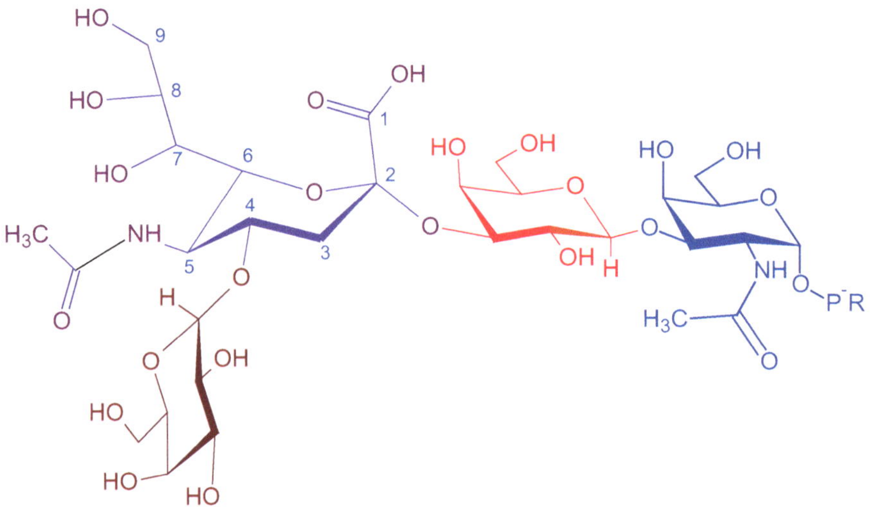

Fig. 2 Structure of the ECO104 repeating unit linear oligosaccharide linked to diphosphate-lipid (P-R). Gal α1-4 (brown), sialic acid α2-3 (purple), Gal β1-3 (red), GalNAc α (blue)

4 Notes

1. There are a number of options to obtain plasmids containing the gene of interest. Before ordering the plasmid, the protein structure has to be assessed by bioinformatics methods (*see* Chapter 11). WbwC, WbwA and WbwB appear to be soluble proteins with short hydrophobic stretches that may be responsible for membrane association.
2. While the BL21 expression strain is suitable, other *E. coli* strains may be preferred as they may have a better ability to express recombinant foreign proteins.
3. For this protein, the Co-resin has shown to be superior to the Ni-resin.
4. The solution for preparing bacterial homogenates can be buffer with or without EDTA or PBS. Sucrose has been a suitable solution to keep the enzymes active.
5. WbwB has been shown to be highly active with low concentrations of the acceptor.
6. HPLC standards with known concentration and high purity are desirable.
7. Procedures for SDS-PAGE and Western blots have been described previously [12].
8. Enzymes in homogenates often have high activities. Only a small fraction of the total enzyme content is usually solubilized

and can be purified. Purified enzymes will have less hydrolytic enzyme contaminations.

9. In the radioactive small-scale assay that is done in parallel to the large-scale assay, radioactive CMP-sialic acid is used to ensure radioactive products contain sialic acid and they can be quantified since only a single sugar is radioactive per molecule. Alternatively, UDP-[^{3}H]Gal can be used and two molecules of [^{3}H] Gal will be incorporated into the final repeating unit.
10. The addition of a sugar residue to the growing GalNAc-PP-PhU oligosaccharide will significantly decrease its elution time. Therefore, the acetonitrile concentration has to be decreased to achieve a good separation. The diphosphate group at pH 7 is partially in the salt form and therefore, the same compound may elute in two or more peaks. MS analysis can resolve this.
11. The mass spectra (ESI-MS in the negative ion mode) of oligosaccharides show a major peak of $[M–H]^-$ and minor ions with one and two Na^+ ions. For GalNAc-PP-PhU we expect a main $[M–H]^-$ peak of m/z 626. For Gal-GalNAc-PP-PhU we expect m/z 788. For $[M–H]^-$ of Sialyl-Gal-GalNAc-PP-PhU m/z is 1079 and for $[M–H]^-$ of Gal-sialyl-Gal-GalNAc-PP-PhU m/z is 1041.

    MALDI analysis is also suitable to assess purity and mass of the oligosaccharide.
12. For 2D-NMR analysis, the repeating unit-linked to PP-PhU needs to be reasonably pure and at least 500 μmol should be obtained. ROESY or NOESY analyses are usually able to confirm the linkages [12, 19, 20].
13. Exo-glycosidases can confirm the anomeric linkages but did not work for confirmation of the Galα1-4sialic acid linkage.

References

1. Muniesa M, Hammeri J, Hertwig S et al (2012) Shiga toxin-producing Escherichia coli O104:H4: a new challenge for microbiology. Appl Environ Microbiol 78:4065–4073
2. Raetz CRH, Whitfield C (2002) Lipopolysaccharide endotoxins. Annu Rev Biochem 71:635–700
3. Gamian A, Romanowska E, Ulrich J et al (1992) The structure of the sialic acid-containing *Escherichia coli* O104 O-specific polysaccharide and its linkage to the core region in lipopolysaccharide. Carbohydr Res 236:195–208
4. Romanow A, Haselhorst T, Stummeyer K et al (2013) Biochemical and biophysical characterization of the sialyl-/hexosyltransferase synthesizing the meningococcal serogroup W135 heteropolysaccharide capsule. J Biol Chem 288:11718–11730
5. Whitfield C, Trent MS (2014) Biosynthesis and export of bacterial lipopolysaccharides. Annu Rev Biochem 83:99–128
6. Wang L, Briggs CE, Rothemund D et al (2001) Sequence of the *E. coli* O104 antigen gene cluster and identification of O104 specific get al enes. Gene 270:231–236
7. Samuel G, Reeves P (2003) Biosynthesis of O-antigens: genes, and pathways involved in nucleotide sugar precursor synthesis and O-antigen assembly. Carbohydr Res 338:2503–2519
8. Lehrer J, Vigeant KA, Tatar LD et al (2007) Functional characterization and membrane topology of *Escherichia coli* WecA, a sugar-

phosphate transferase initiating the biosynthesis of enterobacterial common antigen and O-antigen lipopolysaccharide. J Bacteriol 189:2618–2628

9. Rush JS, Alaimo C, Robbiani R et al (2010) A novel epimerase that converts GlcNAc-P-P-undecaprenol to GalNAc-P-P-undecaprenol in *Escherichia coli* O157. J Biol Chem 285:1671–1680
10. Gao Y, Liu B, Strum S et al (2012) Biochemical characterization of WbdN, a beta1,3-glucosyltransferase involved in O-antigen synthesis in enterohemorrhagic *Escherichia coli* O157. Glycobiology 22:1092–1102
11. Wang S, Hao Y, Lam JS et al (2015) Biosynthesis of the common polysaccharide antigen of *Pseudomonas aeruginosa* PAO1: characterization and role of GDP-D-rhamnose: GlcNAc/GalNAc-diphosphate-lipid alpha1,3-D-rhamnosyltransferase WbpZ. J Bacteriol 197:2012–2019
12. Wang S, Czuchry D, Liu B et al (2014) Characterization of two UDP-gal: GalNAc-diphosphate-lipid β1,3-galactosyltransferases WbwC from *Escherichia coli* serotypes O104 and O5. J Bacteriol 196:3122–3133
13. Riley JG, Menggad M, Montoya-Peleaz PJ et al (2005) The wbbD gene of *E. coli* strain VW187 (O7:K1) encodes a UDP-gal: GlcNAc{alpha}-pyrophosphate-R beta1,3-galactosyltransferase involved in the biosynthesis of O7-specific lipopolysaccharide. Glycobiology 15:605–613
14. Xu C, Liu B, Hu B et al (2011) Biochemical characterization of UDP-gal:GlcNAc- pyrophosphate-lipid-1,4-Galactosyltransferase WfeD, a new enzyme from *Shigella boydii* type 14 that catalyzes the second step in O-antigen repeating-unit synthesis. J Bacteriol 193:449–459
15. Brockhausen I, Hu B, Liu B et al (2008) Characterization of two beta-1,3-glucosyltransferases from *Escherichia coli* serotypes O56 and O152. J Bacteriol 190:4922–4932
16. Chen C, Liu B, Xu Y et al (2016) Biochemical characterization of the novel α-1, 3-galactosyltransferase WclR from *Escherichia coli* O3. Carbohydr Res 430:36–43
17. Yi W, Shao J, Zhu L et al (2005) *Escherichia coli* O86 O-antigen biosynthetic gene cluster and stepwise enzymatic synthesis of human blood group B antigen tetrasaccharide. J Am Chem Soc 127:2040–2041
18. Hanisch FG, Baldus SE (1997) The Thomsen-Friedenreich (TF) antigen: a critical review on the structural, biosynthetic and histochemical aspects of a pancarcinoma-associated antigen. Histol Histopathol 12:263–281
19. Czuchry D, Desormeaux P, Stuart M et al (2015) Identification and biochemical characterization of the novel alpha2,3-sialyltransferase WbwA from the pathogenic *Escherichia coli* serotype O104. J Bacteriol 197:3760–3768
20. Czuchry D, Szarek WA, Brockhausen I (2017) Identification and biochemical characterization of WbwB, a novel UDP-gal: Neu5Ac-R alpha1,4-galactosyltransferase from the intestinal pathogen *Escherichia coli* serotype O104. Glycoconj J 35:65–76
21. Si A, Misra AK (2016) Expedient synthesis of the pentasaccharide repeating unit of the polysaccharide O-antigen of *Escherichia coli* O11. Chemistry Open 5:47–50
22. Bi N, Xiong C, Jin G et al (2017) Synthesis of a trisaccharide repeating unit of the O-antigen from *Burkholderia cenocepacia* and its dimer. Carbohydr Res 451:1–11
23. Bhaumik I, Kar RK, Bhunia A et al (2016) Expedient synthesis of the pentasaccharide repeating unit of the O-antigen of *Escherichia coli* O86 and its conformational analysis. Glycoconj J 33:887–896
24. Mitra A, Mukhopadhyay B (2016) Linear synthesis of the hexasaccharide related to the repeating unit of the O-antigen from Shigella flexneri serotype 1d (I: 7,8). Carbohydr Res 426:1–8
25. Mandal PK (2014) Synthesis of the pentasaccharide repeating unit of the O-antigen of E. coli O117:K98:H4. Beilstein J Org Chem 10:2724–2728

Chapter 16

Sulfhydryl Labeling as a Tool to Investigate the Topology of Membrane Proteins Involved in Lipopolysaccharide Biosynthesis

Faviola Tavares-Carreón, Xiang Ruan, Amy Ford, and Miguel A. Valvano

Abstract

Establishing the topology of membrane proteins, especially when their tridimensional structures are unavailable, is critical to identify functional regions, delimit the protein orientation in the membrane, the number of transmembrane segments, and the position of critical amino acids (whether exposed to the solvent or embedded in the lipid bilayer). Elucidating the topology of bacterial integral membrane proteins typically involves the construction of deletion-fusions whereby regions of the protein are fused to reporters. Although these methods have several advantages, they are also artifact prone. In contrast, methods based on single amino acid substitutions preserve the native protein intact. We describe here an assay to analyze the topology of membrane proteins involved in the biogenesis of bacterial glycoconjugates, which is based on the accessibility of cysteine substitutions at various places in the protein under in vivo and in vitro conditions. Cysteine residues are detected with polyethylene glycol-maleimide (PEG-Mal). This procedure can be applied to crude bacterial cell extracts and does not require protein purification.

Key words Glycosyltransferase, Sulfhydryl labeling, Membrane proteins, O antigen, Lipid A-core oligosaccharide, Substituted cysteine accessibility mutagenesis

1 Introduction

Lipopolysaccharide (LPS) is the major lipoglycan molecule at the cell surface of gram-negative bacteria, which is embedded into the outer leaflet of the outer membrane by the hydrophobic lipid A moiety [1, 2]. External to lipid A is the core oligosaccharide (core), consisting of several sugar residues that vary in number and linkages depending on the bacterial species [2]. The O antigen (OAg) is the outermost component of the LPS, generally attached to a terminal core residue, which displays high structural and immunochemical variability among bacteria [3]. Lipid A-core and OAg are

Faviola Tavares-Carreón and Xiang Ruan contributed equally to this work.

Inka Brockhausen (ed.), *Bacterial Polysaccharides: Methods and Protocols*, Methods in Molecular Biology, vol. 1954,
https://doi.org/10.1007/978-1-4939-9154-9_16,

synthesized by independent pathways. In most bacteria, lipid A is a disaccharide of glucosamine and contains a variable number of acyl chains. Lipid A is also glycosylated at the 6′-position with two residues of 3-deoxy-d-*manno*-oct-2-ulopyranosonic acid (Kdo), and the inner Kdo serves as the point of attachment for the remaining core that is assembled by sequential glycosyltransferase reactions [1, 2, 4]. Both the lipid A and core are assembled on the cytoplasmic side of the inner membrane and translocated across the inner membrane by the ATP-dependent transporter MsbA [1].

The OAg is assembled onto undecaprenyl-phosphate (Und-P), a C_{55}-polyisoprenol, forming an Und-PP linked saccharide [5]. This precursor is also translocated to the periplasmic side of the cell membrane by several different mechanisms. Lipid A-core and OAg synthesis pathways converge by the ligation of the Und-PP-linked OAg glycan onto the lipid A-core, with the release of Und-PP [5] in a reaction that occurs on the periplasmic face of the inner membrane. Newly formed LPS must then be shuttled across the periplasm by a complex multiprotein pathway, designated LPS transport pathway (Lpt), which is also responsible for the insertion of LPS into the outer leaflet of the outer membrane. In addition, the LPS molecule is further modified by reactions that introduce nonstoichiometric modifications of its basic chemical structure, which are often regulated in response to environmental cues and protect bacteria against innate immune responses and other environmental stresses [4, 6].

Several integral membrane proteins participate in various aspects of LPS biogenesis. For example, the initiation, membrane translocation and elongation (polymerization) of the Und-PP-OAg requires membrane proteins with multiple transmembrane helices (TMHs), as well as cytoplasmic- and periplasmic-exposed functional regions, depending on their specific function. For example, the enzymes involved in the initiation of OAg synthesis catalyze the formation of a phosphoanhydride bond between Und-P and the first sugar 1-phosphate of the OAg unit. One of these enzymes is WcaJ, which belongs to the polyisoprenyl-phosphate hexosamine-1-phosphate transferases family (PHPT) [7, 8]. The OAg ligase, WaaL, also has multiple TMHs and a relatively large periplasmic exposed loop presumably required for its catalytic function [9–11]. Similarly, nonstoichiometric modifications introduced in the lipid A-core require membrane embedded enzymes. One of them is ArnT, a glycosyltransferase that catalyzes the addition of 4-amino-4-deoxy-L-arabinose (L-Ara4N) to the lipid A [12–14].

The topology of ArnT, WaaL, and WcaJ was widely studied using different topological methods to provide insight into the organization, structure and nature of functional amino acids [7, 9–11, 15]. Reporter fusions based on β-galactosidase (LacZ) and alkaline phosphatase (PhoA) reporters [16, 17] and the substituted cysteine accessibility method (SCAM) [18, 19] are experimental approaches to probe and elucidate the topology of membrane proteins. However, both methods have various advantages and disadvantages. For example, reporter fusion approaches involve the construction of sequential deletion-fusion derivatives of the native protein. Depending on the location of the fusion endpoint, the reporter can provide inconclusive topological information [16, 17, 20]. SCAM is based on single cysteine replacements followed by methods that take advantage of sulfhydryl chemistry using membrane-permeable and -impermeable thiol reactive agents [18, 19]. These reagents can be used to block solvent-exposed thiol groups and prevent the reaction with another reagent that can be labeled, such as biotin maleimide. The results obtained in the presence and absence of the blocker provide information to deduce the location of the residue. The main advantage of SCAM is the preservation of the native protein structure–function. However, one absolute requirement for SCAM is the construction of a cysteine-less derivative of the native protein, which must be proven to remain functional before further manipulations can take place. In our experience, this is not a problem with the proteins we have investigated by SCAM including the initiation enzymes WecA [21] and WcaJ [7], the WaaL ligase [22], the Wzx OAg flippase [23], and ArnT [13, 14]. One potential drawback of the classical SCAM method is the use of biotin maleimide since the biotinylated protein must be purified prior to detection with streptavidin [21]. Purification is critical to separate the modified protein from the background of other native proteins in the membrane that became biotinylated in their native cysteines. Because the membrane enzymes involved in LPS biogenesis are typically poorly expressed, purification of the protein is not always successful even if they are constructed with tags to facilitate this step (e.g., His10× tag). Therefore, the classical SCAM method based on biotinylation can give considerable background and is highly labor intensive [21, 23].

In contrast, sulfhydryl labeling using the covalent attachment of polyethylene glycol-maleimide (PEG-Mal) does not require protein purification [24]. The PEG-Mal label is large and membrane-impermeable, but when used in combination with EDTA (to permeabilize the outer membrane), it can react with the periplasmic-exposed cysteine residues of the protein in intact cells. Labeling can be demonstrated by the increased apparent molecular mass of the protein by Western blot using antibodies against the protein or a short epitope tag fused to the N- or C-terminus of the native protein, as needed. To assess if a cysteine residue is exposed to the cytoplasm, it is necessary to lyse the bacterial cells and obtain membrane pellets. Total membrane preparations result in inverted vesicles where cytoplasmic cysteines are exposed and therefore available to react with PEG-Mal. In this chapter, we describe methodological details for sulfhydryl labeling of membrane proteins using ArnT, WaaL, and WcaJ, as examples of membrane proteins involved in LPS biogenesis. We believe that the method can be easily applied to explore the topology of any bacterial integral membrane protein for which native cysteine residues can be removed without any effect on its structure–function.

2 Materials

2.1 Recombinant Genes Expressing Cysteine-Less Proteins

1. A version of the gene of interest lacking cysteine codons must be constructed by substitution of native cysteine residues with alanine. The plasmid containing the cysteine-less gene is used as template for single cysteine codon substitution with primers containing the desired mutations. Cysteine substitutions are introduced by site-directed mutagenesis in the cloning vector pBAD24. Using this vector allows us to place the gene under the control of the BAD arabinose inducible promoter, which reduces a potential toxic effect of overexpressing a membrane protein. Other plasmid vectors can also be used, as appropriate, if so desired.

2.2 Growth Medium and Strains

2.2.1 Growth Medium

1. Bacterial cells containing arabinose-inducible plasmids are grown in Luria-Bertani (LB) medium.
2. ArnT, WaaL and WcaJ are constructed with a C-terminal FLAG epitope, so that the proteins can be detected by immunoblot analysis.
3. Grow bacteria overnight in LB containing the appropriate antibiotics (100 μg/mL ampicillin, 40 μg/mL kanamycin, and 20 μg/mL tetracycline) as required.
4. L-arabinose: stock solution of 20% (w/v) in deionized water, filtered through a 0.22-μm filter.
5. Cell disruptor (Constant Systems Ltd.).

2.2.2 Strains

1. *E. coli* DH5a strain (F^{-} *ϕ80lacZ*M15 *endA recA hsdR*(r^{-}_{K} m^{-}_{K}) *supE thi gyrA relA Δ(lacZYA-argF) U169*) is transformed with pFT1 [14], pXR1 [9, 22], and pLA3 [7] plasmids, which are pBAD24 derivatives encoding $ArnT_{FLAG\text{-}10\times His}$, $WaaL_{FLAG\text{-}10\times His}$, and $_{FLAG}WcaJ$, respectively.

2.3 Sulfhydryl Labeling Components

1. Resuspension buffer: 50 mM 4-(2-hydroxyethyl)-1-piperazineethanesulfonic acid (HEPES) pH 6.8, 5 mM $MgCl_2$.
2. 500 mM ethylenediaminetetraacetic acid (EDTA), pH 8.0.
3. 10% (w/v) sodium dodecylsulfate (SDS).
4. 500 mM dithiothreitol (DTT).

5. 10 mM methoxypolyethylene glycol maleimide (PEG-Mal) (PEG ~5000 Da; Sigma-Aldrich).
6. 50 mM N-ethylmaleimide (NEM).
7. 25× protease inhibitor: dissolve 1 tablet of complete, EDTA free Protease Inhibitor Cocktail Tablets (Roche Diagnostics) in 2 mL deionized water.
8. 3× dye buffer: 0.825 M Tris–HCl, pH 6.8, 6% (w/v) SDS, 30% (v/v) glycerol, 15% (v/v) β-mercaptoethanol, and 0.03% (w/v) bromophenol blue.

2.4 SDS-Polyacrylamide Gel Electrophoresis

1. Acrylamide stock solution (30% T, 2.6% C). Add 29.22 g acrylamide and 0.78 g bisacrylamide to 100 mL H_2O. Filter the stock solution through Whatman filter paper and store at 4 °C.
2. Running gel: 14% ABA 44:0.8, 0.2% SDS, 0.375 M Tris–HCl, pH 8.0, 0.025% ammonium persulfate (APS), 0.025% N,N,N′, N′-tetramethylethylenediamine (TEMED).
3. Stacking gel: 4.5% ABA 44:0.8, 0.1% SDS, 0.125 M Tris–HCl, pH 6.8, 0.025% APS, 0.025% TEMED.
4. 1× SDS buffer (25 mM Tris, pH 7.6, 200 mM glycine, 2.3 mM SDS).
5. 3× dye buffer (50 mM Tris–HCl, pH 6.8, 2% SDS, 10% glycerol, and 0.1% bromophenol blue).
6. SDS-PAGE is performed using the Mini-Protean 3 Cell (Bio-Rad) at 110–120 V until the dye front runs through the stacking gel, and then increased to 130–150 V until the dye front reaches the bottom of the gel.

2.5 Detection of Sulfhydryl Labeling

1. Nitrocellulose membrane (Whatman, GE Health Care Life Sciences).
2. TBS (Tri-buffered saline): 50 mM Tris, 150 mM NaCl, pH 7.5.
3. 10% Western Blocking Solution: dilute the Western Blocking Solution (Roche Diagnostics) with TBS to 10% (v/v).
4. The primary antibody, 4.6 mg/mL anti-FLAG M2 monoclonal antibody (Sigma-Aldrich), was diluted to 1: 10,000 and applied for 1–2 h.
5. The secondary antibody consists of 2 mg/mL goat anti-mouse Alexa fluor 680 IgG (ThermoFisher Scientific) is diluted to 1: 20,000 and applied for 1 h.
6. Odyssey infrared imaging system (Licor Biosciences).

3 Methods

3.1 Sulfhydryl Labeling EDTA-Permeabilized Intact Cells Preparation

1. From overnight cultures, dilute cells to an optical density at 600 nm (OD_{600}) of 0.2 in 5 to 10 mL LB broth containing the respective antibiotics.
2. Incubate the culture with shaking at 37 °C until reaching an OD_{600} of 0.5–0.7, and induce gene expression with 0.2% (w/v) L-arabinose for 3 h at 37 °C.
3. Harvest the bacterial cells by centrifugation at 9300 × *g* for 10 min at 4 °C.
4. Wash the pellet twice with 2 mL resuspension buffer.
5. Resuspend the pellet in 1 mL resuspension buffer and take 330 μL aliquots of the cell suspension into clean microcentrifuge tubes. One aliquot is for the whole-cell PEG-Mal treatment; the second aliquot is the nontreatment control, and the last one is for the PEG-Mal treatment of lysed cells.

   For the PEG-Mal treatment of whole bacterial cells, permeabilize cells with 5 mM EDTA for 10 min at room temperature. For the nontreatment control, incubate cells at room temperature for 1 h with resuspension buffer. For the PEG-Mal treatment of lysed cells incubate lysate with 2% SDS for 5 min (*see* **Note 1**). Then, add 1 mM of PEG-Mal (*see* **Note 2)** and incubate at room temperature for 1 h.
6. Quench the reactions with 45 mM DTT for 10 min at room temperature after PEG-Mal treatment.
7. Add 0.1 mL 3× dye buffer to each sample and incubate for 30 min at 45 °C (*see* **Note 3**) before loading the sample into the polyacrylamide gel to analyzed by immunoblotting or store pellet at −20 °C (*see* **Note 4**).

3.2 Labeling from Total Membrane Fraction

1. From overnight cultures, inoculate 50 mL LB broth to an OD_{600} of 0.2 in LB broth containing the appropriate antibiotics.
2. Incubate the culture at 37 °C with shaking until reaching an OD_{600} of 0.5–0.7, and induce with 0.2% (w/v) L-arabinose for 3 h at 37 °C.
3. Harvest cells by centrifugation 8000 × *g* and wash twice with 5 mL resuspension buffer (*see* **Note 5**).
4. Resuspend bacterial pellet in 1 mL resuspension buffer with protease inhibitors and lysed at 10,000 PSI with a cell disruptor.
5. Pellet cell debris at 27,216 × *g* and transfer the supernatant, containing the total membranes, to a fresh tube.

6. To isolate total membranes, centrifuge the supernatant in microfuge tubes at 39,191 × *g* for 40 min at 4 °C. Resuspend the pellet in 50 μL resuspension buffer with 1× protease inhibitors (*see* **Note 6**).
7. Protein concentrations were determined by Bradford protein assay (Bio-Rad).
8. Sulfhydryl labeling of crude membranes: 10 μg of total membranes are used for the following conditions: (a) PEG-Mal treatment of whole cells, (b) nontreatment condition, and (c) PEG-Mal treatment of bacterial cell membranes. For PEG-Mal treatment of intact cells add 1 mM PEG-Mal and incubate at room temperature for 1 h. For the nontreatment control, incubate membranes at room temperature for 1 h with 20 μL cold resuspension buffer. PEG-Mal treatment for the membrane pellet: incubate with 2% SDS for 5 min at room temperature. After that, add 1 mM of PEG-Mal and incubate for 1 h at room temperature.
9. Quench sulfhydryl samples with 45 mM DTT for 10 min at room temperature. Continue with immunoblotting protocol or store pellet at −20 °C (*see* **Note 4**).

3.3 Thiol-Specific Chemical Blocking with N-Ethylmaleimide (NEM)

1. Resuspend whole cell or crude membrane fraction with 0.1 mL resuspension buffer. Add 5 mM NEM for 30 min at room temperature, and centrifuged at 16,000 × *g* for 1 min.
2. Wash the pellets twice with 1 mL ice-cold resuspension buffer.
3. Resuspend the samples in 200 μL resuspension buffer, and incubate with 2% SDS at room temperature.
4. Add 1 mM PEG-Mal and incubate for 1 h at room temperature.
5. After incubation with PEG-Mal quench sulfhydryl labeling by adding sufficient DDT to reach a final concentration of 45 mM and incubate for 10 min at room temperature. Continue with immunoblotting protocol or store pellet at −20 °C.

3.4 Detection of Sulfhydryl Labeling Proteins

1. Sulfhydryl labeling of intact cells and total membrane preparations is analyzed by Immunoblotting.
2. Perform the immunoblotting by separating intact cells and crude membrane preparations by 14% SDS-PAGE and transferring to nitrocellulose membrane.
3. Block the membranes overnight in 5% Western blocking reagent and TBS.
4. Incubate the membrane with properly diluted antibody for the appropriate time required for each specific protein.
5. Sulfhydryl labeling of the protein results in a band shift of approximately 10 to 25 kDa gain in apparent molecular mass

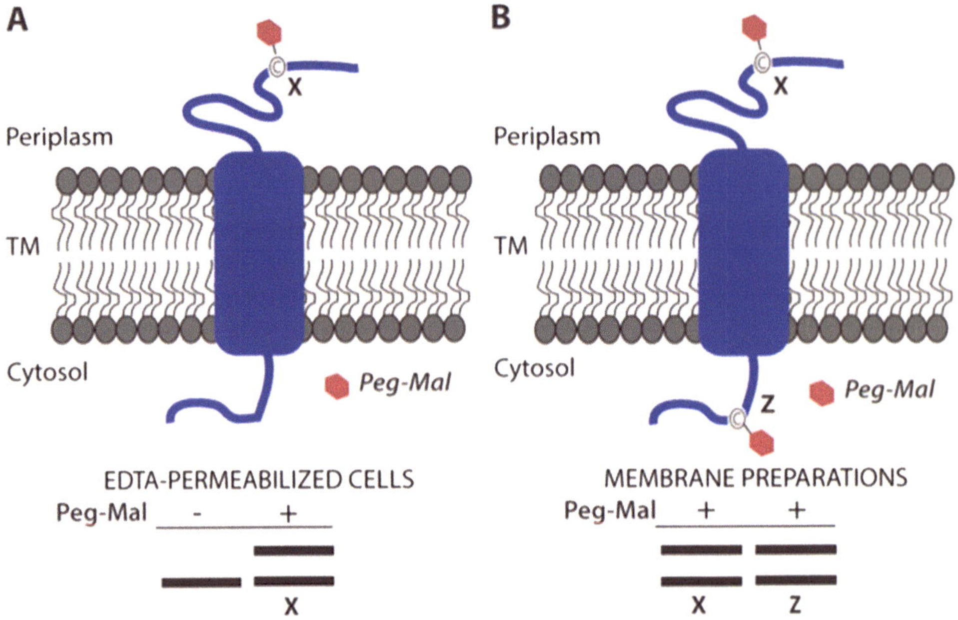

Fig. 1 Schematic illustration of sulfhydryl labeling and the theoretical immunoblot based on the labeling pattern. (**a**) In EDTA-permeabilized cells, only the periplasmic cysteine residue X can react with PEG-Mal molecule (+) and produce a band-shift observable on the immunoblot. (**b**) In membrane crude preparations, cysteines X and Z on either side of the inner membrane, respectively, can react with PEG-Mal (+) and produce a band-shift. Cysteines located into the lipid bilayer (TM) do not react with PEG-Mal

relative to that of the parental protein. This is visualizing by a characteristic double-band in the in the Western blot (*see* Figs. 1 and 2 and **Notes 7–9**).

4 Notes

1. To enhance cell lysis by SDS, heat the sample for 5 up to min at 37 °C. Longer time is not recommended.
2. PEG-mal, NEM and DTT should always be prepared fresh before the experiment. PEG-Mal should be dissolved in dimethyl sulfoxide (DMSO) ultra-pure grade. Similarly, DTT should be dissolved in fresh water, and NEM in ethanol first and then in fresh water (For example, to prepare 50 mM NEM, dissolve 5 μmol NEM in ethanol, bubble with nitrogen gas to half-dry, and add 100 μL water). Once dissolved, storage is not recommended. HEPES (pH 6.8) buffer is also recommended to be freshly prepared.
3. Mild denaturation conditions work best for membrane proteins [9], and ArnT, WaaL and WcaJ detection by Western blot is not possible is the proteins are denatured at 100 °C.

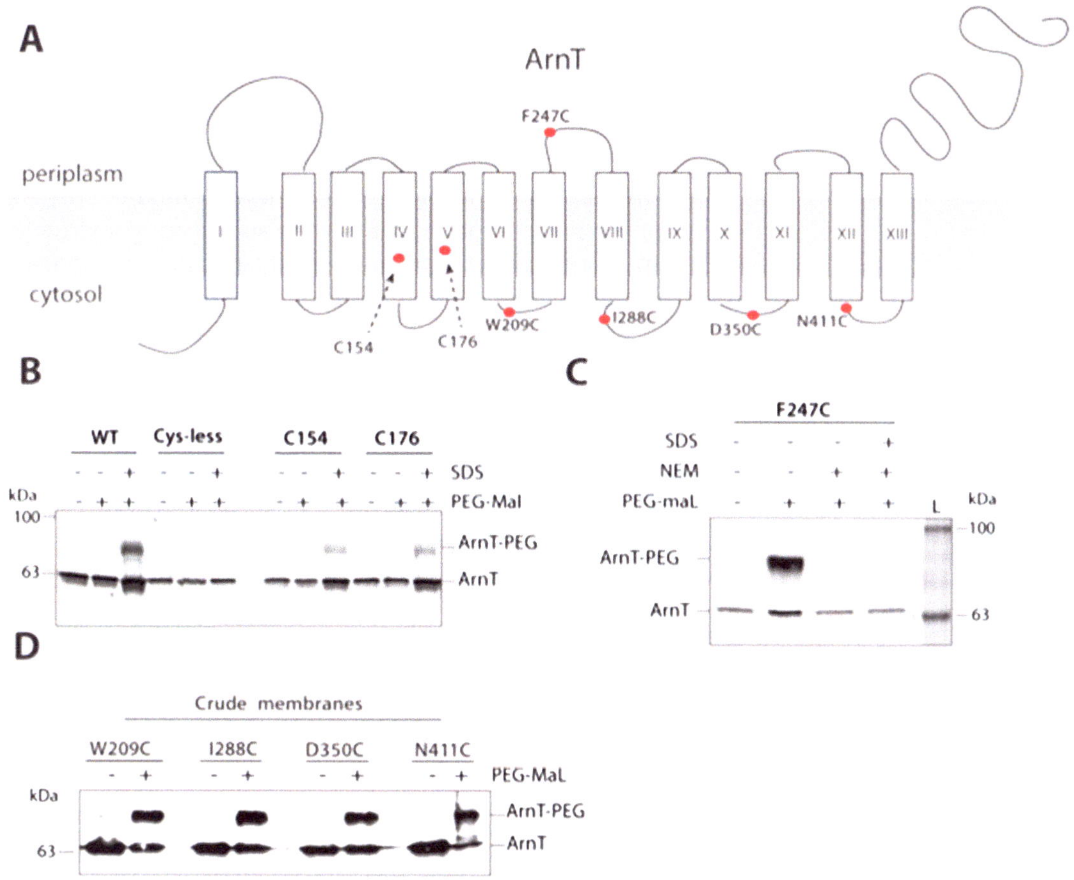

Fig. 2 Example of SCAM analysis by PEGylation using ArnT. (**a**) Topological representation of the ArnT from *Burkholderia cenocepacia*. Roman numerals indicate the TM helices. Red circles indicate cysteine-substitutions for sulfhydryl labeling assay. (**b**) Analyses of sulfhydryl labeling assays (*modified from ref.* 14). *E. coli* DH5α was transformed with plasmids expressing parental ArnT ($ArnT_{WT}$), its cysteine-less version ($ArnT_{Cys\text{-}less}$), and various single cysteine replacement derivatives. $ArnT_{C154A}$ and $ArnT_{C176A}$ were only PEGylated in detergent (2% SDS), while $ArnT_{Cys\text{-}less}$ was not PEGylated under any condition. (**c**) The results with $ArnT_{F247C}$ indicate the residue at position 247 is exposed to the periplasm since it could be PEGylated in the absence of NEM. (**d**) Cysteine replacements of selected cytosolic residues were PEGylated in membrane preparations only in presences of PEG-Mal since membrane fractions allow the exposure of cytosolic residues due to formation of inverted vesicles [25]. All proteins were separated by SDS-PAGE and ArnT derivatives were detected with anti-FLAG

However, some proteins may also not tolerate even 45 °C. Therefore, the denaturation step can also be performed at room temperature before loading onto the SDS-PAGE gel.

4. Short-term storage at −20 °C with protease inhibitor is usually optimal and the PEGylated proteins may remain stable in these conditions for 1 week. Long-term storage may alter protein stability.

5. The bacterial cell pellets can be stored at −20 °C for 2–3 days before being processed.
6. Total membrane pellets will appear amber to brown color and may be difficult to resuspend. Gentle pipetting is necessary to homogeneously resuspend the pellet.
7. The introduction of novel cysteine residues can affect the stability of the protein, particularly when substituting a structurally important residue. This may result in a poorly expressed protein and/or inconsistent labeling results. If this occurs, it may be better to substitute neighboring residues to build a more robust topological model.
8. Membrane proteins are differentially expressed. For this reason, it may be useful to optimize the experimental protocol to avoid either overexpression that may lead to smearing and a large amount of unlabeled protein or underexpression resulting in inconclusive results. This may be achieved a number of ways by altering the protocol at the initial experimental stage including adjusting the number of cells used, the induction time, and the growth temperature.
9. Residues at the membrane borders may appear to be partially PEGylated/partially blocked by NEM. Again, to eliminate any uncertainty of the positioning of the residue, it is advisable to look at neighboring residues.

Acknowledgments

This work was supported by grants from the Canadian Institutes of Health Research and Cystic Fibrosis Canada to M.A.V. A.F. was supported by a Doctoral Research Scholarship from the Department of Employment and Learning, Northern Ireland.

References

1. Raetz RH, Whitfield C (2002) Lipopolysaccharide endotoxins. Annu Rev Biochem 71:635–700
2. Whitfield C, Trent MS (2014) Biosynthesis and export of bacterial lipopolysaccharides. Annu Rev Biochem 83:99–128
3. Samuel S, Reeves P (2003) Biosynthesis of O-antigens: genes and pathways involved in nucleotide sugar precursor synthesis and O-antigen assembly. Carbohydr Res 338:2503–2519
4. Raetz CR, Reynolds CM, Trent MS et al (2007) Lipid A modification systems in gram-negative bacteria. Annu Rev Biochem 76:295–329
5. Valvano MA (2011) Common themes in glycoconjugate assembly using the biogenesis of O-antigen lipopolysaccharide as a model system. Biochemistry-Moscow 76:729–735
6. Needham BD, Trent MS (2013) Fortifying the barrier: the impact of lipid A remodelling on bacterial pathogenesis. Nat Rev Microbiol 11:467–481
7. Furlong SE, Ford A, Albarnez-Rodriguez L et al (2015) Topological analysis of the *Escherichia coli* WcaJ protein reveals a new conserved configuration for the polyisoprenyl-phosphate hexose-1-phosphate transferase family. Sci Rep 5:9178

8. Lukose V, Walvoort MTC, Imperiali B (2017) Bacterial phosphoglycosyl transferases: initiators of glycan biosynthesis at the membrane interface. Glycobiology 27:820–833
9. Ruan X, Pérez JM, Marolda CL et al (2012) The WaaL O-antigen lipopolysaccharide ligase has features in common with metal ion-independent inverting glycosyltransferases. Glycobiology 22:288–299
10. Pan Y, Ruan X, Valvano MA et al (2012) Validation of protein topology models by oxidative labeling and mass spectrometry. J Am Soc Mass Spectr 23:889–898
11. Pérez JM, McGarry MA, Marolda CL et al (2008) Functional analysis of the large periplasmic loop of the *Escherichia coli* K-12 WaaL O-antigen ligase. Mol Microbiol 70:1424–1440
12. Trent MS, Ribeiro AA, Lin S et al (2001) An inner membrane enzyme in *Salmonella* and *Escherichia coli* that transfers 4-amino-4-deoxy-L-arabinose to lipid A: induction on polymyxin-resistant mutants and role of a novel lipid-linked donor. J Biol Chem 276:43122–43131
13. Tavares-Carreon F, Fathy Mohamed Y, Andrade A et al (2016) ArnT proteins that catalyze the glycosylation of lipopolysaccharide share common features with bacterial N-oligosaccharyltransferases. Glycobiology 26:286–300
14. Tavares-Carreón F, Patel KB, Valvano MA (2015) *Burkholderia cenocepacia* and *Salmonella enterica* ArnT proteins that transfer 4-amino-4-deoxy-L-arabinose to lipopolysaccharide share membrane topology and functional amino acids. Sci Rep 5:10773. accepted
15. Islam ST, Taylor VL, Qi M et al (2010) Membrane topology mapping of the O-antigen flippase (Wzx), polymerase (Wzy), and ligase (WaaL) from *Pseudomonas aeruginosa* PAO1 reveals novel domain architectures. mBio 1: e00189–00110–e00189–00119
16. Haardt M, Bremer E (1996) Use of *phoA* and *lacZ* fusions to study the membrane topology of ProW, a component of the osmoregulated ProU transport system of *Escherichia coli*. J Bacteriol 178:5370–5381
17. Manoil C (1991) Analysis of membrane protein topology using alkaline phosphatase and beta-galactosidase gene fusions. Methods Cell Biol 34:61–75
18. Bogdanov M, Heacock PN, Dowhan W (2010) Study of polytopic membrane protein topological organization as a function of membrane lipid composition. Methods Mol Biol 619:79–101
19. Bogdanov M, Zhang W, Xie J et al (2005) Transmembrane protein topology mapping by the substituted cysteine accessibility method (SCAM™): application to lipid-specific membrane protein topogenesis. Methods 36:148–171
20. Bogdanov M, Heacock PN, Dowhan W (2002) A polytopic membrane protein displays a reversible topology dependent on membrane lipid composition. EMBO J 21:2107–2116
21. Lehrer J, Vigeant KA, Tatar LD et al (2007) Functional characterization and membrane topology of *Escherichia coli* WecA, a sugar-phosphate transferase initiating the biosynthesis of enterobacterial common antigen and O antigen lipopolysaccharide. J Bacteriol 189:2618–2628
22. Ruan V, Monjarás Feria J, Hamad M, Valvano MA (2018) Escherichia coliand Pseudomonas aeruginosalipopolysaccharide O-antigen ligases share similar membrane topology and biochemical properties. Molecular Microbiology 110(1):95–113. https://doi.org/10.1111/mmi.14085
23. Marolda CL, Li B, Lung M et al (2010) Membrane topology and identification of critical amino acid residues in the Wzx O-antigen translocase from *Escherichia coli* O157:H4. J Bacteriol 192:6160–6171
24. Koch S, Fritsch MJ, Buchanan G et al (2012) *Escherichia coli* TatA and TatB proteins have N-out, C-in topology in intact cells. J Biol Chem 287:14420–14431
25. Rivera-Ordaz A, Bracher S, Sarrach S et al (2013) The sodium/proline transporter PutP of *Helicobacter pylori*. PLoS One 8:e83576

Chapter 17

Assay Methods for the Glycosyltransferases Involved in Synthesis of Bacterial Polysaccharides

Tasnim Abukar, Nakita Buenbrazo, Bettina Janesch, Laura Kell, and Warren Wakarchuk

Abstract

Glycans play many important roles in bacterial biology and the complexity of the glycan structures requires biochemical assays in place to help characterize the biosynthetic pathways. Our focus has been on the use of enzymes from pathogens which make molecular mimics of host glycans. We have been examining glycosyltransferases that make strategic linkages in biologically active glycans which can be also exploited for potential therapeutic glycoconjugate synthesis. This chapter will provide details on assays for a variety of bacterial glycosyltransferases that we and others have used for the characterization of pathogen glycoconjugate biosynthetic pathways, and for the in vitro synthesis of human-like glycans produced by bacterial pathogens. The methods presented here should enable other assays to be developed for new pathway characterization.

Key words Lipopolysaccharide, Sialyltransferase, BODIPY, Enzyme assay, Synthetic acceptor

1 Introduction

Bacterial surface carbohydrates are the first line molecules when interacting with hosts for colonization and or infection. It is not surprising that human pathogens have developed the ability to exploit these molecular recognition events to evade detection by the host defenses, and to use host receptors to dampen the immune response [1] or to aid in colonization [2]. The study of bacterial lipopolysaccharide and capsule biosynthesis has been explored for many bacterial species by both genetic and biochemical means. The development of biochemical tools has helped clarify biosynthetic pathways and has led to the discovery of glycosyltransferases with potential for synthesis of bioactive glycoconjugates. The species which have received the most attention are the human pathogens, where the structures being produced mimic those found in the host: (1) head groups of common glycolipids (e.g., lacto-*N*-neotetraose, and gangliosides [3–6]); (2) glycosaminoglycans

Inka Brockhausen (ed.), *Bacterial Polysaccharides: Methods and Protocols*, Methods in Molecular Biology, vol. 1954,
https://doi.org/10.1007/978-1-4939-9154-9_17,

(hyaluronic acid and unsulfated chondroitin [7]; (3) Lewis blood group antigens [8] and polysialic acid [9].

An excellent and very detailed review of glycosyltransferase structure and function was published by Lairson et al. which outlines many biochemical features of transferases from various sources [10]. Carbohydrate active enzymes have been classified into families in the sequence database for carbohydrate active enzymes, CAZy [11]. At present this database contains 105 families of glycosyltransferases, and 152 families of glycoside hydrolases obtained from the analysis of over 9200 bacterial genomes (as well as ~800 other organisms). Synthetic assay substrates have been used to characterize many bacterial glycosyltransferases from LPS core, LPS O-Chain, and capsule biosynthetic pathways [12–15]. Many of the synthetically useful bacterial glycosyltransferase enzymes are found in enzyme families having only bacterial members (e.g., CAZy GT-42 and GT-52), while a few are found in very large families which have representation from all branches of the tree of life (e.g., CAZy GT-2,GT-4, GT-8).

In this chapter, we outline steps to synthesize several fluorescent acceptor molecules to probe various enzyme activities from bacteria which make human-like glycan linkages. The general principals can be applied to a variety of structures and enzyme families provided a suitable sugar acceptor can be found as a starting material. The assays are sensitive enough to probe crude lysates of bacteria to see if the enzyme activity is present. The assay designs use commercial starting materials and can be adapted to produce a wide variety of assay substrates. We provided examples of simple substrate synthesis, or how sequential enzyme reactions can be used to broaden the utility of the substrates.

A number of pathogens make *N*-acetyllactosamine, which we have illustrated with the conversion of BODIPY-β-GlcNAc to BODIPY-LacNAc (Fig. 1) using the *Helicobacter* β-1,4-galactosyltransferase (HP0826). Once the fluorescent boron-dipyrromethene (BODIPY) dye is reacted with the sugar acceptor all of these transformations can be followed by thin layer chromatography. A facile reaction to reduce azides to amines is the Staudinger reaction using triphenylphosphine and water. This gentle reaction (Fig. 2) is easily performed in a biochemistry lab setting. *N*-Hydroxylsuccinimidyl (NHS) ester reaction uses the newly formed primary amine with an activated BODIPY dye under mild alkaline conditions to produce a stable amide bonded labeled product (Fig. 3). The lactose substrate is a convenient acceptor for most bacterial sialyltransferases; this is illustrated with the conversion of BODIPY-lactose (BDP-Lac) to BODIPY-α2,3-sialyl lactose (BODIPY-GM3) (Fig. 4) using the *Campylobacter jejuni* α2,3-sialyltransferase (CST-I). The BDP-GM3 substrate can be used for other enzyme reactions, illustrated with the conversion of BODIPY-GM3 to BODIPY-α2,8-α2,3-sialyl lactose

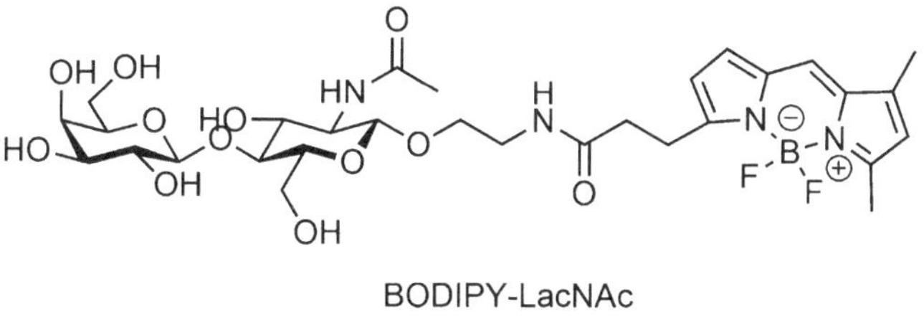

Fig. 1 Chemical structure of BODIPY-LacNAc (BDP-LacNAc). This acceptor can be further used to assay for GT-52 sialyltransferases from *Neisseria*, as well as GT-42 sialyltransferases from *Helicobacter* species

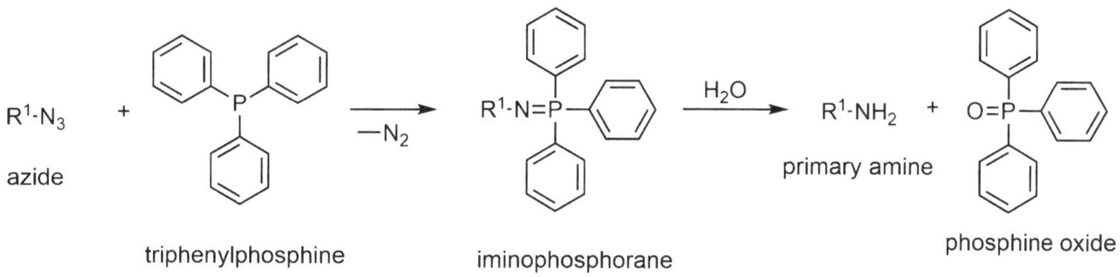

Fig. 2 Staudinger ligation reaction. The carbohydrates used were R^1 = carbohydrate (i.e., azidoethyl-lactose, azidoethyl-β-D-GlcNAc, azidoethyl-β-D-GalNAc)

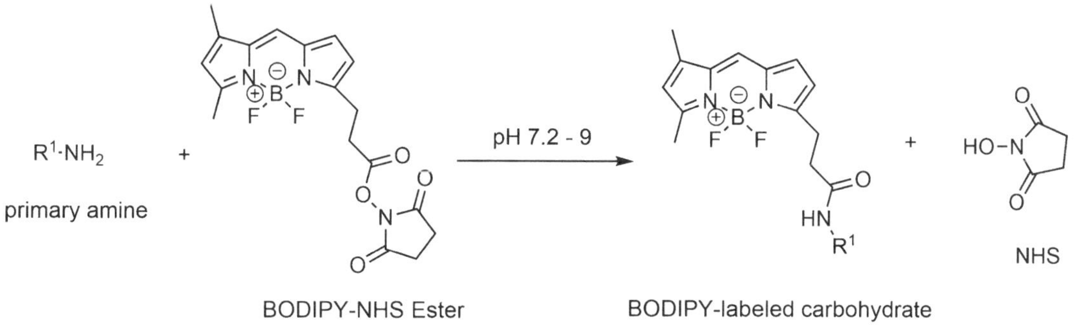

Fig. 3 *N*-Hydroxysuccinimidyl (NHS) ester reaction. R^1 = carbohydrate (i.e., aminoethyl-lactose/GlcNAc/GalNAc). The reaction is monitored by TLC

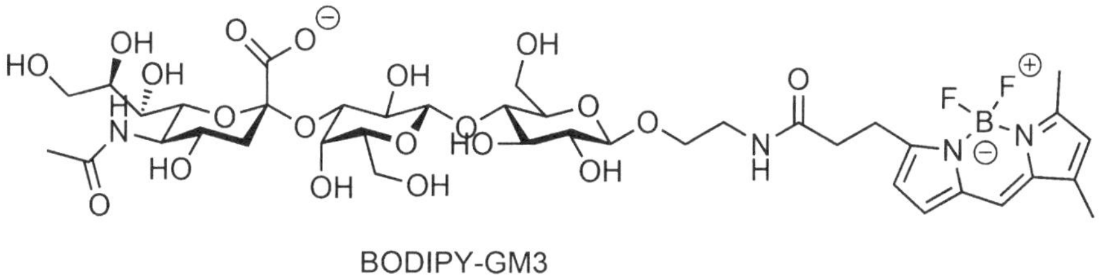

Fig. 4 Chemical structure of BODIPY-α2,3-sialyl lactose (BDP-GM3)

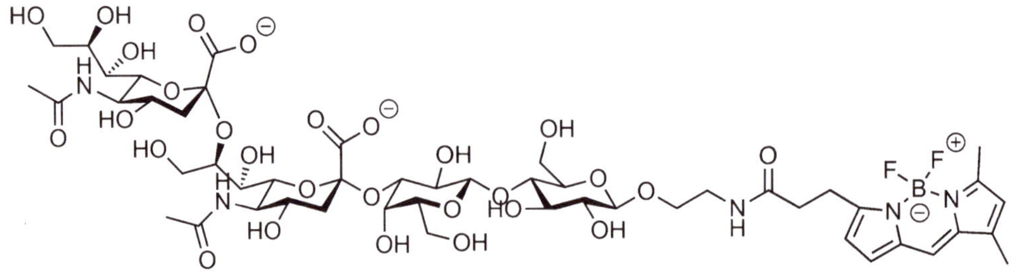

Fig. 5 Chemical structure of BODIPY-α2,8-α2,3-sialyl lactose (BDP-GD3)

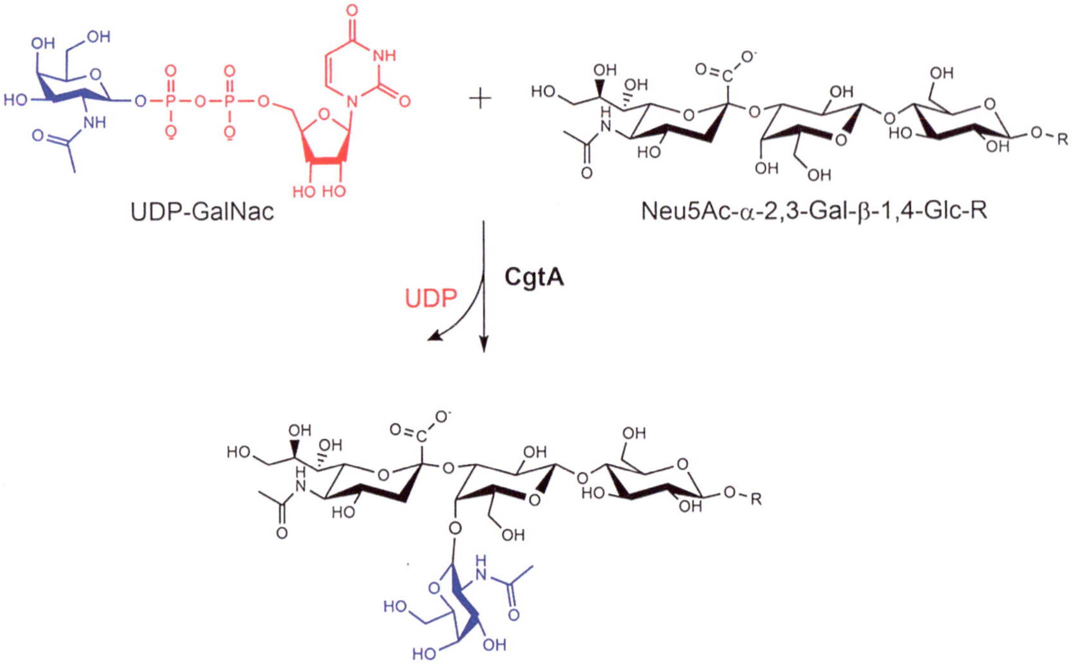

Fig. 6 The reaction scheme for CgtA addition of N-acetylgalactosamine to a GM3-like substrate

(BODIPY-GD3) (Fig. 5) using the *Campylobacter jejuni* α2,8-sialyltransferase (CST-II).

The glycosyltransferase β-1,4-*N*-acetylgalactosaminlytransferase (CgtA) transfers an *N*-acetylgalactosamine (GalNAc) molecule to the galactose (Gal) in the αNeu5Ac(2–3)βGal disaccharide to make a GM2 mimic (Fig. 6). Discovered originally by Gilbert et al. [16] to be one of the four enzymes responsible for the biosynthesis of the GT1a ganglioside mimic in the lipooligosaccharide of the gram negative bacterial pathogen *Campylobacter jejuni* OH4384, an engineered version of the enzyme can be produced in recombinant *E. coli*. CgtA can be used on small molecules, and glycolipids to build oligosaccharide chains, which mimic the gangliosides GM2 or GD2 [16].

Sialic acid terminates oligosaccharide chains on microbial cell surfaces, playing critical roles in host recognition and adherence. The enzymes that transfer the sialic acid moiety from cytidine-5′-monophospho-*N*-acetyl-neuraminic acid (CMP-NeuAc) to the terminal positions of these key glycoconjugates (Fig. 7) are known as sialyltransferases. The bacterial sialyltransferases are found in four distinct protein families, GT38/42/52/80,100. The GT42 enzymes CstI and CstII have been well studied and have both been examined by X-ray crystallography [17, 18]. Polysialic acid capsules are important virulence factors for a handful of pathogens.

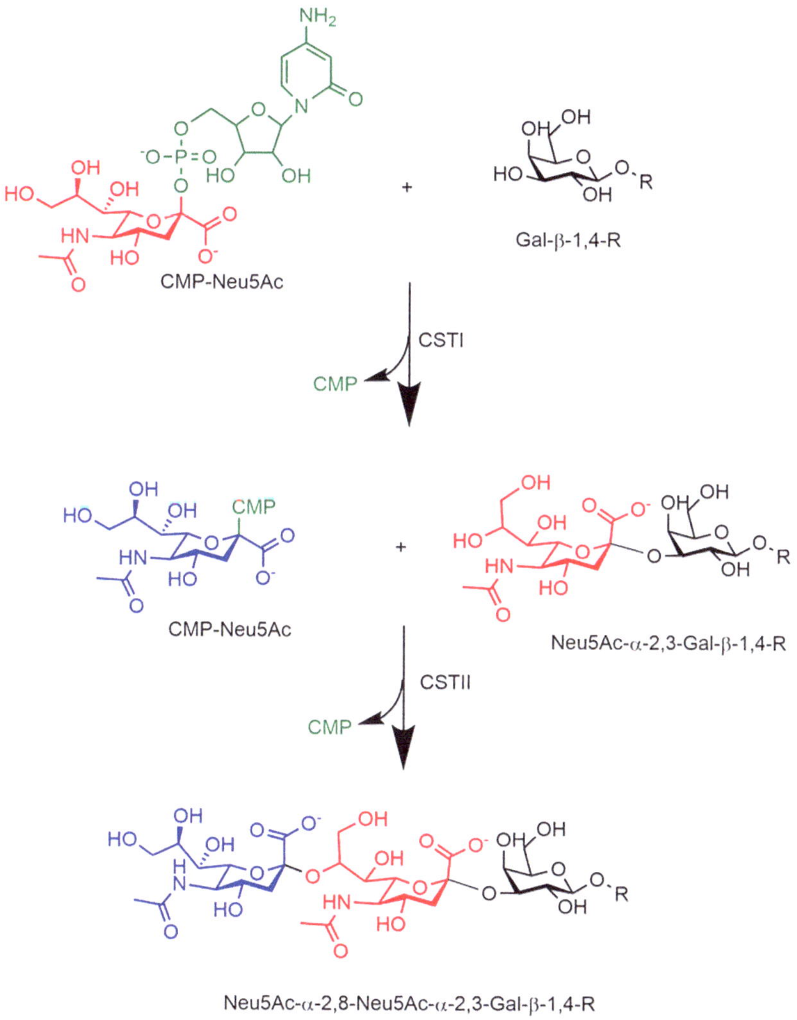

Fig. 7 The reaction scheme for Cst-II addition of sialic acid to a GM3-like substrate to generate GD3-mimic structures

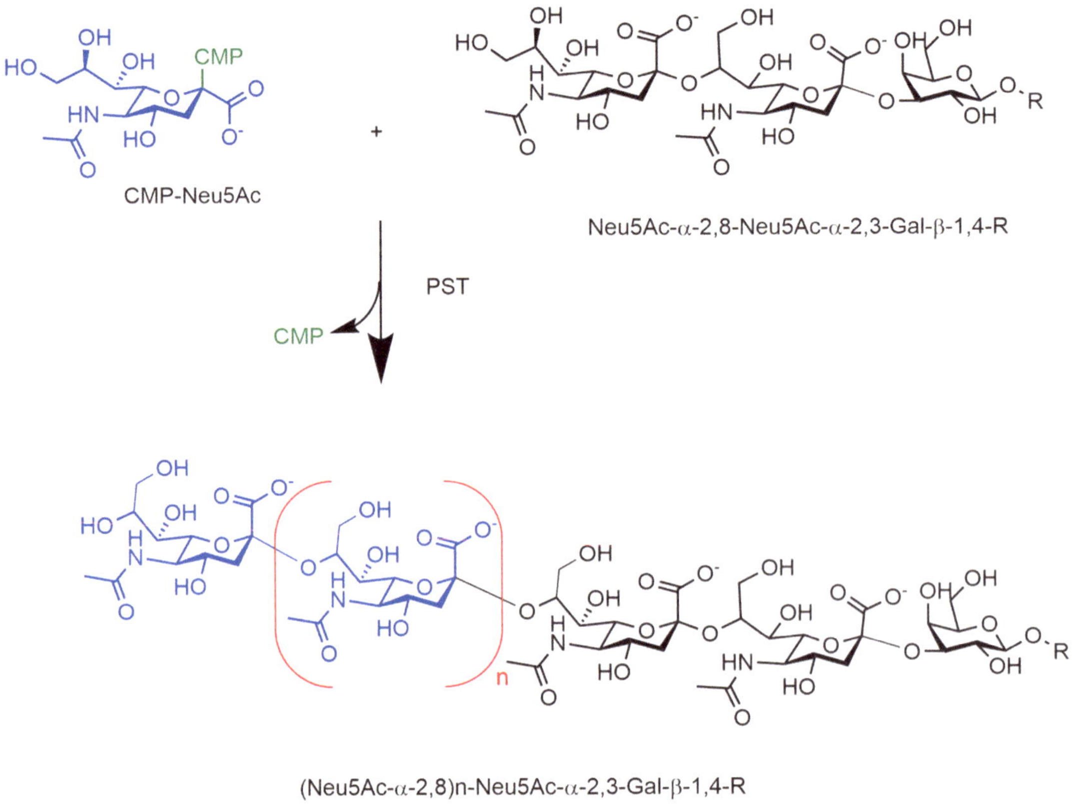

Fig. 8 The reaction scheme for polysialyltransferase (PST) reactions on GD3-mimic structures

This homopolymeric capsular polysaccharide has been studied in bacterial pathogens such as *Mannheimia haemolytica* serotype A2, *Neisseria meningitidis* group B/C, and *Escherichia coli* K1/K92 [19]. This modification is also found in the human host—but so far only on a small group of proteins. The role of this capsule in virulence is multifactorial, including resistance to antimicrobial peptides, evasion from phagocytosis, and escape from an endosome if engulfed [20]. These enzymes normally transfer to a glycolipid acceptor which has a sialic acid/ketodeoxyoctulosonic acid terminus, which means that in vitro we need to have at least two sialic acids on the acceptor before the reaction will proceed [15] (Fig. 8).

2 Materials

2.1 BODIPY-Labeled Carbohydrates

1. 25 mM sodium borate, pH 8.5.
2. C-18 silica gel columns for concentration and desalting of products. These can be self-packed or purchased from a variety of sources (Sigma-Aldrich).

3. 200 mM and 500 mM HEPES, pH 7.5.
4. 100 mM $MgCl_2$.
5. 100 mM CMP-Neu5Ac: (dissolved in 200 mM HEPES).
6. 10 mM UDP-gal.
7. Azido-ethyl carbohydrates: lactose, β-D-*N*-acetylglucosamine (β-GlcNAc) or α-*N*-acetylgalactosamine (GalNAc) (Sussex Research Chemicals, Ottawa).
8. Triphenylphosphine.
9. Carbohydrate alkylamines.
10. Tetrahydrofuran (THF).
11. Dimethylformamide (DMF).

2.2 Glycosyltransferases

2.2.1 General Materials for Protein Expression

1. 2YT media or components thereof.
2. Baffled Erlenmeyer flask(s).
3. Isopropyl β-D-1-thiogalactopyranoside (IPTG) (Sigma).

2.2.2 General Materials for Enzyme Purification

1. DNase A (Roche).
2. Protease inhibitors (Sigma).
3. Lysozyme (BioShop).
4. Emulsiflex (Avestin).
5. Purification buffer(s), vacuum filter sterilized (see specific materials below).
6. Loose resin or prepacked columns (see specific materials below).

2.2.3 Enzyme-Specific Materials: CgtA

1. 50 mM HEPES, 150 NaCl, 10% glycerol, pH 7.2.

2.2.4 Enzyme-Specific Materials: MBP-CstI

1. *Buffer A*: 20 mM HEPES, 200 mM NaCl, 1 mM EDTA, pH 7.2.
2. *Buffer B*: 20 mM HEPES, 200 mM NaCl, 1 mM EDTA, 10 mM maltose, pH 7.2.
3. *Dialysis Buffer*: 20 mM HEPES, 100 mM NaCl, 10% glycerol, pH 7.2.
4. Amylose/Dextrose prepacked column for MBP-Trap purification (GE Healthcare).

2.2.5 Enzyme-Specific Materials: CstII

1. *Buffer A*: 20 mM Tris, pH 8.3.
2. *Buffer B*: 20 mM Tris, 1 M NaCl, pH 8.3.
3. *Buffer C*: 20 mM Tris, 500 mM NaCl, pH 8.3

4. Q Sepharose prepacked column for anion exchange chromatography (GE Healthcare).
5. S-200 Sephacryl HR gel column for size exclusion chromatography (GE Healthcare).

2.2.6 Enzyme-Specific Materials: MBP-PolyST

1. *Buffer A*: PBS buffer, pH 7.4.
2. *Buffer B*: PBS buffer, 2 M NaCl, pH 7.4.
3. Heparin Sepharose prepacked column for affinity chromatography (GE Healthcare).

2.3 Small Molecule Activity Tests and Determination of Specific Activity

All enzyme activity tests and assays require five key components: acceptor substrate, activated donor, metal cofactor, enzyme, and buffer. All reactions are carried out in 50 mM HEPES Buffer, pH 7.2. Certain enzymes require additives, such as detergents, for optimization. Table 1 summarizes reaction components for specific enzymes.

1. Chemical solvents (Caledon, all HPLC grade).
2. Acetonitrile (ACN).
3. Ethyl acetate (EtOAc).
4. Methanol (MeOH).
5. Acetic acid (HOAc).
6. Precoated silica, plastic backed analytical TLC sheets (Millipore Sigma).

2.4 Analysis by TLC or HPLC

1. Plastic-backed silica TLC plate.
2. TLC running solvent: 4:2:1:0.1 or 3:2:1:0.1 (ethyl acetate–methanol–water–acetic acid).
3. Proteomix WAX-NP3 column (Sepax).
4. 10 mM NH_4HCO_3, 20% methanol.
5. 500 mM NH_4HCO_3, 20% methanol.

Table 1
Enzyme reaction components

Enzyme	Substrate	Activated donor	Metal cofactor	Additional components
CgtA	BDP-GM3	UDP-GalNAc	$MnCl_2$	Dithiothreitol (DTT) *Optional: Methanol*[a]
MBP-CstI CstII MBP-PST	BDP-Lac BDP-GM3 BDP-GD3	CMP-Neu5Ac	$MgCl_2$	Alkaline Phosphatase (Sigma)

[a]Methanol can be used in this reaction (up to 20% volume) in case of a substrate with low solubility

6. Dionex HPLC system: DNAPac PA-100 guard column (Dionex).
7. Acetonitrile.
8. 2 M ammonium acetate pH 7.0.

3 Methods

3.1 BODIPY-Labeled Carbohydrates

3.1.1 Purification Techniques for BODIPY-Labeled carbohydrates Using C-18 Silica Columns

1. Wash 2 × 1 mL C-18 columns by running 20 mL of methanol (MeOH) and 20 mL of water through the columns, successively.
2. Slowly load reaction mixture onto the C-18 column. BODIPY-glycoside should be retained on the column. Wash this column with 10 mL of water.
3. Elute BODIPY-glycoside with approximately 4 mL of MeOH.

3.1.2 Purification Using Preparative Glass-Backed TLC Plates

1. Reduce the total volume of the BODIPY-labeled mixture to approximately 80 μL using a vacuum concentrator.
2. The concentrated BODIPY-labeled mixture is then layered onto a preparative glass-backed TLC plate (20 cm × 20 cm × 1 mm thick) and allowed to completely dry (*see* **Note 1**).
3. Develop TLC plate using the corresponding solvent mix (Table 2).
4. Allow the developed TLC plate to completely dry in the dark at room temperature (RT) for 10–24 h.
5. Once TLC plate is completely dry, scrape off the BODIPY-labeled band of interest (the mobility of products is slightly different based on it being a monosaccharide or disaccharide) carefully with a Scoopula and transfer into a 50 mL conical tube (Table 3).
6. Wash the silica with 6 × 14 mL of warm water (40 °C) or until it is completely clear. Centrifuge at 1751 × *g* at RT for 5 min

Table 2
TLC solvent mix ratios used for each BODIPY-labeled substrate

Substrate	TLC solvent mix (EtOAc–MeOH–H_2O–HOAc) ratios
BODIPY-Lactose, BODIPY-GalNAc, and BODIPY-GlcNAc	4:2:1:0.1
BODIPY-GM3, BODIPY-GD3, and BODIPY-GT3	3:2:1:0.2
BODIPY-LacNAc	7:2:1:0.1

Table 3
R_f values of BODIPY-labeled carbohydrates in different solvents

	R_f values in TLC solvent mix: (EtOAc–MeOH–H_2O–HOAc)		
Substrate	**4:2:1:0.1 ratio**	**3:2:1:0.2 ratio**	**7:2:1:0.1 ratio**
BDP-Lactose	0.53	0.42	–
BDP-GlcNAc	0.61	–	0.56
BDP-GalNAc	0.63	–	–
BDP-GM3	–	0.33	–
BDP-GD3	–	0.16	–
BDP-GT3	–	0.07	–
BDP-LacNAc	–	–	0.39

and gently decant the supernatant without including silica. Keep the supernatant as it contains your sample.

7. Purify supernatant by following the method described in the Subheading 3.1.1.
8. Determine the absorbance at 504 nm (ABS_{504nm}). Use MeOH to dilute sample and as a blank. Using ABS_{504nm}, calculate the final concentration of the sample (Extinction coefficient = 80,000 cm^{-1} M^{-1}).
9. Completely dry down sample using a vacuum concentrator; store the final product at −20 °C.

3.1.3 Reduction of the Azide to an Amine and Labeling the Amine with BODIPY-NHS

1. In a microcentrifuge tube, add 10 mg of azido-ethyl carbohydrate (lactose, β-GlcNAc or GalNAc) and dissolve in 500 μL of tetrahydrofuran (THF) and sonicate for 15 min to help dissolve mixture.
2. Add 10 mg of triphenylphosphine into the mixture and sonicate for 1 h. Add 50 μL of water to the mixture to completely dissolve everything.
3. Leave reaction mixture in the dark constantly rotating for 24 h at RT.
4. Manual reverse phase purification of reaction mixture: Wash 2 × 1 mL C-18 silica columns by running 20 mL of methanol (MeOH) and 20 mL of water through the columns, successively. Slowly load reaction mixture onto the C-18 column and then elute with water. The flow-through contains the primary amine product. Collect four 500 μL water fractions in microcentrifuge tubes. The primary amine should be in the flow-through, and some of the early water wash fractions.

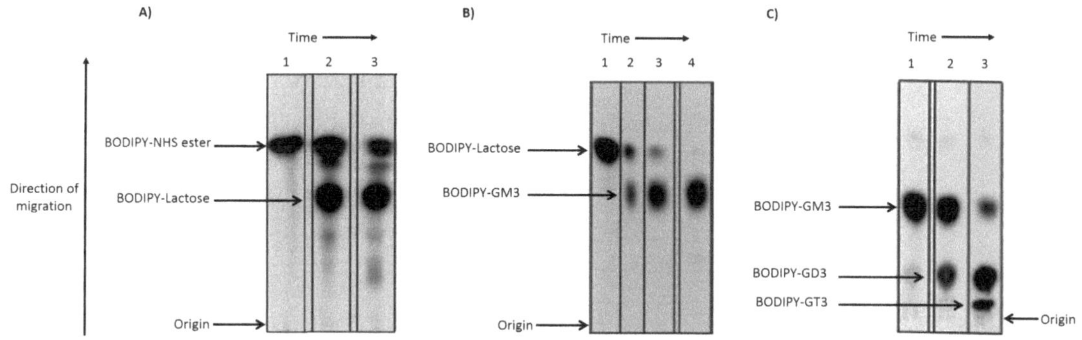

Fig. 9 TLC analysis of various reactions with BDP-glycosides. (**a**) BDP-Lac synthesis from the aminoethyl-lactose made by a Staudinger reaction. Lane 1: starting material, BODIPY-NHS; Lanes 2–3, NHS reaction. During the reaction progress, aminoethyl-lactose is labeled with BODIPY. Lanes 2–3 time points: 55 min and 100 min (**b**) BDP-GM3 preparative synthesis from BDP-Lac. During the reaction progress, a sialic acid moiety is transferred onto the BDP-Lac resulting in the gradual accumulation of BDP-GM3. Lanes 1–4 time points: 0 min, 15 min, 35 min, and 80 min. (**c**) BDP-GD3 preparative synthesis from BDP-GM3: During the reaction progress, a sialic acid moiety is transferred onto the BDP-GM3 resulting in the gradual accumulation of BDP-GD3 (target substrate) and BDP-GT3. Lanes 1–3 time points: 5 min, 15 min, and 50 min

5. Using a vacuum concentrator, the fractions are taken to dryness (*see* **Note 2**).
6. Pool the dried down fractions of the carbohydrate alkylamine by resuspension in 500 μL of sodium borate (25 mM, pH 8.5) and 220 μL of dimethylformamide (DMF).
7. In a separate microcentrifuge tube dissolve 5 mg of BODIPY-NHS ester in 300 μL of DMF.
8. Mix components from **items 6** and 7. In a single microcentrifuge tube and incubate at RT while constantly rotating the mixture for 1–2 h.
9. Monitor the progress of the reaction using thin layer chromatography on plastic-backed TLC plates (silica 60) and develop using the solvent mix (Table 2, Fig. 9). The product will have a lower mobility than the free BODIPY compound. Refer to Table 3 for approximate RF values of BODIPY-β-Lactose, BODIPY-α-GalNAc and BODIPY-β-GlcNAc.
10. Purify BODIPY-labeled carbohydrates following the method described in Subheading 3.1.2, **steps7–9**.

3.1.4 BODIPY-α2,3-Sialyl Lactose (BODIPY-GM3)

1. For a 1 mL reaction, mix the following final concentrations in a microcentrifuge tube: 2 mM BODIPY-lactose; 50 mM HEPES, pH 7.5; 10 mM $MgCl_2$; 3 mM CMP-Neu5Ac; 0.067 mg/mL CSTI enzyme; and water to make up the final 1 mL volume (*see* **Note 3**).
2. Incubate reaction mixture at 37 °C.

3. Monitor the progress of the reaction using plastic-backed TLC plates and develop using the solvent mix (Table 3). Usually after 2 h, reaction is ~98% complete (*see* **Note 4**).
4. Apply reaction mixture to Sep-Pak columns by following the method described in Subheading 3.1.1. BDP-GM3 will elute with MeOH.
5. Purify BDP-labeled carbohydrates following the method described in Subheading 3.1.2.

3.1.5 BODIPY-α2,8-α2,3-Sialyl Lactose (BDP-GD3)

1. For a 1 mL reaction, mix ingredients to achieve the following final concentrations in a microcentrifuge tube: 2 mM BDP-GM3, 50 mM HEPES, pH 7.5; 10 mM $MgCl_2$; 2 mM CMP-Neu5Ac; 0.1 mg/mL CST-II enzyme; and water to make up the final 1 mL volume (*see* **Note 5**).
2. Follow **steps 2–5**. From Subheading 3.1.4 (BODIPY-GM3).

3.1.6 BODIPY-N-Acetyllactosamine (BDP-LacNAc)

1. For a 1 mL reaction, mix ingredients to achieve the following final concentrations in a microcentrifuge tube: 0.5 mM BDP-β-GlcNAc, 2 mM UDP-Gal, 50 mM HEPES, pH 7.5; 2 mM $MgCl_2$; 0.75 mg/mL β-1,4-galactosyltransferase (HP0826) crude lysate [21]; and water to make up the final 1 mL volume (*see* **Note 6**).
2. Incubate reaction mixture at 30 °C.
3. Monitor the progress of the reaction using plastic-backed TLC plates and develop using the solvent mix; EtOAc–MeOH–-H_2O–HOAc (7:2:1:0.1). Usually after 1.5 h, reaction is ~100% complete. Refer to Table 3 for approximate R_f values.
4. Purify reaction mixture using C-18 silica columns following the method described in Subheading 3.1.1.
5. Determine the final concentration and store the purified sample following the method described in **steps 7–9**. Of Subheading 3.1.2.

3.2 Glycosyltransferases

All enzymes listed are produced recombinantly in *E. coli* cells. Plasmid constructs and preferred cell type/strain for production can be found in published literature [17, 19, 22].

3.2.1 Production of Recombinant Glycosyltransferases

1. Using a liquid starter culture, inoculate 250 mL sterile 2YT media in a 1-L baffled Erlenmeyer flask to an OD_{600} of 0.01.
2. Grow cells at 37 °C until the culture reaches an OD_{600} between 0.3 and 0.6 or is entering exponential phase of growth.
3. Induce the culture with IPTG to a final concentration of 0.5 mM.

4. Return flask to incubator and reduce the temperature accordingly: CgtA (25 °C), MBP-CstI (25 °C),CstII (30 °C), MBP-PST (20 °C).
5. Continue culture growth for 16–20 h.
6. Harvest cells by centrifugation at 5000 × *g* for 20 min at 4 °C, decant the supernatant and store the cell pellet at −20 °C until needed.

3.2.2 Purification of Recombinant Glycosyltransferases

Purification of the glycosyltransferases is completed efficiently to preserve the specific activity of the target enzyme. It is vital to do the lysis of the cells and subsequent purification on the same day, and to keep the samples on ice or cold during these processes. Purification buffers and resins are different for each enzyme; however, the general method of purification is similar. In the case of CgtA, this enzyme is used as a clarified lysate and has been found to be inactivated during purification.

1. Resuspend the cells in 15 mL/g of cells with appropriate buffer (*see* Subheadings 3.1.3–3.1.5). Commercial additives such as DNase, RNase A, protease inhibitors, and/or lysozyme can be added according to the supplier's directions. We do recommend protease inhibitors for bacterial extracts.
2. Lyse cells with an Emulsiflex at 15000 psi.
3. Centrifuge lysate at 17000 × *g* for 30 min at 4 °C.
4. Decant the supernatant and ultracentrifuge at 215619 × *g* for 1 h at 4 °C.
5. Filter supernatant through 0.2 μm syringe filter.

3.2.3 CgtA Purification

1. Aliquot clarified lysate and store at −80 °C, use as needed.

3.2.4 MBP-CstI Purification

1. Purify the enzyme on a 5-mL MBP-trap with a linear gradient of 0–100% *Buffer B* over 5 column volumes (CV).
2. Dialyze the peak fractions at 4 °C and concentrate as needed.
3. Aliquot and store the purified enzyme at −80 °C, use as needed.

3.2.5 CstII Purification

1. Purify using 5 mL Q Sepharose column (anion-exchange) with a linear gradient of 0–50% *Buffer B* over 5 CV.
2. Peak fractions (fractions showing activity) are combined and concentrated to <5 mL, then purified using size-exclusion chromatography on a 125 mL Sephacryl S-200 HR gel filtration column with a *Buffer C* isocratic gradient.
3. Peak fractions are collected and concentrated as needed.
4. Aliquot and store the purified enzyme −80 °C, use as needed.

3.2.6 MBP-PST Purification

1. Purify the enzyme on a 5 mL heparin Sepharose column with a linear gradient of 0–70% *Buffer B* over 5 CV.
2. Peak fractions (fractions showing activity) are combined and concentrated to a max of 2 mg/mL (if needed).
3. For long term storage add DTT and glycerol to a final concentration of 1 mM DTT and 10% glycerol.
4. Aliquot and store the purified enzyme −80 °C, use as needed.

3.3 Small Molecule Activity Tests and Determination of Specific Activity

Activity tests can be completed as needed throughout the lysis/purification protocols. It may be useful to first test the activity of crude lysates before purification, to ensure an active enzyme was produced. After purification activity tests can be completed on individual fractions, subsequently these "peak fractions" (fractions with high activity) can be combined to achieve a maximum specific activity of the enzyme prep. For specific activity determination, the concentration of the enzyme should be known.

3.3.1 Enzyme Activity Tests

All enzyme reactions have a similar setup (Table 4), using the components listed in Table 1. Read the enzyme specific notes (Table 5) to ensure the reaction is tailored to the target enzyme. Complete the following steps to confirm enzyme activity. A negative control tube should always be prepared with all components except the activated donor.

1. Add all components, except the enzyme, listed in Table 4 to a small Eppendorf tube.
2. If suggested, add additive components from Table 5.
3. Mix components well, spin tube at low speed to collect liquid.
4. Add suggested amount of enzyme (Table 4) to mixture and stir gently with pipette tip.
5. Place tube in incubator at 30 °C for 30 min (*see* **Note** 7).

Table 4
General set-up for test reactions

Reaction component	Final concentration
Substrate	0.1 mM to 2 mM
Activated donor	0.2–3 mM
Metal cofactor	10 mM
Buffer	50 mM
Enzyme (crude lysate or peak fractions)	2–5 μL
Sterile H_2O	Fill up to 20 μL

Table 5
Suggested additive components

Enzyme	Additive component	Final concentration
CgtA	DTT *Methanol (optional)*	1 mM *Up to 20%*
MBP-CstI CstII	Alkaline phosphatase	Package directions
MBP-PST	N/A	

6. Quench the reaction with a 1:1 volume ratio of 80% ACN. Spin tube at low speed to collect liquid.
7. Spot all reactions (including negative control) on TLC sheet and resolve in solvent containing 4:2:1:0.1 ratio of EtOAc–MeOH–H_2O–HOAc.
8. Image plate with UV or Blue light transilluminator.

3.3.2 Determination of Specific Activity

Specific activity is more accurately calculated with pure enzyme after the concentration of the enzyme is determined. It can be calculated crudely using the TLC plates and Image software such as Image Lab (Bio-Rad), which measures conversion rate by densitometry of the reactant and product "spots" (more qualitative measure) but is best calculated by HPLC on an appropriate column, which measures conversion rate by integration of the reactant and product peaks on a chromatogram (more quantitative measure). *For more information see section TLC and HPLC Analysis.* Reactions that have a conversion rate in the linear range for the enzyme, usually between 30 and 60%, give a more precise value for specific activity. As such two experiments should be conducted to determine the optimal variables for the calculation. First, the correct dilution or concentration of the enzyme should be defined (this is a value within the linear range of conversion). Second, a time course should be completed to achieve a linear range of conversion (0–100%) over a period of 0–180 min. Analysis of TLC plates or HPLC chromatograms will yield conversion rates (ideally <35%) which can substitute into the equation below to determine the specific activity of the enzyme. The average specific activities of CSTI, CSTII and MBP-PST are listed in Table 6 (*see* **Note 8**).

3.3.3 Equation to Determine Enzyme activity and Specific Activity of an Enzyme

$$\text{Enzyme activity (units/mL)} = \frac{(\%\ \text{conversion})(\text{mM substrate})(\mu L\ \text{reaction volume})}{(\text{min of reaction})(\mu L\ \text{enzyme})} \times 1000.$$

Table 6
Average specific activity values for each enzyme discussed

Enzyme	Avg. specific activity (units mg^{-1})
MBP-CstI	0.5
CstII	0.6
MBP-PST	0.3–0.7

Table 7
Enzyme dilution series for Experiment 1

Tube #	Enzyme dilution
1 (negative control)	0 (no enzyme)
2	1 in 40
3	1 in 20
4	1 in 10
5	1 in 5
6	1 in 2.5

$$\text{Specific activity (units/mg)} = \frac{\text{Enzyme activity } \left(\text{units mL}^{-1}\right)}{\text{Enzyme concentration } \left(\text{mg mL}^{-1}\right)}$$

3.4 Experiments

3.4.1 Experiment 1, Enzyme Dilution Series, Fixed Time Point

1. Follow **steps 1–8** from above (Enzyme Activity Tests), in 6 tubes using the 30 min fixed time point analysis. Reactions in each tube will contain a different dilution of enzyme (the negative control will be absent of enzyme), outlined in Table 7 below.
2. Conversion % of each tube can be determined by TLC or HPLC.
3. Plot the results, dilutions expressed as reciprocals on the *x* axis and conversion % on the *y* axis.
4. Choose a dilution within the linear range of the curve and determine the new concentration (mg mL^{-1}). If dilutions do not fall within a linear range, increase or decrease dilutions as needed.

3.4.2 Experiment 2: Linear Range of Conversion, Time Variable

1. Follow **steps 1–8** from above (Enzyme Activity Tests), in 7 tubes. Each tube will be quenched at a different time point (disregard the fixed time point in **step 5**). The negative control will be the time zero 0 time point. It is helpful to also prepare a donor-less negative control. See time points outlined in Table 8.
2. Substrate conversion % of each tube can be determined by TLC or HPLC.
3. Plot the results, time in minutes on the x axis and conversion % on the y axis (*see* **Notes 9** and **10**).
4. Choose a time point within the linear range of conversion and resulting in <35% conversion, substitute the parameters from this reaction into the equation to determine enzyme activity and specific activity using the equation above (*see* **Note 8**).

3.5 Analysis by TLC or HPLC

3.5.1 Analysis of Small Molecule tests Using TLC

1. Spot 0.5 μL of the stopped reaction onto the TLC plate.
2. Wait for spots to dry completely; a blow-dryer can be used to quickly dry spots.
3. Develop by placing the plate within a chamber with solvent covering the bottom. The solvent for separating BDP-Lac, BDP-GM3, BDP-GM2, BDP-GD3 and BDP-GT3 is 4:2:1:0.1 (ethyl acetate–methanol–water–acetic acid). The solvent for separating BDP-GD3 and BDP-Sia oligomers is 3:2:1:0.1 (ethyl acetate–methanol–water–acetic acid).
   Spots on TLC plates are visualized on a Bio-Rad EZDOC gel scanner, using either the Blue light or the UV transilluminator tray (*see* Figs. 10 and 11).

Table 8
Time point series for Experiment 2

Tube #	Time point (min)
1 (negative control)	0
2	5
3	10
4	20
5	40
6	80
7	160

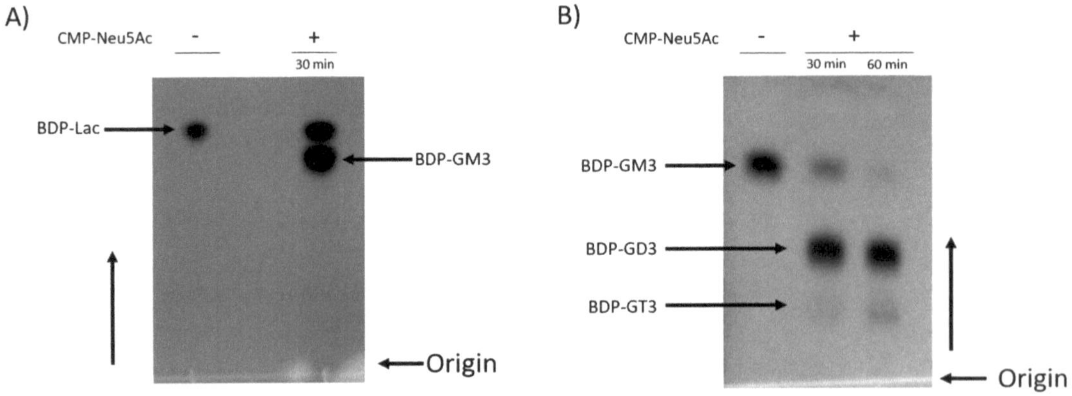

Fig. 10 Thin layer chromatography of a polysialyltransferase (PST) in vitro enzyme assay. A) Synthesis of BDP-GM3 with CstI; B) Synthesis of BDP-GD3 with CstII

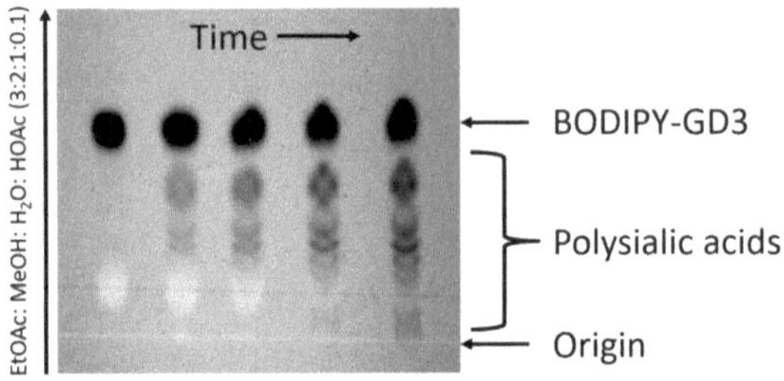

Fig. 11 Thin layer chromatography of a polysialyltransferase in vitro enzyme assay using BODIPY-GD3 as the substrate. An increase in chain length can be seen during the process of polysialylation over time

3.5.2 Analysis of BDP-Glycoside Assays Using HPLC with Fluorescence Detection

1. Use a Proteomix WAX-NP3 at 42 °C on an HPLC.
2. Monitor at excitation 504 nm and emission 514 nm (for BDP-labeled compounds).
3. Use the following buffers: A, 10 mM NH_4HCO_3 + 20% methanol; B, 500 mM NH_4HCO_3, 20% methanol.
4. Dilute samples with Buffer A until peaks are within range of the detector (5–15 pmol).
5. Products are separated by a gradient from 0–100% B at 0.3 mL/min over 20 min.

3.6 Analysis of Polysialylated Compounds Using HPLC with Fluorescence Detection

1. Dilute the reactions 1:100 with 12.5% acetonitrile.
2. Apply 10 μL of stopped reaction sample, corresponding to 25 pmol of acceptor, to a DNAPac PA-100 guard column (Strong anion exchanger) for separation.
3. Chromatography is conducted at 0.5 mL min^{-1} with a column temperature of 40 °C, with the mobile phases consisting of

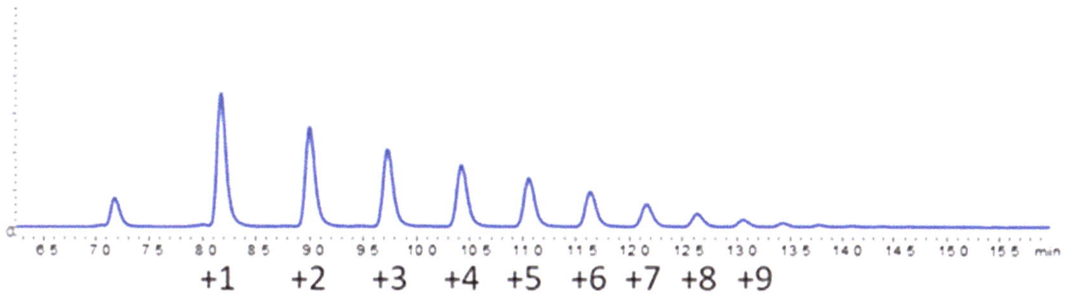

Fig. 12 HPLC strong anion exchanger chromatography of a polysialyltransferase in vitro enzyme assay. BDP-GD3 was modified with one or more additional sialic acids. An increase in chain length can be seen during the process of polysialylation, and the individual peaks are the addition of single monomers to the growing polymer

acetonitrile (M1) and 2 M ammonium acetate pH 7.0 (M2). Products are separated using an elution gradient from 0 to 50% M2 over 15 min (Fig. 12).

4 Notes

1. Make sure sample is completely dried on TLC before continuing, or else you will not get a clear separation.
2. Dried down fractions of the carbohydrate alkyl amine should be stored in −20 °C freezer if not in use.
3. BODIPY-β-lactose must be completely dried down before use in this reaction. A speed vacuum concentrator can be used to dry down the substrate.
4. Do not incubate this reaction at 37 °C for very long periods of time as the CST-I enzyme will convert GM3 back to lactose through the reverse reaction.
5. BODIPY-GM3 must be completely dried down before use in the reaction.
6. BDP-β-GlcNAc must be completely dried down before use in the reaction.
7. To visualize conversion rate over a period of time, 1–2 μL samples can be removed from tubes and quenched (e.g., 5, 10, 20, 40, 60 min).
8. Specific activity is usually not determined for the CgtA enzyme since it is used as a crude clarified lysate and the exact concentration of the enzyme in unknown.
9. This plot should have a linear range. If it is not linear please see **Note 10**; troubleshoot and then repeat the experiment.

10. Factors that affect linear range of conversion are (1) too low/too high or fluctuating substrate concentration; (2) too low/too high or fluctuating incubator temperature; (3) inconsistent reaction volume; (4) inconsistency between tubes (pipetting, timing, order of addition).

Acknowledgements

We thank Dr. Hong-Ming Chen, University of British Columbia, for the gift of BODIPY-NHS. We thank Sussex Research Chemicals, for the gift of azido-sugars. The work was supported by grants to WW from GlycoNet doi: https://doi.org/10.13039/501100009056, and BJ was supported by an Erwin Schrödinger Fellowship.

References

1. Carlin AF, Uchiyama S, Chang Y-C et al (2009) Molecular mimicry of host sialylated glycans allows a bacterial pathogen to engage neutrophil Siglec-9 and dampen the innate immune response. Blood 113:3333–3336
2. Harvey HA, Jennings MP, Campbell CA et al (2001) Receptor-mediated endocytosis of *Neisseria gonorrhoeae* into primary human urethral epithelial cells: the role of the asialoglycoprotein receptor. Mol Microbiol 42:659–672
3. Pavliak V, Brisson JR, Michon F et al (1993) Structure of the sialylated L3 lipopolysaccharide of *Neisseria meningitidis*. J Biol Chem 268:14146–14152
4. Penner JL, Aspinall GO (1997) Diversity of lipopolysaccharide structures in *Campylobacter jejuni*. J Infect Dis 176:S135–S138
5. Hood DW, Cox AD, Wakarchuk WW et al (2001) Genetic basis for expression of the major globotetraose-containing lipopolysaccharide from *H. influenzae* strain Rd (RM118). Glycobiology 11:957–967
6. Fox KL, Cox AD, Gilbert M et al (2006) Identification of a bifunctional lipopolysaccharide sialyltransferase in *Haemophilus influenzae*: incorporation of disialic acid. J Biol Chem 281:40024–40032
7. DeAngelis PL (2002) Microbial glycosaminoglycan glycosyltransferases. Glycobiology 12:9R–16R
8. Monteiro MA (2001) Helicobacter pylori: a wolf in sheep's clothing: the glycotype families of *Helicobacter pylori* lipopolysaccharides expressing histo-blood groups: structure, biosynthesis, and role in pathogenesis. Adv Carbohydr Chem Biochem 57:99–158
9. Bhattacharjee AK, Jennings HJ, Kenny CP et al (1975) Structural determination of the sialic acid polysaccharide antigens of *Neisseria meningitidis* serogroups B and C with carbon 13 nuclear magnetic resonance. JBiolChem 250:1926–1932
10. Lairson LL, Henrissat B, Davies GJ et al (2008) Glycosyltransferases: structures, functions, and mechanisms. Annu Rev Biochem 77:521–555
11. Cantarel BL, Coutinho PM, Rancurel C et al (2009) The Carbohydrate-Active EnZymes database (CAZy): an expert resource for glycogenomics. Nucleic Acids Res 37:D233–D238
12. Greenfield LK, Richards MR, Li J et al (2012) Biosynthesis of the polymannose lipopolysaccharide O-antigens from *Escherichia coli* serotypes O8 and O9a requires a unique combination of single- and multiple-active site mannosyltransferases. J Biol Chem 287:35078–35091
13. Kolkman MA, Wakarchuk W, Nuijten PJ et al (1997) Capsular polysaccharide synthesis in Streptococcus pneumoniae serotype 14: molecular analysis of the complete cps locus and identification of genes encoding glycosyltransferases required for the biosynthesis of the tetrasaccharide subunit. Mol Microbiol 26:197–208
14. Wakarchuk W, Martin A, Jennings MP et al (1996) Functional relationships of the genetic locus encoding the glycosyltransferase enzymes involved in expression of the lacto-N-neotetraose terminal lipopolysaccharide structure in

Neisseria meningitidis. J Biol Chem 271:19166–19173
15. Willis LM, Stupak J, Richards MR et al (2013) Conserved glycolipid termini in capsular polysaccharides synthesized by ATP-binding cassette transporter-dependent pathways in gram-negative pathogens. Proc Natl Acad Sci U S A 110:7868–7873
16. Gilbert M, Karwaski MF, Bernatchez S et al (2002) The genetic bases for the variation in the lipo-oligosaccharide of the mucosal pathogen, *Campylobacter jejuni*. Biosynthesis of sialylated ganglioside mimics in the core oligosaccharide. J Biol Chem 277:327–337
17. Chiu CP, Watts AG, Lairson LL et al (2004) Structural analysis of the sialyltransferase CstII from *Campylobacter jejuni* in complex with a substrate analog. Nat Struct Mol Biol 11:163–170
18. Chiu CP, Lairson LL, Gilbert M et al (2007) Structural analysis of the alpha-2,3-sialyltransferase Cst-I from *Campylobacter jejuni* in apo and substrate-analogue bound forms. Biochemistry 46:7196–7204
19. Lindhout T, Bainbridge CR, Costain WJ et al (2013) Biochemical characterization of a polysialyltransferase from *Mannheimia haemolytica* A2 and comparison to other bacterial polysialyltransferases. PLoS One 8:e69888
20. Willis LM, Whitfield C (2013) Structure, biosynthesis, and function of bacterial capsular polysaccharides synthesized by ABC transporter-dependent pathways. Carbohydr Res 378:35–44
21. Namdjou DJ, Chen HM, Vinogradov E et al (2008) A beta-1,4-galactosyltransferase from *Helicobacter pylori* is an efficient and versatile biocatalyst displaying a novel activity for thioglycoside synthesis. Chembiochem 9:1632–1640
22. Gilbert M, Brisson JR, Karwaski MF et al (2000) Biosynthesis of ganglioside mimics in *Campylobacter jejuni* OH4384. Identification of the glycosyltransferase genes, enzymatic synthesis of model compounds, and characterization of nanomole amounts by 600-mhz (1) h and(13)c NMR analysis. J Biol Chem 275 (6):3896–3906

Chapter 18

Glycosyltransferase Activity Assay Using Colorimetric Methods

M. Shafiqur Rahman and Wensheng Qin

Abstract

The glycosyltransferases (GTs) are an important subclass of enzymes that catalyze the biosynthesis of glycosidic bonds in oligosaccharides, polysaccharides and glycoconjugates by transferring a sugar residue from a donor substrate to an acceptor substrate. The membrane-associated GTs play a vital role in the biosynthesis of bacterial cell-wall polysaccharides. Characterization and quantification of GT activities is important for studies of biosynthesis of polysaccharides, drug target development, and production of bacterial products. In this chapter, colorimetric assays for the measurement of GT activities will be presented. Assays for GTs acting on monosaccharide-derivatives are based on the cleavage of unreacted glycosyl-*p*-nitrophenol acceptors followed by detection of *p*-nitrophenolate. GT reactions coupled with phosphatases and detection of inorganic phosphate are suitable for most GTs. These assays permit convenient quantification of GT activities and kinetics without the use of radioactive sugars.

Key words Bacterial glycosyltransferases, Substrate specificity, Polysaccharide glycosyltransferases, Colorimetric assay, Phosphate-coupled assay

1 Introduction

Microbial polysaccharides are extremely abundant and varied in structure. Monosaccharides and oligosaccharides in the form of glycoconjugates are widespread in microorganisms, and play substantial roles in cell wall structure, antigenicity, and resistance to antibiotics [1–4]. Glycosidic bonds are synthesized by a vast number of glycosyltransferases (GTs). GTs have been classified into more than 100 families in the Carbohydrate-Active Enzyme (CAZy) data base [5]. In addition, CAZy lists glycohydrolases that break down complex carbohydrates and glycoconjugates and also play important biological roles [6]. GTs are the enzymes responsible for the biosynthesis of simple and complex glycans by transferring a sugar residue from a specific phospho-activated sugar donor to an acceptor substrate [7–9]. Usually, GTs are highly specific for both, their donor and acceptor substrates. Depending

Inka Brockhausen (ed.), *Bacterial Polysaccharides: Methods and Protocols*, Methods in Molecular Biology, vol. 1954, https://doi.org/10.1007/978-1-4939-9154-9_18,

on the GT specificity, monosaccharides, disaccharides, and oligosaccharides, as well as proteins, lipids, DNAs, and numerous small molecules and their derivatives are acceptor substrates [9]. Consequently, GTs play a crucial role in biosynthesis pathways of oligosaccharides, polysaccharides and glycosylated proteins and lipids.

Notwithstanding their abundance and important functions, assaying GTs remains challenging for glycoscientists. The scientific hurdle has been the lack of availability of defined acceptor substrates and complex donor substrates required for bacterial GTs. Synthetic compounds help to overcome this hurdle (*see* Chapters 12, 13, 15, 17, 19). GTs can be evaluated by the transfer of a radiolabeled sugar from a donor to an acceptor substrate [10, 11] followed by the separation and quantification of radiolabeled reaction products using chromatography methods such as column or thin layer chromatography. Several nonradioactive methods have also been reported for assaying GTs [12–15].

In this chapter, we describe nonradioactive colorimetric methods for assaying GTs. The first colorimetric assay method is suitable for GTs that require only one sugar in the acceptor substrate. As an example, the glycosyl acceptor for certain Gal-transferases is GlcNAcβ-4-nitrophenol (GlcNAcβ-*p*NP), and for certain GlcNAc-transferases is Galβ-*p*NP. While the GT product is resistant to hydrolysis by exoglycosidases, the monosaccharide acceptor is hydrolyzed by a specific exoglycosidase to release *p*NP which is quantified by measuring its visible absorption. The enzymatic reactions for the GT reaction and release of *p*NP from the acceptor substrate are shown in Fig. 1. This GT activity assay is comparatively easy, simple, sensitive, and quantitative. However, the method is limited to enzymes that can utilize glycosyl-*p*NP as acceptor substrate such as GlcNAc-transferase LgtA and Gal-transferases LgtB and LgtC from *Neisseria meningitides* [14]. Other bacterial GTs may require more complex oligosaccharide acceptor substrates [10, 11].

A method that is broadly applicable to most GTs is a Phosphatase-coupled assay [15]. One of the GT reaction products is a nucleotide (e.g., UDP, GDP, or CMP). The terminal phosphate of this reaction product can be acted on by a phosphatase, thus releasing inorganic phosphate that can be quantified with a color reagent. The enzymatic reactions for releasing free phosphate are shown in Fig. 2. This method is unrelated to the structure of the acceptor substrate and thus is applicable to most GTs, including LgtA, LgtB and LgtC. However, it is recommended that purified enzyme preparations are used that do not have contaminating phosphatases.

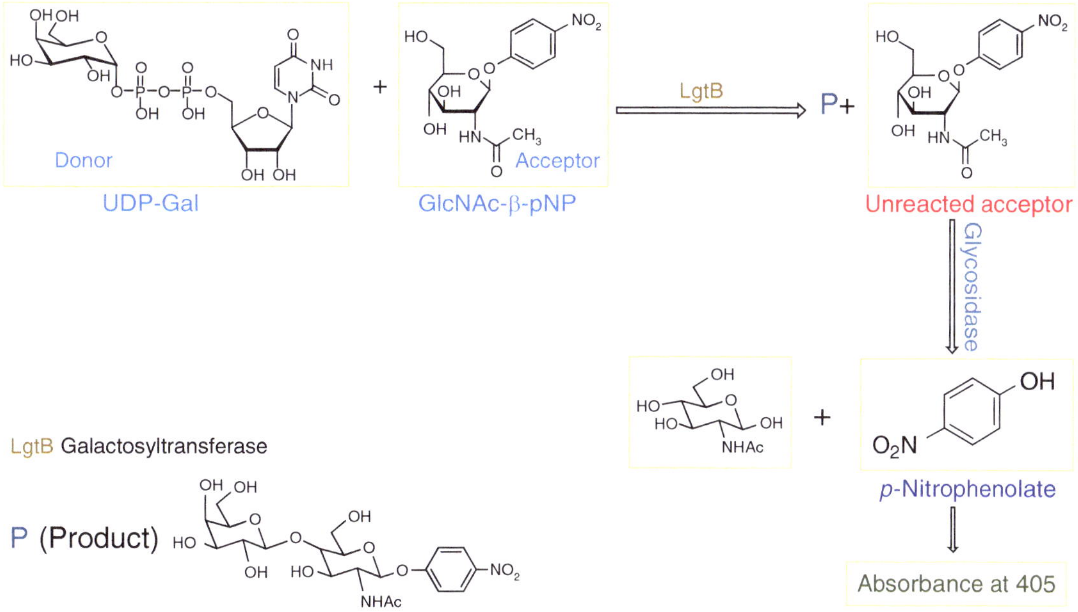

Fig. 1 Colorimetric screening method for glycosyltransferase activity. After transfer of the Gal residue to GlcNAcβ-*p*NP by Gal-transferase LgtB the unreacted acceptor is cleaved by exo-β-*N*-acetyl-hexosaminidase, releasing *p*-Nitrophenolate which is detected by measuring the change of visible absorption at 405 nm

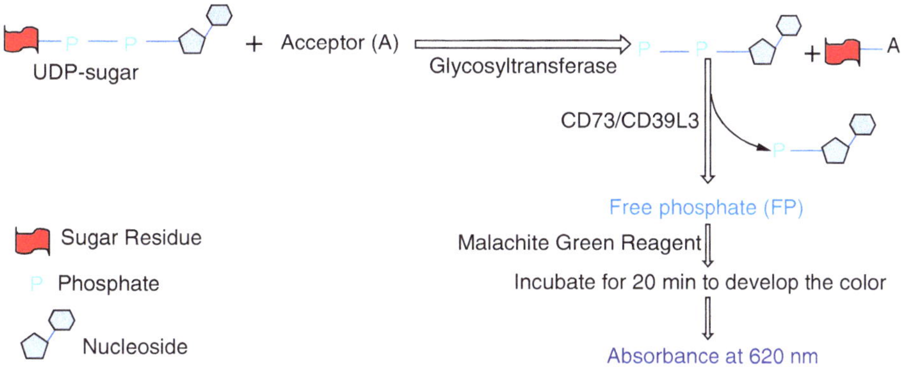

Fig. 2 Phosphatase-coupled screening for glycosyltransferase activity. This approach can be applied to any glycosyltransferase reaction with a removable phosphate-containing leaving group. In the glycosyltransferase reaction mixture, a phosphonucleotide reaction product is acted upon by a coupling phosphatase CD73 or CD39L3. The inorganic phosphate is detected using phosphate detection reagents. Quantification is done from a standard curve with various concentrations of phosphate

2 Materials

All the reagents and microbial cultures (if needed) are prepared under aseptic conditions (*see* **Note 1**). The reagents and buffers are prepared by using deionized, doubly distilled or milli-Q water at room temperature and the reagents are stored at room temperature (unless indicated otherwise). Always use analytical or reagent grade chemicals.

2.1 Colorimetric Assay for Gal-transferases (See Note 2)

1. 0.5 M 4-(2-hydroxyethyl)-1-piperazine-ethanesulfonic acid (Hepes) buffer: Dissolve 11.92 g Hepes in water, adjust to pH 7.2 and add water to make up to 100 mL.
2. 0.1 M 4-nitrophenyl-*N*-acetyl-β-D-glucosaminide (GlcNAc-β-*p*NP) solution (Sigma).
3. 10 mM uridine diphosphate galactose (Sigma).
4. 0.1 M $MnCl_2$.
5. As an example, recombinant Gal-transferase LgtB from *Neisseria meningitides* is expressed in *E. coli* BL21 and purified [14]. It is then dialyzed against 20 mM phosphate-buffered saline (PBS) pH 7.5 (*see* **Note 3**). The enzyme is stored in 50% glycerol in 10 mM PBS at −20 °C.
6. Bovine serum albumin (BSA).
7. 0.5 M Na_2CO_3.
8. Exo-*N*-acetylglucosaminidase (Sigma).
9. Microplate reader.

2.2 Phosphatase-Coupled Glycosyltransferase Assay

1. Reaction buffer: 0.1 M Tris–HCl, pH 7.5.
2. 0.15 M NaCl.
3. 0.1 M $MgCl_2$.
4. 0.1 M $MnCl_2$.
5. 10 mM UDP-Gal.
6. Acceptor substrate: 0.1 M GlcNAcβ–benzyl (Sigma).
7. Purified Gal-transferase LgtB [14] (*see* **Note 4**).
8. Coupling phosphatase (ecto-5′-nucleotidase CD73) solution: Dissolve 1.0 mg CD73 (Sigma) in 1.0 mL reaction mixture.
9. Malachite Green Phosphate Detection Kit (R&D Systems, Minneapolis, USA).
10. Microplate reader.

3 Methods

3.1 Colorimetric Assay for Gal-transferases

1. Prepare a reaction mixture on 96 well microplates in a total volume of 100 μL by adding 10 μL 0.5 M Hepes buffer (pH 7.2), 1 μL 0.1 M GlcNAc-β-*p*NP, 5 μL 0.1 M $MnCl_2$, 4 μL 10 mM UDP-Gal, 10 μL purified glycosyltransferase (in 10 mM PBS/50% glycerol), 10 μL BSA (0.1 mg/mL) (*see* **Note 5**) and 60 μL water. All assays are done in triplicate and the average values will be determined. Control assays lack the Gal-transferase but have 10 mM PBS/50% glycerol instead. Another set of control assays lack UDP-Gal.
2. Mix the solution gently.
3. Incubate at 37 °C for 30 min.
4. Add an excess (1 unit) of exo-N-acetylhexosaminidase to cleave unreacted GlcNAcβ-*p*NP and quench the reaction.
5. Incubate at 37 °C for 30 min.
6. Adjust the final pH to 9.6 by adding 0.5 M Na_2CO_3.
7. Monitor the *p*-nitrophenolate by measuring absorbance at 405 (A_{405}). The absorbance of the *p*-nitrophenolate is directly correlated to the amount of unreacted GlcNAcβ-*p*NP in the GT reaction mixture and to the GT activity. For comparison, a standard curve is made using different concentrations of *p*NP under the assay conditions but lacking enzymes and donor substrates. Subtract values from control assays and calculate activity (μmol × min^{-1} × mg $protein^{-1}$).

3.2 Phosphatase-Coupled Glycosyltransferase Assay

1. In a 96-well plate, prepare a reaction mixture in a total volume of 50 μL: Add 12 μL 0.1 M Tris–HCl, pH 7.5, 10 μL 150 mM NaCl, 2 μL 0.1 M $MgCl_2$, 2 μL 0.1 M $MnCl_2$, 2 μL 0.1 M GlcNAcβ-benzyl acceptor (*see* **Note 6**), 10 μL purified LgtB in 10 mM PBS/50% glycerol. Vortex the mixture gently. Initiate reaction with 5 μL 10 mM UDP-Gal and 0.1 μg coupling phosphatase CD73. Add water to make up to 50 μL (*see* **Note 7**).
2. Incubate for 30 min at 37 °C.
3. Quench the reaction by adding 30 μL of Malachite Reagent A and 100 μL of water to each well.
4. Add 30 μL of Malachite Green Reagent B to develop color, followed by gentle mixing.
5. Prepare control assays under the same conditions but without Gal-transferase.
6. Perform all measurements in triplicate.
7. Incubate at room temperature for 20 min.

8. Read the absorbance at 620 nm with a plate reader.
9. Average the values and subtract values from control assays.
10. Perform a phosphate standard curve to determine the conversion factor between the absorbance and the inorganic phosphate contents. The concentrations of phosphate groups in the assay mixtures are determined by the standard curve using different concentrations of phosphate. The GT activity is then determined (μmol phosphate produced x min^{-1} x mg GT $protein^{-1}$).

4 Notes

1. Aseptic technique is a set of principles and practices used for bacterial culture preparation and transferring culture to reduce undesirable microbes.
2. This protocol is presented to assay a Gal-transferase that acts on GlcNAcβ-*p*NP as acceptor substrate but other Gal-transferases may also be assayed with this method. The GT should be purified but bacterial homogenates may also have significant GT activity. Other glycosyl-*p*NP acceptors could be used (e.g., Galβ-*p*NP together with UDP-GlcNAc donor substrate to assay GlcNAc-transferase LgtA from *Neisseria meningitides*) [14]. In this case, an exo-β-galactosidase is used to liberate *p*NP.
3. Other GTs can be expressed in BL21 cells, but the purification procedure may be adjusted according to the affinity tag they carry. It is important that they can utilize Glycosyl-*p*NP as an acceptor substrate. The assays are not suitable for GTs that require a glycosyl-diphosphate-lipid or complex oligosaccharide as acceptors.
4. To obtain kinetic parameters for UDP-Gal the concentration should be varied from 0.005 to 1 mM and a fixed saturating concentration of GlcNAcβ-*p*NP (2 mM).
5. BSA is added to stabilize highly purified GTs.
6. Many other acceptors are appropriate, depending on the specificity of the GT, such as GlcNAcβ-*p*NP or more complex oligosaccharide derivatives.
7. This method is a general method for GTs and a variety of donor substrates. Thus, GT enzymes and their donor and acceptor substrates could be changed.

Acknowledgments

This work was supported by Natural Sciences and Engineering Research Council of Canada (RGPIN-2017-05366) to W. Q.

References

1. Rademacher TW, Parekh RB, Dwek RA (1988) Glycobiology. Annu Rev Biochem 57:785–838
2. Varki A (1993) Biological roles of oligosaccharides: All of the theories are correct. Glycobiology 3:97–130
3. Tzeng YL, Datta A, Kolli VK et al (2002) Endotoxin of *Neisseria meningitidis* composed only of intact lipid A: Inactivation of the meningococcal 3-deoxy-D-manno-octulosonic acid transferase. J Bacteriol 184:2379–2388
4. Miyake K, Iijima S (2004) Bacterial capsular polysaccharide and sugar transferases. Adv Biochem Eng Biotechnol 90:89–111
5. Cantarel BL, Coutinho PM, Rancurel C et al (2009) The Carbohydrate-Active EnZymes database (CAZy): an expert resource for Glycogenomics. Nucleic Acids Res 37: D233–D238
6. Yip VL, Withers SG (2006) Breakdown of oligosaccharides by the process of elimination. Curr Opin Chem Biol 10:147–155
7. Brockhausen I (2014) Crossroads between bacterial and mammalian glycosyltransferases. Front Immunol 5(492):1–21
8. Gilbert HJ, Davies G, Henrissat B et al (1999) Recent advances in carbohydrate bioengineering Cambridge. The Royal Society of Chemistry, Cambridge, pp 3–12
9. Lairson LL, Henrissat B, Davies GJ et al (2008) Glycosyltransferases: Structures, functions, and mechanisms. Annu Rev Biochem 77:521–555
10. Wang S, Hao Y, Lam JS et al (2015) Biosynthesis of the Common Polysaccharide Antigen of *Pseudomonas aeruginosa* PAO1: Characterization and role of WbpZ – a GDP-D-rhamnose: GlcNAc/GalNAc-diphosphate-lipid α1,3-D-rhamnosyltransferase WbpZ. J Bacteriol 197:2012–2019
11. Wang S, Czuchry D, Liu B et al (2014) Characterization of two UDP-Gal: GalNAc-diphosphate-lipid β1,3-galactosyltransferases WbwC from *Escherichia coli* serotypes O104 and O5. J Bacteriol 196:3122–3133
12. Hang HC, Yu C, Pratt MR et al (2004) Probing glycosyltransferase activities with the Staudinger ligation. J Am Chem Soc 126:6–7
13. Wongkongkatep J, Miyahara Y, Ojida A et al (2006) Label-free, real-time glycosyltransferase assay based on a fluorescent artificial chemosensor. Angew Chem Int Ed Engl 45:665–668
14. Shen S, Wang S, Ma X et al (2010) An easy colorimetric assay for glycosyltransferases. Biochemist 75(7):944–950
15. Wu ZL, Ethen CM, Prather B et al (2011) Universal phosphatase-coupled glycosyltransferase assay. Glycobiology 21(6):727–733

Chapter 19

In Vitro Characterization of a Multidomain Glycosyltransferase Using Fluorescently Tagged Synthetic Acceptors

Danielle M. Williams, Olga G. Ovchinnikova, and Chris Whitfield

Abstract

In vitro assays using fluorescently tagged sugar residues can facilitate the characterization of glycosyltransferase function. Here we describe the use of in vitro assays to characterize the three glycosyltransferase modules of the protein designated WbbB from *Klebsiella pneumoniae* O12. This protein combines key activities necessary to synthesize the O antigenic polysaccharide portion of lipopolysaccharide. The specificities of the three glycosyltransferases were investigated in vitro, using purified proteins, the activated donor sugars (dTDP-Rha, UDP-GlcNAc and CMP-β-Kdo) and synthetic acceptors terminating in either α1,3-linked Rha or β1,4-linked GlcNAc. The reaction products were verified by mass spectrometry and nuclear magnetic resonance methods.

Key words Microbial glycobiology, Lipopolysaccharide, O antigen, Glycosyltransferase, In vitro biochemical reaction, Fluorescently tagged synthetic acceptors

1 Introduction

In vitro assays have become a well-established method to characterize enzyme function and are commonly used in the fields of molecular biology and biochemistry. To investigate some biological systems, natural or synthetic acceptors are common components of in vitro reactions and have allowed for a more diverse investigation of enzyme catalysis. A more recent advance is the use of synthetic acceptor molecules attached to a fluorophore [1]. The fluorophore can be easily visualized under UV light which simplifies detection of reaction progress following thin layer or other chromatography separations [1–3]. Further structural analysis of the reaction products can then be carried out using mass spectrometry and nuclear magnetic resonance techniques to determine enzyme function [2, 3]. Here, we demonstrate the use of these techniques to characterize the polymerase domain of the multidomain glycosyltransferase, WbbB, from *Klebsiella pneumoniae* O12 [3].

Inka Brockhausen (ed.), *Bacterial Polysaccharides: Methods and Protocols*, Methods in Molecular Biology, vol. 1954, https://doi.org/10.1007/978-1-4939-9154-9_19,

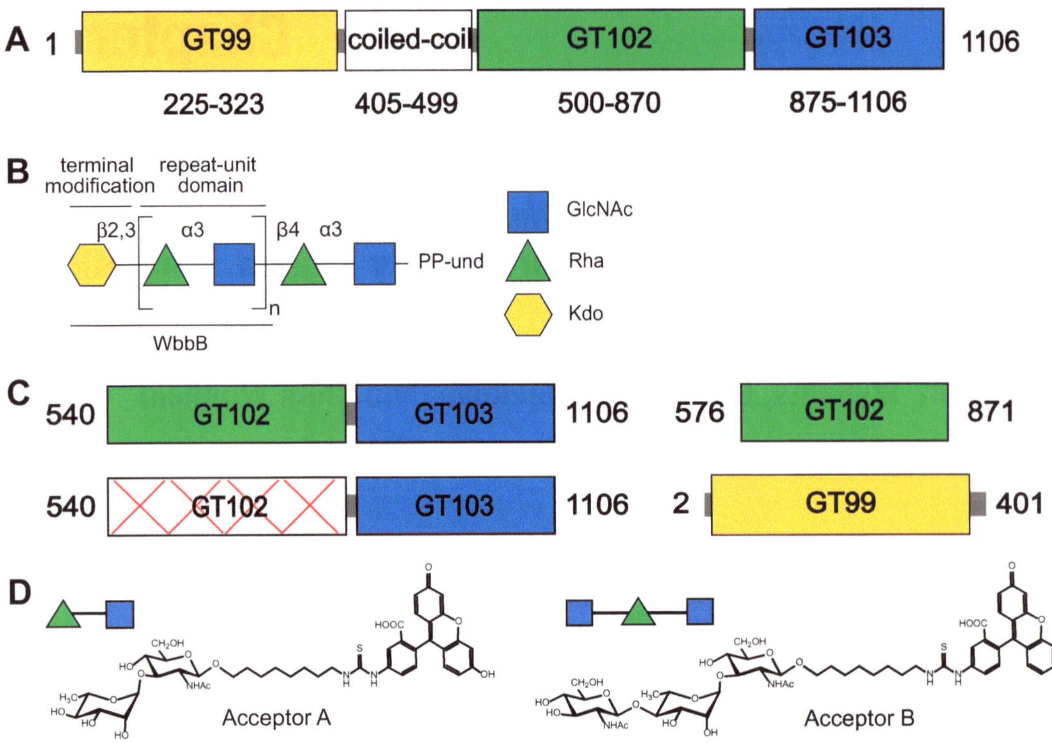

Fig. 1 Structure and biochemical characterization of WbbB in *Klebsiella pneumoniae* 012 0 antigenic polysaccharide. (**a**) Predicted location of WbbB glycosyltransferase modules. (**b**) Structure of the *K. pneumoniae* 012 OPS. (**c**) WbbB constructs used to biochemically characterize WbbB glycosyltransferase modules. (**d**) Structures of the synthetic acceptors. This research was originally published in Proceeding of the National Academy of Sciences of the United States of America [3]

O antigenic polysaccharide (OPS) is an essential component of lipopolysaccharide (LPS), an outer membrane glycolipid present in most Gram-negative bacteria [4, 5]. OPS consists of long chains of repeating sugar units which often provide resistance to opsonophagocytosis and complement-mediated killing [6, 7]. WbbB, a protein found in *Klebsiella pneumoniae* O12 and *Raoultella terrigena* 33257, contains three separate glycosyltransferase modules which combine several activities necessary for biosynthesis of a shared OPS structure (Fig. 1a, b) [3]. The O12 OPS repeat unit is a disaccharide of N-acetylglucosamine (GlcNAc) and rhamnose (Rha) [8], which is synthesized by the polymerase domain of WbbB [3] which consists of modules belonging to the GT102 and GT103 families in the CAZy database [9]. The OPS chain is then terminated with a β-linked 3-deoxy-D-*manno*-oct-2-ulosonic acid (Kdo) residue (Fig. 1b), which prevents further chain elongation and serves as a signal for export of the complete OPS prior to its transfer to the lipid A-core portion of LPS and translocation to the cell surface [4, 5]. This terminal β-Kdo is added by the GT99 module. The functions of the three GT modules were established using in vitro assays requiring purified proteins containing different active modules alone or in combination (Fig. 1c), activated precursors (dTDP-Rha, UDP-GlcNAc, and CMP-β-Kdo) and synthetic acceptors representing short fragments of the backbone structure terminating in either α1,3-linked Rha (acceptor A) or β1,4-linked GlcNAc (acceptor B) (Fig. 1d). A single polypeptide ($WbbB^{540-1106}$) which included the GT102 and GT103 modules was confirmed as the polymerase; this construct generated native OPS polymer by extending either of the synthetic acceptors. Specificities of the GT102 and GT103 enzymes were determined using the same in vitro approach. The GT102 ($WbbB^{576-871}$) module possessed rhamnosyltransferase activity and catalyzed the transfer of a single Rha residue to acceptor B. The GT103 module was inactive in the absence of the GT102 domain, so a construct containing GT103 and a catalytically inactive version of GT102 ($WbbB^{R738A/R739A/K740A}$) was constructed. The GT103 module in the mutant protein was confirmed as the N-acetylglucosaminyltransferase since it was able to add a single GlcNAc residue to acceptor A [3]. A construct containing the isolated GT99 ($WbbB^{2-401}$) was confirmed as the β-Kdo transferase and was able to add a single β-Kdo residue to acceptor A recapitulating the authentic terminus of the natural OPS product [2]. Together these experiments confirm that WbbB possessed three novel GT modules which perform all the functions necessary for OPS chain elongation and termination.

2 Materials

Prepare all solutions using ultrapure water and analytical grade reagents. Prepare and store all reagents at room temperature, unless otherwise indicated. Diligently follow all waste disposal regulations when disposing of waste materials.

2.1 WbbB Terminator Domain Storage Buffer

1. 50 mM Tris–HCl, 250 mM NaCl, pH 7.4: add 6.06 g of Tris base and 14.61 g of NaCl. Add water to a volume of 900 mL. Mix and adjust pH using concentrated HCl. Bring to a volume of 1 L with water (*see* **Note 1**).

2.2 KdsB Storage Buffer

1. 50 mM Tris–HCl, 250 mM NaCl, pH 8: add 6.06 g of Tris base and 14.61 g of NaCl. Add water to a volume of 900 mL. Mix and adjust pH using concentrated HCl. Bring to a volume of 1 L with water (*see* **Notes 1** and **2**).

2.3 WbbB Polymerase Domain Storage Buffer

1. 0.5 M Na_2HPO_4 (dibasic anhydrous): add 70.9 g Na_2HPO_4 to water and bring to a volume of 1 L.

2. 1 M NaH_2PO_4 (monobasic): add 119.9 g of NaH_2PO_4 to water and bring to a volume of 1 L.
3. 0.1 M sodium phosphate Buffer, 0.25 M NaCl, pH 7.4: combine 154.8 mL of 0.5 M Na_2HPO_4 with 22.6 mL of 1 M NaH_2PO_4 and bring to 500 mL with water. Add 14.6 g of NaCl, mix and bring to 1 L with water. Confirm the buffer has the desired pH. (*see* **Note 1**).

2.4 In Vitro Reaction Components

1. 1 M HEPES, pH 7.5 Stock: add 23.83 g HEPES to water and bring to a volume of 100 mL. Bring to desired pH using concentrated NaOH.
2. 0.5 M HEPES Buffer pH 7.5: mix 5 mL of water with 5 mL of 1 M HEPES. Store at −20 °C.
3. 100 mM $MgCl_2$: add 0.95 g $MgCl_2$ to water and bring to a volume of 100 mL. Store 1 mL aliquot of this buffer at −20 °C for use in in vitro assay.
4. 40 mM Kdo: dissolve 5 mg Kdo in 524.79 μL of water. Store in aliquots at −20 °C.
5. 100 mM CTP: add 161.6 mg CTP to water and bring to a volume of 5 mL with water. Store in aliquots at −20 °C.
6. Synthetic Acceptor: 3.17 mM stock (*see* **Notes 3–5**).
7. dTDP-L-Rha: 27 mM stock (*see* **Note 6**).
8. 50 mM UDP-GlcNAc stock: add 0.033 g UDP-GlcNAc sodium salt to 1 mL of water. Store at −20 °C (*see* **Note 7**).

2.5 Thin-Layer Chromatography (TLC)

1. TLC Solvent for WbbB Terminator Domain: In a fume hood, measure out and mix thoroughly chloroform, methanol, water, and acetic acid in a ratio of 25:15:4:2, respectively. Can be stored in a tightly sealed container in the fume hood.
2. TLC Solvent for WbbB Polymerase Domain: In a fume hood mix ethyl acetate, 1-butanol, water, and acetic acid in a ratio of 5:4:4:2.5, respectively. Can be stored in a tightly sealed container in the fume hood.
3. Aluminum Foil Silica Gel 60 F_{254} TLC plate.

3 Methods

3.1 In Vitro Reactions for WbbB Terminator Domain (Small Scale)

1. Combine components of reaction into PCR tube on ice to a final volume of 19 μL. Components include 50 mM HEPES, 10 mM $MgCl_2$, 2 mM Kdo, 5 mM CTP, 0.1 mM acceptor, and 10 μg of WbbB.
2. Finally, add 1 μg of KdsB to a final reaction volume of 20 μL which will initiate CMP-Kdo synthesis. Immediately incubate

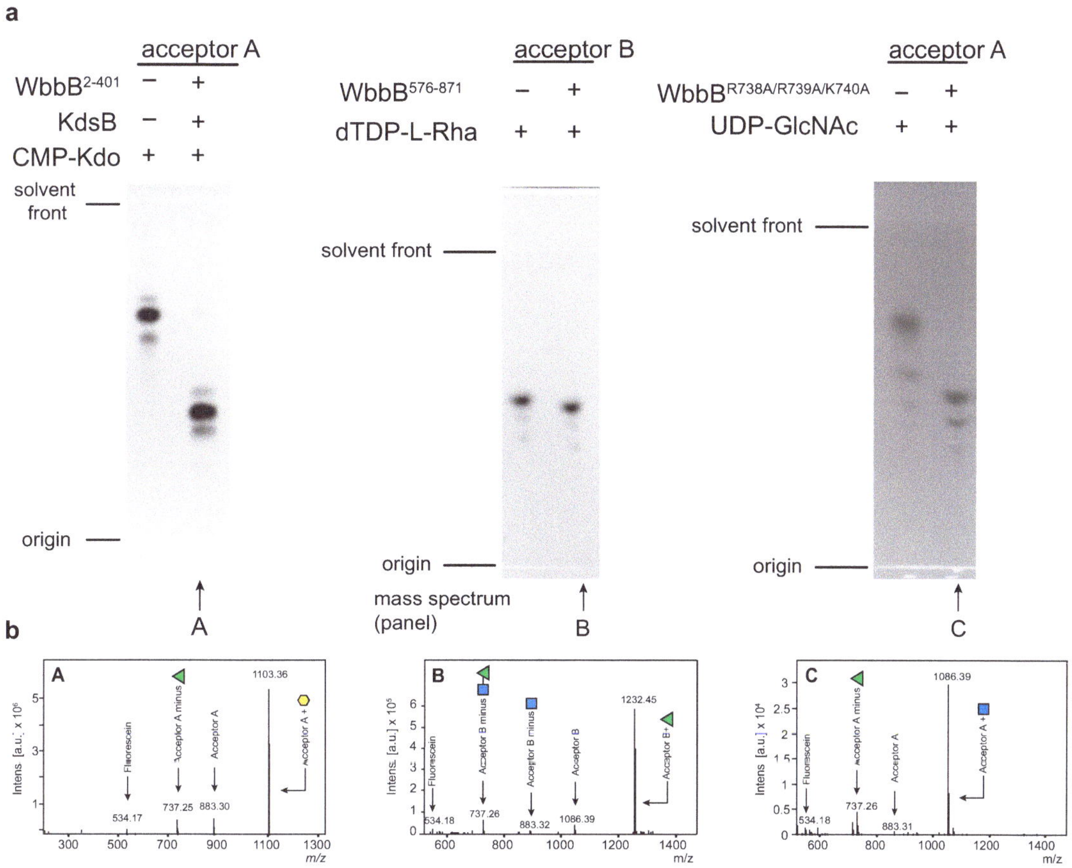

Fig. 2 In vitro activity of the individual glycosyltransferase modules of WbbB. (**a**) TLC analysis of the reaction mixtures obtained after incubating the indicated enzyme with the appropriate acceptor and donor sugar. (**b**) ESI mass spectrum of the indicated reaction mixture from panel A. This research was originally published in Proceeding of the National Academy of Sciences of the United States of America [2, 3]

reaction at 37 °C for 30 min in thermal cycler. Set the lid temperature to 50 °C to prevent condensation.

3. Follow TLC analysis and mass spectrometry protocols listed below to examine reaction progress (Fig. 2).

3.2 In Vitro Reactions for WbbB Polymerase Domain or Individual Modules (Small Scale)

1. Combine components of reaction into PCR tube on ice to a final volume of 20 μL. Components include 0.2 mM synthetic acceptor, 50 mM of HEPES buffer, 5 mM dTDP-Rha, and 5 mM UDP-GlcNAc. Finally add 10 μg of WbbB to begin the reaction.
2. Incubate PCR tube at 37 °C for 30 min in a thermal cycler. Set the lid temperature to 50 °C to prevent condensation.
3. Follow TLC analysis and mass spectrometry protocols listed below to examine reaction progress (Figs. 2 and 3).

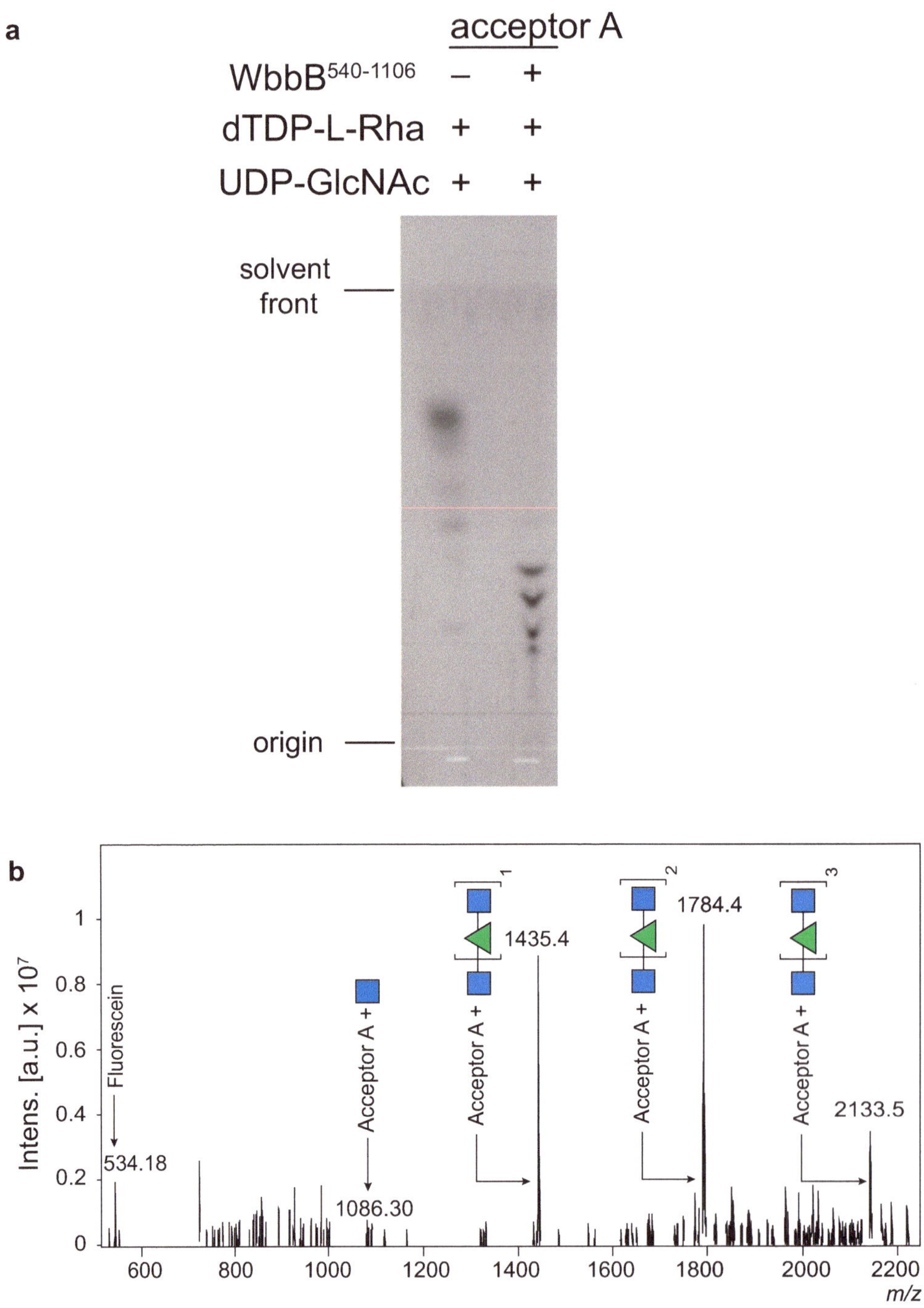

Fig. 3 In vitro activity of the polymerase domain of WbbB. (**a**) TLC analysis of the reaction mixtures obtained after incubating WbbB$^{540-1106}$ with acceptor A and both donor sugars. (**b**) ESI mass spectrum of the indicated reaction mixture from panel A. This research was originally published in Proceeding of the National Academy of Sciences of the United States of America [3]

3.3 In Vitro Reactions for WbbB Terminator Domain (Large Scale)

1. Combined components of reaction into Eppendorf tube on ice to a final volume of 96 μL. Components include 50 mM HEPES, 10 mM $MgCl_2$, 2 mM Kdo, 5 mM CTP, 0.34 mM acceptor A, and 8 μg of WbbB. (*see* **Note 8**).
2. Finally, add 4 μg of KdsB to a final reaction volume of 100 μL which will initiate CMP-Kdo synthesis. Immediately incubate reaction mixture at 37 °C for 30 min.
3. Follow TLC analysis, mass spectrometry and nuclear magnetic resonance spectroscopy protocols listed below to determine the structure of reaction products.

3.4 In Vitro Reactions for WbbB Polymerase Domain (Large Scale)

1. Combine components of reaction into Eppendorf tube on ice to a final volume of 500 μL. Components include 0.5 mM synthetic acceptor, 50 mM of HEPES buffer, 5 mM dTDP-Rha, and 5 mM UDP-GlcNAc. Finally add 100 μg of WbbB to begin the reaction.
2. Incubate Eppendorf tube at 37 °C for 1 h.
3. Follow TLC analysis, mass spectrometry and nuclear magnetic resonance spectroscopy protocols listed below to determine the structure of reaction products.

3.5 TLC Analysis

1. Cut TLC plate to appropriate size based on the number of reactions (Fig. 2 in Chapter 12). Leave at least 0.5 cm between each sample and at each side. TLC plate should be at least 6 cm long and the samples should be spotted 1.5 cm above the base of the plate.
2. After 30 min incubation spot 1 μL of reaction mixture on TLC plate along with a synthetic acceptor standard.
3. Cover TLC plate and allow to dry for 10 to 15 min.
4. In fume hood pour appropriate TLC solvent into chamber until the bottom of the chamber is completely covered. Place lid on chamber and equilibrate for 10 to 15 min.
5. Place dry TLC plate into chamber and cover with the lid. Allow the solvent to run up the TLC until approximately 1 cm from the top of the plate.
6. Remove TLC plate from chamber and while covered allow to dry.
7. Image TLC plate using UV light.

3.6 Mass Spectrometry Sample Preparation

1. Centrifuge the samples at 13,000 × *g* for 5 min.
2. Take a 10 μL aliquot of the reaction mixture and place in a sample vial for MS analysis (*see* **Note 9**).

3.7 Nuclear Magnetic Resonance Spectroscopy Sample Preparation

1. For structural analysis by NMR reactions were scaled up to 500 μL volume to generate enough product.
2. Purify the reaction mixture using a Sep-Pak C18 cartridge.
3. Condition the cartridge using 10 mL of 100% acetonitrile.
4. Equilibrate the cartridge with 20 mL of water.
5. Dilute the reaction mixture in 10 mL of water and load onto the Sep-Pak C18 cartridge.
6. Wash the Sep-Pak cartridge with 10 mL of water.
7. Elute the product in 6 mL of 60% (vol/vol) aqueous acetonitrile.
8. Dry the sample completely using a SpeedVac.
9. Resuspend sample in 99.9% D_2O.
10. Dry sample completely using SpeedVac.
11. Repeat **steps 9** and **10**.
12. Finally resuspend dried sample in 250 μL 99.96% D_2O (*see* **Note 10**).

4 Notes

1. To store WbbB protein long term (months to years) add glycerol to a final concentration of 10% and place at −80 °C.
2. The nucleotide sugar donor, CMP-β-Kdo, is inherently unstable and therefore must be synthesized in situ. This synthesis is accomplished via addition of CTP and Kdo along with a purified CMP-Kdo synthetase, KdsB, from *E. coli.*
3. Acceptors should be aliquoted into ~20 μL aliquots to limit freeze–thaw cycles.
4. When in use wrap synthetic acceptor aliquots in foil to protect from light.
5. Synthetic acceptor was synthesized in the Lab of Dr. Todd Lowary at the University of Alberta Department of Chemistry according to the protocol found in Ovchinnikova et al. [2] and Williams et al. [3].
6. dTDP-Rha synthesis is carried out following the procedure in Graninger et al. [10].
7. For in vitro reactions, protein, acceptor, and sugar nucleotide concentrations must be optimized based on enzyme activity.
8. For this upscale reaction nine 100 μL reactions are performed and then combined for structural analysis.

9. Mass Spectra are analyzed using Bruker Data Analysis 4.0 program.
10. NMR data are acquired and processed using the Bruker Top-Spin 2.1 program.

Acknowledgments

We thank Akihiko Koizumi, Iain L. Mainprize, Matthew S. Kimber, and Todd L. Lowary for their contributions to the projects originally published as refs. 2 and 3. This work was supported by the Natural Sciences and Engineering Research Council of Canada. C.W. acknowledges additional support from the Canada Research Chairs program.

References

1. Clarke BR, Richards MR, Greenfield LK et al (2011) *In vitro* reconstruction of the chain termination reaction in biosynthesis of the *Escherichia coli* O9a O-polysaccharide: the chain-length regulator, WbdD, catalyzes the addition of methyl phosphate to the non-reducing terminus of the growing glycan. J Biol Chem 286:41391–41401
2. Ovchinnikova OG, Mallette E, Koizumi A et al (2016) Bacterial β-Kdo glycosyltransferases represent a new glycosyltransferase family (GT99). Proc Natl Acad Sci U S A 113:3120–3129
3. Williams DM, Ovchinnikova OG, Koizumi A et al (2017) Single polysaccharide assembly protein that integrates polymerization, termination, and chain-length quality control. Proc Natl Acad Sci U S A 114:E1215–E1223
4. Raetz CRH, Whitfield C (2002) Lipopolysaccharide endotoxins. Annu Rev Biochem 71:635–700
5. Whitfield C, Trent MS (2014) Biosynthesis and export of bacterial lipopolysaccharides. Annu Rev Biochem 83:99–128
6. Cress BF, Englaender JA, He W et al (2014) Masquerading microbial pathogens: capsular polysaccharides mimic host-tissue molecules. FEMS Microbiol Rev 38:660–697
7. Lerouge I, Vanderleyden J (2002) O-antigen structural variation: mechanisms and possible roles in animal/plant-microbe interactions. FEMS Microbiol Rev 26:17–47
8. Vinogradov E, Frirdich E, MacLean LL et al (2002) Structures of lipopolysaccharides from *Klebsiella pneumoniae*. Eluicidation of the structure of the linkage region between core and polysaccharide O chain and identification of the residues at the non-reducing termini of the O chains. J Biol Chem 277:25070–25081
9. Lombard V, Golaconda Ramulu H, Drula E et al (2014) The carbohydrate-active enzymes database (CAZy) in 2013. Nucleic Acids Res 42:490–495
10. Graninger M, Nidetzky B, Heinrichs DE et al (1999) Characterization of dTDP-4-dehydrorhamnose 3,5-Epimerase and dTDP-4-dehydrorhamnose Reductase, Required for dTDP-L-rhamnose Biosynthesis in *Salmonella enterica* Serovar Typhimurium LT2*. J Biol Chem 274:25069–25077

Chapter 20

Glycosyltransferase-Coupled Assays for 4-Epimerase WbpP from *Pseudomonas aeruginosa*

Sulav Sharma, Carole Creuzenet, Kenneth F. Jarrell, and Inka Brockhausen

Abstract

The donor substrates for the biosynthesis of bacterial polysaccharides include UDP-Glc/Gal and UDP-GlcNAc/GalNAc. The conversion of these nucleotide sugars is catalyzed by 4-epimerases. The *wbpP* gene of *Pseudomonas aeruginosa* encodes a 4-epimerase that has a preference for UDP-GlcNAc/GalNAc as substrates. Other 4-epimerases have broad specificities or preference for UDP-Glc/Gal. We have developed coupled assays where the 4-epimerase product is used as a donor substrate for glycosyltransferases that are highly specific for the nucleotide sugar structure. We describe here a method for the study of substrate specificity of WbpP, using coupled assays employing four different glycosyltransferases. These protocols can be applied to the identification and characterization of novel 4-epimerases and to determine their substrate specificities.

Key words Glycosyltransferases, 4-Epimerase, WbpP, UDP-Glc, UDP-Gal, UDP-GlcNAc, UDP-GalNAc

1 Introduction

UDP-hexose/hexNAc 4-epimerases catalyze the interconversion of UDP-Glc to UDP-Gal and/or UDP-GlcNAc to UDP-GalNAc [1]. These enzymes can be classified into different groups based on their substrate specificities. The 4-epimerases belonging to Group 1 preferably act on UDP-Glc and UDP-Gal. Group 2 4-epimerases can accommodate UDP-GlcNAc and UDP-GalNAc as well as UDP-Glc and UDP-Gal while Group 3 4-epimerases show a strong preference toward UDP-GlcNAc and UDP-GalNAc [1, 2] (Table 1 [3–5]). The WbpP 4-epimerase from *P. aeruginosa* can be classified into Group 3. Another type of 4-epimerase belonging to Group 4 only acts on GlcNAc and GalNAc linked to undecaprenyl-pyrophosphate (PP-Und) but does not act on nucleotide sugars [6]. Furthermore, there are 4-epimerases that act on UDP-pentoses and UDP-sugar acids [7].

Inka Brockhausen (ed.), *Bacterial Polysaccharides: Methods and Protocols*, Methods in Molecular Biology, vol. 1954, https://doi.org/10.1007/978-1-4939-9154-9_20, © Springer Science+Business Media, LLC, part of Springer Nature 2019

Table 1
The different 4-epimerase groups

Group	Species (examples)	Abbreviated name	Preferred substrate	Reference
1	*E.coli*	eGalE	UDP-Gal/Glc	[3]
2	Human	hGalE	UDP-Gal/Glc,UDP-GalNAc/GlcNAc	[4]
3	*P. aeruginosa*	WbpP	UDP-GalNAc/GlcNAc	[2, 5]
4	*E.coli O157*	GnU	GlcNAc-PP-Und	[6]
5	*P. aeruginosa*	GnaB	UDP-GlcNAcA	[7]

WbpP in the opportunistic pathogen *P. aeruginosa* serotype O6 is essential for the production of lipopolysaccharides (LPS) capped with an O antigen called O-specific antigen (OSA) [2, 8]. The O6 OSA contains galacturonic acid derivatives that are thought to be derived by 4-epimerization of UDP-GlcNAc. The *wbpP* gene is found in the O6 antigen biosynthesis gene cluster, indicating its likely involvement in O6 OSA repeating unit synthesis. Although WbpP can accommodate UDP-Glc and UDP-Gal as substrates, it facilitates the interconversion of UDP-GlcNAc and UDP-GalNAc at a much higher catalytic efficiency [2]. A number of methods are available to measure the activity of 4-epimerases [9–13]. WbpP has been assayed spectrophotometrically by the *p*-dimethylaminobenzaldehyde reaction method, and also by separating the nucleotide sugars by capillary electrophoresis [2]. However, most of the known methods to assay 4-epimerases are limited toward the substrates they utilize or require expensive equipment that is not always available.

As with other 4-epimerases, the cofactor NAD^+ does not have to be added in the assay as it is tightly bound to the enzyme, and metal ion cofactors are not required. The reversible nature of 4-epimerase reactions makes it difficult to quantitatively assess the activity of these enzymes. We have developed 4-epimerase-glycosyltransferase (GT)-coupled assays that take advantage of the high specificity of GTs toward their nucleotide sugar donor substrates. If 4-epimerase and GT are assayed in the same reaction mixture, the 4-epimerase product is used as a donor substrate by the GT, thus preventing the reverse reaction of 4-epimerases and resulting in a high conversion of substrate to product. This method allows the study of 4-epimerase specificity and characterization of its properties. Radioactive assays provide high sensitivity and can be coupled with chromatographic methods to quantify the specific nucleotide sugars present in the assay mixtures.

In this protocol, we describe assays of 4-epimerase WbpP, using as substrate (a) UDP-Glc, coupled with bovine β1,4-Gal-transferase; (b) UDP-Gal, coupled with β1,3-Glc-transferase WbdN from *Escherichia coli* O157; (c) UDP-GlcNAc, coupled with human polypeptide GalNAc-transferase and (d) UDP-GalNAc, coupled with human GlcNAc-transferase I. These enzymes were chosen because they have high activities and have been characterized. The results from these assays confirmed that WbpP from *P. aeruginosa* is a Group 3 4-epimerase that prefers UDP-GlcNAc/GalNAc. The assays can be adapted to use other GTs to specifically identify the reaction products, according to the availability of GTs. The quantification of the exact amount of nucleotide sugar produced by the 4-epimerase is done by constructing a standard curve for the specific GT used [13]. The 4-epimerases can be identified by signature amino acid sequences GxxGxxG and YxxxK. However, biochemical assays are essential to determine 4-epimerase function. The described protocol is useful for studying differences in specificities among homologues of WbpP, or for the identification of novel 4-epimerases.

2 Materials

2.1 Expression and Purification of WbpP

1. *Escherichia coli* BL21 harboring a pET23 vector derivative containing *wbpP* encoding His-tagged WbpP [2] (*see* **Note 1**).
2. Luria-Bertani (LB) medium.
3. Ampicillin (Sigma).
4. Isopropyl-β-D-thiogalactoside (IPTG) (Sigma).
5. Incubator shaker.
6. Sterile flasks, pipette tips, and culture tubes.
7. Centrifuge.
8. Sonicator.
9. Buffer A: 5 mM imidazole, 20 mM Tris–HCl, pH 8, 0.1 M NaCl.
10. Wash Buffer B: 20 mM imidazole, 20 mM Tris–HCl, pH 8, 0.1 M NaCl.
11. Elution Buffer C: 1 M imidazole, 20 mM Tris–HCl, pH 8, 0.1 M NaCl.
12. Dialysis Buffer D: 50 mM Tris–HCl, pH 8.
13. Ni^{2+}-NTA Agarose (Qiagen).
14. Dialysis tubing (3500 Dalton cutoff) and clips.

2.2 WbpP-β1,4-Gal-Transferase (β4GalT)-Coupled Assay

1. Water bath set to 37 °C.
2. MES buffer: 0.5 M 2-(N-morpholino)ethanesulfonic acid (MES) buffer, pH 7.5.
3. UDP-[^{3}H]Glc (American Radiolabeled Chemicals): dry 10 μCi under nitrogen, add 1 mL 5 mM UDP-Glc (Abcam) to yield a concentration of 5 mM and 2 μCi/μmol, store at −20 °C.

4. UDP-[^{3}H]Gal (American Radiolabeled Chemicals), dry 10 μCi under nitrogen, add 0.5 mL 10 mM UDP-Gal (MP Biomedicals) to yield a concentration of 10 mM and 2 μCi/μmol, store at −20 °C.
5. Purified WbpP.
6. Bovine milk β1,4Gal-transferase (β4GalT) (Sigma): Add 1 mL water to 5 units, store solution in aliquots at −80 °C. The enzyme is also stable for several months at −20 °C (*see* **Note 2**).
7. Acceptor substrate for β4GalT: GlcNAcβ-Bn (Sigma) (*see* **Note 3**).
8. $MnCl_2$.
9. Dowex 1 × 8, chloride form, 200–400 mesh (Sigma) (*see* **Note 4**).
10. Scintillation counter (Beckmann).
11. Scintillation vials, 7 mL (Thermo-Fisher).
12. Scintillation Cocktail (MP Biomedicals).

2.3 WbpP-β1,3-Glc-Transferase (WbdN)-Coupled Assay

1. 0.5 M MES buffer, pH 7.5
2. UDP-[^{3}H]Gal: 10 mM, 2 μCi/μmol.
3. UDP-[^{3}H]Glc: 5 mM, 2 μCi/μmol.
4. Purified WbpP.
5. WbdN β1,3-Glc-transferase: Homogenates of BL21 expressing WbdN from *E.coli* O157 prepared by sonication of bacteria in 50 mM sucrose on ice [14] (*see* **Note 5**).
6. Acceptor substrate for WbdN: 2 mM GalNAcα–PO_3–PO_3–$(CH_2)_{11}$-O-Phenyl (GalNAc-PP-PhU) [15] (*see* **Note 6,** *see* Chapter 13).
7. 0.1 M $MnCl_2$.
8. Methanol.
9. Sep-Pak C18 cartridge and 3 mL syringe (Waters) (*see* **Note 7**).
10. Scintillation counter, 7 mL scintillation vials, scintillation cocktail.

2.4 WbpP-Polypeptide-GalNAc-Transferase (ppGalNAcT)-Coupled Assay

1. Water bath set to 37 °C.
2. 0.5 M MES buffer, pH 7.5.
3. 0.1 M $MnCl_2$.
4. UDP-[^{3}H]GlcNAc: 10 mM, 2 μCi/μmol.
5. UDP-[^{3}H]GalNAc: 10 mM, 2 μCi/μmol.
6. Purified WbpP.
7. Acceptor substrate for ppGalNAcT: 10 mM AQPTPPP (Queen's University) (*see* **Note 8**).

8. ppGalNAcT: purified bovine ppGalNAcT 1 [16] (*see* **Note 9**).
9. Dowex 1 × 8, chloride form, 200–400 mesh (Sigma) (*see* **Note 10**).
10. Scintillation counter, 7 mL scintillation vials, scintillation cocktail.

2.5 WbpP-GlcNAc-Transferase I (GlcNAcTI)-Coupled Assay

1. 0.5 M MES buffer, pH 7.5.
2. UDP-[^{3}H]GalNAc: 10 mM, 2 μCi/μmol.
3. UDP-[^{3}H]GlcNAc: 10 mM, 2 μCi/μmol.
4. Purified WbpP.
5. GlcNAcTI: insect cell supernatant containing the soluble form of GlcNAcT1 [17] (*see* **Note 11**).
6. Acceptor substrate for GlcNAcT1: Manα1-6(Manα1-3)Man-β-octyl (Toronto Research Chemicals) (*see* **Note 12**).
7. 0.1 M $MnCl_2$.
8. Dowex 1 × 8, chloride form, 200–400 mesh (Sigma).
9. Scintillation counter, 7 mL scintillation vials, scintillation cocktail.

3 Methods

3.1 Expression and Purification of WbpP

1. Prepare an overnight culture by transferring *E. coli* BL21 bacteria harboring the pET23 vector containing *wbpP* from a glycerol stock at −80 °C into a sterile culture tube containing 5 mL sterile LB broth and 5 μL of 100 mg/mL stock solution of ampicillin. Incubate the mixture in an incubator shaker set to 37 °C and shake at 200 rpm overnight (*see* **Note 1**).
2. Transfer the 5 mL overnight culture to an Erlenmeyer flask containing 100 mL sterile LB broth containing 100 μg/mL ampicillin. Incubate in the shaker at 30 °C until the optical density at 600 nm reaches 0.6.
3. Add 150 μL of 100 mM IPTG to the bacterial culture to obtain a final concentration of 0.15 mM. Incubate at 30 °C for 5–6 h to induce protein expression.
4. Harvest cells by centrifugation at 5000 × *g* for 15 min at 4 °C.
5. Resuspend bacterial pellet with 10 mL Buffer A.
6. Sonicate cells for 2 min with intervals of 5 s-on and 5 s-off on ice.
7. Centrifuge the sonicate at 13,000 × *g* for 15 min at 4 °C.
8. Load the supernatant onto a 3 mL Ni^{2+}-NTA Agarose column equilibrated with Buffer A.
9. Wash the column with 30 mL Buffer A and 15 mL Buffer B.

10. Elute bound proteins with 9 mL Buffer C, collect 3 mL fractions.
11. Analyze fractions by SDS-PAGE and pool fractions containing WbpP (38 kDa) [2].
12. Dialyze the pooled fractions against Buffer D at 4 °C overnight.
13. Store the purified enzyme in aliquots at −20 °C in 25% glycerol for use in the GT-coupled assays.

3.2 WbpP-β1,4-Gal-Transferase (β4GalT)-Coupled Assay to Test Specificity Toward UDP-Glc

All WbpP/GT-coupled assays are carried out in duplicates for each condition. Background control assays have water replacing WbpP or replacing acceptor substrate for the GT, or replacing both. Therefore, the minimum number of assay tubes is 8 (Table 2) (*see* **Note 13**). The coupled 4-epimerase WbpP-β4-Gal-transferase reactions are:

$$\text{UDP-}[^3\text{H}]\text{Glc} \xrightarrow{\textit{WbpP}} \text{UDP-}[^3\text{H}]\text{Gal} \xrightarrow{\textit{β4-Gal-transferase} + \text{GlcNAcβ-Bn}} [^3\text{H}]\text{Galβ1-4GlcNAcβ-Bn}$$

1. Prepare the WbpP assay mixtures for the 4-epimerase reaction in 20 μL total volume: add 6 μL 0.5 M MES buffer, pH 7.5, 3 μL 5 mM UDP-[^{3}H]Glc (3000 cpm/nmol), 10 μL purified WbpP and 1 μL H_2O (*see* **Note 13**). Control assays lack WbpP (*see* Table 1).

Table 2
Step 1: 4-epimerase assay incubation protocol. Prepare 9 tubes. Cocktail tube contains sufficient volume for one extra assay. Add 10 μL of cocktail to each tube, then add 10 μL purified WbpP (or water). Then incubate for 10 min at 37 °C

Assay tube	0.5 M MES, pH 7.5	5 mM UDP-[^{3}H]Glc (3000 cpm/nmol)	WbpP or H_2O	H_2O
1	6 μL	3 μL	10 μL	1 μL
2	6 μL	3 μL	10 μL	1 μL
3	6 μL	3 μL	10 μL	1 μL
4	6 μL	3 μL	10 μL	1 μL
5	6 μL	3 μL	10 μL H_2O	1 μL
6	6 μL	3 μL	10 μL H_2O	1 μL
7	6 μL	3 μL	10 μL H_2O	1 μL
8	6 μL	3 μL	10 μL H_2O	1 μL
WbpP Cocktail (9 assays)	54 μL	27 μL	–	9 μL

2. Incubate mixtures for 10 min at 37 °C, keep on ice until the GT assay mixtures are added.
3. Prepare Dowex 1 × 8 columns: plug short Pasteur pipettes with Nylon wool and add 0.4 mL Dowex 1 × 8. Prewash columns three times with water.
4. Confirm the concentration of the UDP-[^{3}H]Glc (5 mM) and UDP-[^{3}H]Gal (10 mM) solutions by measuring the absorbance at 262 nm of a 1:1000 diluted solution. The Molar extinction coefficient of UDP is 10,000. Thus, a 10 mM solution (1:1000) should have an absorbance of 0.1, and a 5 mM solution (1:1000) has an absorbance of 0.05. Then measure the cpm of 2 μL of UDP-[^{3}H]Glc and UDP-[^{3}H]Gal, and calculate the specific radioactivity (cpm/nmol) (*see* **Note 14**).
5. Prepare the β4GalT assay mixture by mixing 8 μL 5 mM GlcNAcβ-Bn (or water for control assays), 6 μL 0.5 M mM MES buffer, pH 7.5, 2 μL 0.1 M $MnCl_2$ and 4 μL containing 1.8 μg β4GalT. Prepare assay cocktail (*see* **Note 13**).
6. Add the β4GalT assay mixture to the 20 μL 4-epimerase reaction mixture from the first step containing UDP-[^{3}H]Glc/Gal (final volume is 40 μL) and vortex for 2 s.
7. Incubate mixtures for 30 min at 37 °C (*see* **Note 15**).
8. Quench reactions by adding 0.6 mL ice cold water and freezing.
9. Apply mixture to a Dowex 1 × 8 column and collect the flow-through in a scintillation vial (*see* **Note 4**).
10. Wash the column two times with 0.6 mL H_2O and collect the flow-through in the same scintillation vial.
11. Add 4.5 mL Scintillation Cocktail to the scintillation vial and quantify the β4GalT crude product by scintillation counting (*see* **Note 16**).
12. Calculate the (pure) cpm β4GalT product by subtracting the cpm obtained from assays without acceptor substrate from the cpm of crude product.
13. To construct a standard curve for β4GalT activity, six different concentrations of UDP-[^{3}H]Gal should be tested in duplicates. For each concentration of UDP-Gal two control assays lacking acceptor substrate are necessary under the same conditions (Table 3). The cpm obtained from assays lacking GlcNAcβ-Bn acceptor are subtracted from cpm of assays with acceptor to obtain cpm of β4GalT product. Plot cpm of product versus UDP-Gal concentration (Fig. 1). The cpm obtained from 4-epimerase assays are used to determine the UDP-Gal concentration in 40 μL assay volume. This allows the

Table 3
Step 2, glycosyltransferase-coupled assays. Prepare nine tubes. Cocktail tube contains sufficient for one or two extra assays. Add 20 μL of cocktail to each tube, add to 20 μL of WbpP assay mixtures from step 1. Then incubate 30 min at 37 °C

Assay tube	β4GalT	5 mM GlcNAc-β-Bn	0.5 M MES, pH 7	0.1 M $MnCl_2$	4-epimerase mixture from step 1
1a	4 μL	8 μL H_2O	6 μL	2 μL	20 μL
2a	4 μL	8 μL H_2O	6 μL	2 μL	20 μL
3a	4 μL	8 μL	6 μL	2 μL	20 μL
4a	4 μL	8 μL	6 μL	2 μL	20 μL
5a	4 μL	8 μL H_2O	6 μL	2 μL	20 μL
6a	4 μL	8 μL H_2O	6 μL	2 μL	20 μL
7a	4 μL	8 μL	6 μL	2 μL	20 μL
8a	4 μL	8 μL	6 μL	2 μL	20 μL
GT cocktail (9 assays)	36 μL	–	54 μL	18 μL	–

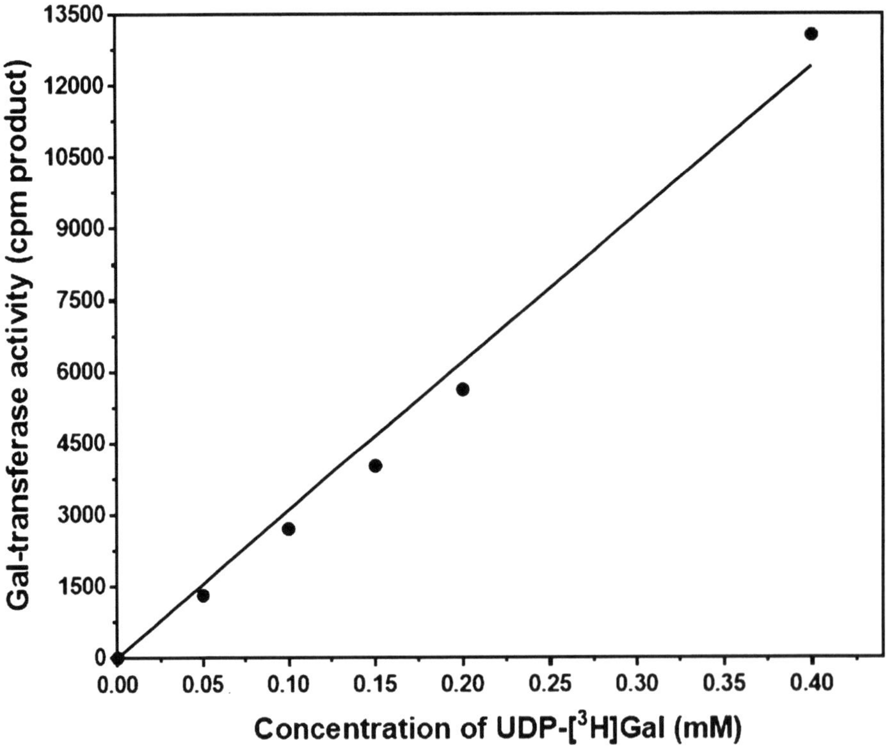

Fig. 1 Standard curve for β1,4-Gal-transferase to quantify the amount of 4-epimerase product UDP-Gal from UDP-Glc. The concentration of UDP-Gal (mM) in the Gal-transferase assay was related to the amount of total Gal-transferase reaction product (cpm). The cpm of 4-epimerase-Gal-transferase reaction product is determined by the standard curve to quantify the concentration and thus the amount of UDP-Gal formed by 4-epimerase during the given conditions

calculation of the total amount of UDP-Gal formed during the 4-epimerase incubation.

14. To determine the specific activity of 4-epimerase (nmol/h/mg) toward UDP-Glc, calculate the nmol UDP-Gal formed during the assay, divided by the incubation time (h) × amount of enzyme protein per assay (mg) (*see* **Note 17**).

3.3 WbpP-β1,3-Glc-Transferase (WbdN)-Coupled Assay to Test Specificity Toward UDP-Gal

The coupled WbpP-WbdN-reactions are:

$$\text{UDP-}[^3\text{H}]\text{Gal} \xrightarrow{WbpP} \text{UDP-}[^3\text{H}]\text{Glc} \xrightarrow{WbdN\,+\,\text{GalNAc-PP-PhU}} [^3\text{H}]\text{Glc}\beta 1\text{-3GalNAc-PP-PhU}$$

1. Prepare the WbpP assay mixtures for the 4-epimerase reaction in 20 μL total volume: add 6 μL 0.5 M MES buffer, pH 7.5, 3 μL 10 mM UDP-[^{3}H]Gal (3000 cpm/nmol), 10 μL purified WbpP and 1 μL H_2O (*see* **Note 13**). Control assays lack WbpP.
2. Incubate mixtures for 10 min at 37 °C, keep on ice until the GT assay mixtures are added.
3. Prepare the assay mixtures for the second (WbdN) step by mixing 4 μL 2 mM GalNAcα–PO_3–PO_3–$(CH_2)_{11}$–O–Phenyl (or water for control samples), 4 μL 0.5 M MES buffer, pH 7.5, 2 μL 0.1 M $MnCl_2$ and 10 μL WbdN bacterial homogenate (140 μg protein). Add to the 20 μL reaction mixtures from the first step, making the final volume 40 μL.
4. Incubate mixtures for 30 min at 37 °C, keep on ice until the GT assay mixtures are added.
5. Quench the reactions by adding 0.7 mL of ice-cold water and freezing.
6. Recycle cartridges with 3 mL methanol followed by 6 mL water.
7. Apply reaction mixture to a recycled Sep-Pak C18 cartridge and push through the cartridge using a 3 mL syringe very slowly (1 drop per sec). Wash the water-soluble components slowly with 3 × 1 mL of water, collect into scintillation vial (fraction A). Wash cartridge with 1 mL water and collect into another scintillation vial (fraction B). Elute the reaction product with 2 × 1.5 mL methanol, collect two scintillation vials (fractions C, D) (*see* **Note 18**).
8. Add 4 mL of Scintillation Cocktail to scintillation vials and quantify the product by scintillation counting. Substract cpm obtained from control assays to obtain cpm WbdN product.
9. Construct a standard curve for WbdN activity with UDP-[^{3}H]Glc as the donor, as described in Subheading 3.2, **step 13** to obtain the concentration of UDP-[^{3}H]Glc produced by WbdN. Calculate 4 epimerase activity (nmol/h/mg) toward UDP-Gal.

3.4 WbpP: Polypeptide-GalNAc-Transferase (ppGalNAcT)-Coupled Assay to Test Specificity Toward UDP-GlcNAc

The coupled 4-epimerase WbpP-ppGalNAcT- reactions are:

$$\text{UDP-}\left[{}^{3}\text{H}\right]\text{GlcNAc} \xrightarrow{\textit{WbpP}} \text{UDP-}\left[{}^{3}\text{H}\right]\text{GalNAc} \xrightarrow{\textit{ppGalNAcT}\,+\,\text{AQP}\mathbf{T}\text{PPP}} \text{AQP}\left(\left[{}^{3}\text{H}\right]\text{GalNAc-}\right)\mathbf{T}\text{PPP}$$

1. Confirm the exact concentration (10 mM) and specific radioactivity (2 μCi/μmol and cpm/nmol) of the 10 mM UDP-[^{3}H]GlcNAc and UDP-[^{3}H]GalNAc as described in Subheading 3.2, **step 4** (*see* **Note 14**).
2. Prepare the WbpP assay mixtures for the 4-epimerase reaction in 20 μL total volume: add 6 μL 0.5 M MES buffer, pH 7.5, 3 μL 10 mM UDP-[^{3}H]GlcNAc (3000 cpm/nmol), 10 μL purified WbpP and 1 μL H_2O (*see* **Note 13**). Control assays lack WbpP.
3. Incubate mixtures for 10 min at 37 °C, keep on ice until the GT assay mixtures are added.
4. Prepare the assay mixtures for the second (ppGalNAc-T1) step by mixing 5 μL 0.5 M MES, pH 7, 8 μL 10 mM AQTPPP acceptor, 2 μL 5 mM $MnCl_2$, 4 μL (0.033 μg protein) ppGalNAc-T, and 1 μL water. Add to the 20 μL reaction mixtures from the first (4-epimerase) step, making the final volume 40 μL.
5. Incubate mixtures for 1 h at 37 °C.
6. Quench the reactions by adding 600 μL of ice-cold water and freezing.
7. Apply the mixtures to 0.4 mL Dowex 1 × 8 anion exchange columns and quantify cpm product by scintillation counting as described in Subheading 3.2, **steps 9–12**.
8. Construct a standard curve for ppGalNAcT activity with different concentrations of UDP-[^{3}H]GalNAc and determine the amount of UDP-[^{3}H]GalNAc formed by WbpP. Calculate the 4-epimerase activity (nmol/h/mg) toward UDP-GlcNAc.

3.5 WbpP-GlcNAc-Transferase I (GlcNAcTI)-Coupled Assay to Test Specificity Toward UDP-GalNAc

The coupled 4-epimerase WbpP–GlcNAcTI reactions are:

$$\text{UDP-}\left[{}^{3}\text{H}\right]\text{GalNAc} \xrightarrow{\textit{WbpP}} \text{UDP-}\left[{}^{3}\text{H}\right]\text{GlcNAc} \xrightarrow{\textit{GlcNAcTI}\,+\,\text{Man}\alpha 1\text{-}6(\text{Man}\alpha 1\text{-}3)\ \text{Man}\beta\text{-octyl}} \text{Man}\alpha 1\text{-}6\left(\left[{}^{3}\text{H}\right]\text{GlcNAc}\beta 1\text{-}2\text{Man}\alpha 1\text{-}3\right)\text{Man}\beta\text{-octyl}$$

1. Prepare the WbpP assay mixtures for the 4-epimerase reaction in 20 μL total volume: add 6 μL 0.5 M MES buffer, pH 7.5, 3 μL 10 mM UDP-[^{3}H]GalNAc (3000 cpm/nmol), 10 μL purified WbpP and 1 μL H_2O (*see* **Note 13**). Control assays lack WbpP.

2. Incubate mixtures for 10 min at 37 °C, keep on ice until the GT assay mixtures are added.
3. Prepare the assay mixtures for the second (GlcNAc-T1) step by mixing 8 μL 5 mM Manα1–6(Manα1–3)Manβ-octyl, 5 μL 0.5 M MES buffer, pH 7.5, 4 μL 0.1 M $MnCl_2$, and 3 μL (60 μg protein) GlcNAcT1. Add to the 20 μL reaction mixtures from the first step, making the final volume 40 μL [17].
4. Incubate mixtures for 1 h at 37 °C.
5. Terminate the reactions by adding 0.6 mL of ice-cold water and freezing.
6. Apply the mixtures to 0.4 mL Dowex 1 × 8 anion exchange columns and quantify cpm product by scintillation counting as described in Subheading 3.2, **steps 9–12**.
7. Construct a standard curve for GlcNAcTI activity with different concentrations of UDP-[^{3}H]GlcNAc and determine the amount of UDP-[^{3}H]GlcNAc formed by the 4-epimerase. Calculate the 4-epimerase activity (nmol/h/mg) toward UDP-GalNAc.

4 Notes

1. Consult reference [2] for more detailed steps on the expression and purification of WbpP. The bacterial pellet can be stored −20 °C and used later for the protein purification.
2. Bovine milk β1,4-Gal-transferase may be the most economical enzyme to use to detect UDP Gal. Other available α- or β-Gal-transferases are also useful. It is preferable to have a purified enzyme that is not contaminated with 4-epimerases.
3. Bovine milk β1,4-Gal-transferase prefers acceptor substrates that have GlcNAc-β-linked to a hydrophobic group that are not too bulky such as naphthyl. Benzyl aglycone groups are generally good and easy to handle in assays and enzyme product purification by HPLC.
4. Dowex 1 × 8 anion exchanger is equivalent to other anion exchange resins (e.g., AG1 × 8).

   It will bind UDP-sugars and sugar-phosphates. The flow-through contains hydrophilic, neutral breakdown products of donor and acceptor substrates (e.g., free Gal) and if the enzymes are not pure, possibly enzyme products from endogenous substrates. Always do duplicate control assays without the acceptor substrates as a background to assess the β1,4-Gal-transferase product only.
5. WbdN is a very active Glc-transferase, both in bacterial homogenates and as purified enzyme. Other available Glc-transferase should also be useful. Sometimes there is a background

Glc-transferase activity in *E.coli* homogenates that has to be subtracted.

6. The chemical synthesis of GlcNAc–PO_3–PO_3–$(CH_2)_{11}$-O-phenyl has been described in detail [15]. The compound is water or methanol soluble and exists as the free acid, or salt form with one or two Na^{2+} ions (or other counterion). Thus, HPLC analysis will show several peaks.
7. The WbdN product cannot be assayed with the simple Dowex 1 × 8 method since the enzyme product will bind to it due to the two phosphate groups. Thus, the C18 Sep-Pak method is preferred. Alternatively, the assay mixture can be separated directly by HPLC (C18 column, acetonitrile/water = 25/75).
8. Most polypeptide GalNAc-transferases prefer peptides with Thr as the acceptor amino acid. Many other peptides (or glycopeptides) can be successful as acceptor substrates. The sequence -PTPPP- is optimal for polypeptide GalNAc-transferase 1.
9. Any other polypeptide GalNAc-transferase can be used with peptide acceptors. Alternatively, GalNAc-transferases that act on oligosaccharides are useful (with oligosaccharide acceptors) but may not be as active.
10. The Dowex 1 × 8 assay is simple and quick. However, depending on the acceptor substrate structure and charge, other methods for the isolation of enzyme product are available (e.g., the C18 Sep-Pak method or HPLC).
11. GlcNAc-transferase I is a very active enzyme. GlcNAc-transferase II is also active but requires a more complex acceptor Manα1-6(GlcNAcβ1-2Manα1-3)Manβ-octyl that can be synthesized from Manα1-6(Manα1-3)Manβ-octyl with GlcNAcTI and UDP-GlcNAc. Other GlcNAc-transferases can be used but may not be as active.
12. The acceptor Manα1-6(Manα1-3)Manβ-octyl can be produced by chemical synthesis. Hydrophobic aglycone groups other than octyl are also acceptable in active substrates.
13. If many assays are done simultaneously, prepare a cocktail of components common to all assays for two more assays (Tables 2 and 3).
14. The purity of the radioactive nucleotide sugar solution is determined by passing an aliquot through Dowex 1 × 8. This will show how much free sugar is present. HPLC analysis on a C18 column run in 5 mM triethylammonium acetate pH 6 will indicate the amount of sugar-phosphate present (eluting in the early fractions). UDP-sugars are retained by the column.

15. The time and temperature of the incubations can be varied to balance production of sufficient enzyme product with minimal degradation of labile nucleotide sugars.
16. The cpm of radioactive compounds vary slightly with the solvents, ratio of scintillation cocktail to sample, program used and scintillation counter. Always count samples under similar conditions in the same counter, using the same program and use a representative ratio of cpm/nmol for the calculations.
17. The units of activity can be nmol/h/mg but for high activities µmol/min/mg are convenient.
18. The separation by C18 Sep-Pak has to be done very slowly to ensure binding of hydrophobic compounds and proper elution of enzyme product. Make sure no methanol or other hydrophobic solvent is present in the water elution fractions. It is preferred to use new columns for critically important assays but they can be recycled at least ten times.

Acknowledgments

This work was supported by Natural Sciences and Engineering Research Council of Canada Discovery grants to I.B. and to K.J.

References

1. Beerens K, Soetaert W, Desmet T (2015) UDP-hexose 4-epimerases: a view on structure, mechanism and substrate specificity. Carbohydr Res 414:8–14
2. Creuzenet C, Belanger M, Wakarchuk WW et al (2000) Expression, purification, and biochemical characterization of WbpP, a new UDP-GlcNAc C4 epimerase from *Pseudomonas aeruginosa* serotype O6. J Biol Chem 275:19060–19067
3. Guo H, Li L, Wang PG (2006) Biochemical characterization of UDP-GlcNAc/Glc 4-epimerase from *Escherichia coli* O86:B7. Biochemistry 45:13760–13768
4. Schulz JM, Watson AL, Sanders R et al (2004) Determinants of function and substrate specificity in human UDP-galactose 4′-epimerase. J Biol Chem 279:32796–32803
5. Ishiyama N, Creuzenet C, Lam JS et al (2004) Crystal structure of WbpP, a genuine UDP-N-acetylglucosamine 4-epimerase from *Pseudomonas aeruginosa* substrate specificity in UDP-hexose 4-epimerases. J Biol Chem 279:22635–22642
6. Rush JS, Alaimo C, Robbiani R et al (2010) A novel epimerase that converts GlcNAc-P-P-undecaprenol to GalNAc-P-P-undecaprenol in *Escherichia coli* O157. J Biol Chem 285:1671–1680
7. Cunneen MM, Liu B, Wang L et al (2013) Biosynthesis of UDP GlcNAc, UndPP-GlcNAc and UDP-GlcNAcA involves three easily distinguished 4-epimerase enzymes, Gne, Gnu and GnaB. PLoS One 8 (e67646):1–9
8. King JD, Kocíncová D, Westman EL et al (2009) Review: lipopolysaccharide biosynthesis in *Pseudomonas aeruginosa*. Innate Immun 15:261–312
9. Merril CR, Das AK, LaPolla RJ et al (1976) Microassay for UDP-galactose 4-epimerase activity. Anal Biochem 72:606–613
10. Moreno F, Rodicio R, Herrero P (1981) A new colorimetric assay for UDP-glucose 4-epimerase activity. Cell Mol Biol Incl Cyto Enzymol 27:589–592
11. Wilson DB, Hogness DS (1964) The enzymes of the galactose operon in *Escherichia coli*. I. Purification and characterization of uridine diphosphogalactose 4-epimerase. J Biol Chem 239:2469 2481

12. Chen X, Kowal P, Hamad S et al (1999) Cloning, expression and characterization of a UDP-galactose 4-epimerase from *Escherichia coli*. Biotechnol Lett 21:1131–1135
13. Sharma S, Ding Y, Jarrell KF et al (2018) Identification and characterization of the 4-epimerase AglW from the archaeon *Methanococcus maripaludis*. Glycoconj J 35 (6):525–535.
14. Gao Y, Liu B, Strum S et al (2012) Biochemical characterization of WbdN, a beta1,3-glucosyltransferase involved in O-antigen synthesis in enterohemorrhagic *Escherichia coli* O157. Glycobiology 22:1092–1102
15. Wang S, Czuchry D, Liu B et al (2014) Characterization of two UDP-gal: GalNAc-diphosphate-lipid β1,3-galactosyltransferases WbwC from *Escherichia coli* serotypes O104 and O5. J Bacteriol 196:3122–3133
16. Elhammer A, Poorman R, Brown E et al (1993) The specificity of UDP-GalNAc: polypeptide N-acetylgalactosaminyltransferase as inferred from a database of in vivo substrates and from the in vitro glycosylation of proteins and peptides. J Biol Chem 268:10029–10038
17. Reck F, Springer M, Meinjohanns E et al (1995) Synthetic substrate analogues for UDP-GlcNAc: man alpha1-3R beta1-2-N-acetylglucosaminyltransferase I. substrate specificity and inhibitors for the enzyme. Glycoconj J 12:747–754

Chapter 21

Assaying Fucosidase Activity

Zoë Anne Megson, Paul Messner, and Christina Schäffer

Abstract

The characterization of a recombinant glycosidase can be done with commercially available substrates, which enable testing of enzyme functionality and determination of linkage specificity. Colorimetric assays with *p*-nitrophenyl substrates provide a relatively simple and fast way of screening conditions which could affect enzyme activity (buffer, pH, ion dependence, temperature). These substrates are useful for the determination of activity optima and the characterization of basic activity parameters. However, testing for linkage specificity should be performed on more complex sugars presenting a range of different glycosidic bonds and might need more sophisticated methods of analysis. This protocol provides comprehensive instructions on how to perform an initial characterization of your glycosidase using a recombinant α-L-fucosidase as an example.

Key words Fucosidase, Enzymatic activity, Substrate specificity, pH optimum, Cation dependence, HPAEC-PAD, Reaction rate

1 Introduction

The following procedure can be used to determine the glycosidase activity of a purified recombinant enzyme or whole cellular fraction containing various enzymes. Here we describe the steps using a recombinant bacterial fucosidase as an example, but these can be used for all glycosidases alike by adapting the substrates accordingly. When using enzymatic preparations, such as the whole cytoplasmic fraction, nonsoluble contaminants should have been removed by ultracentrifugation and assays should be performed using more of the enzymatic fraction to compensate for the enzyme dilution.

As the unknown/unstudied genes we clone are often annotated in databases as glycosidases based on sequence and structure similarity to other known glycosidases, a first step in the characterization of an enzyme's activity will always be to confirm its proposed function and test the recombinant enzyme's functionality. Here, the enzymatic activity of a purified recombinant fucosidase [1] from a human periodontal pathogen (*Tannerella forsythia*) will be tested

Inka Brockhausen (ed.), *Bacterial Polysaccharides: Methods and Protocols*, Methods in Molecular Biology, vol. 1954, https://doi.org/10.1007/978-1-4939-9154-9_21,

using the standard colorimetric α-fucosidase substrate 4-nitrophenyl-α-L-fucopyranoside (*p*NP-fucose) [2–5] in a range of different pH values. This will confirm its activity as an α-L-fucosidase and establish its pH optimum. By performing the assay at different temperatures and in the presence of $MgCl_2$, KCl, and NaCl, the enzyme's optimum temperature and its cation dependence can be evaluated. Activity on the *p*NP-fucose substrate can also be used for determination of catalytic constants (e.g., K_M and V_{max}) for comparative purposes or studies with inhibitors. However, some fucosidases which are active on larger, more complex glycans are not active on small *p*NP substrates. This should be taken into consideration and testing on larger branched substrates should also be performed.

To determine the enzyme's linkage specificity, the recombinant enzyme is incubated with a set of different fucosylated substrates of defined structure representing, in this specific case, a range of fucose linkages available on host glycoproteins and on oral surfaces. The reaction products obtained after overnight incubation are analyzed using high-performance anion-exchange chromatography with pulsed amperometric detection (HPAEC-PAD) and the release of fucose is confirmed by comparison with the retention time of the standard monosaccharide. In order to obtain accurate activity values on diverse substrates, the K-FUCOSE kit from Megazyme is used according to the manufacture's specifications, with modifications.

2 Materials

Prepare all solutions using ultrapure water and analytical grade reagents. Prepare all reagents at room temperature (22 °C) and store them at the indicated temperatures. Use Milli-Q water (resistivity of 18.2 MΩ·cm at 25 °C; Sigma), where appropriate. Diligently follow all waste disposal regulations when disposing of waste materials. The recombinant hexahistidyl (His_6)-tagged enzyme used here was purified and concentrated yielding a protein concentration of 0.35 mg/mL in 50 mM phosphate buffer and was kept as aliquots frozen at −20 °C [1].

2.1 Enzymatic Characterization with pNP-Substrates

1. Solution of 4-nitrophenyl-α-L-fucopyranoside (*p*NP-fucose; Sigma) at 100 mM (MW 285.25 g/mol) in pure dimethyl sulfoxide (DMSO), aliquot (e.g., 25–50 μL aliquots); store at −20 °C until use.
2. Solution of 4-nitrophenol (Sigma) at 2 mM in water to be used for a standard curve. Weigh 6 mg of 4-nitrophenol (MW 139.11 g/mol) to make 20 mL of solution. In order to solubilize the 4-nitrophenol use a bath sonicator and add a few

drops of 0.1 M NaOH (as solubility is improved at an alkaline pH) until a homogenous solution is obtained.

3. 0.1 M McIlvaine citrate–phosphate buffer [6]: 0.2 M disodium phosphate (Na_2HPO_4) and 0.1 M citric acid, to be used to assay the pH range from 3.0–8.0. These will then be combined at different volumes to make a total of 20 mL of buffer at the required pH. For the 0.2 M disodium phosphate, weigh 28.4 g of Na_2HPO_4 (MW 141.96 g/mol) and dissolve in water to make 1 L of buffer. For the 0.1 M citric acid, weigh 19.21 g of citric acid (MW 192.12 g/mL) and add water to make 1 L of buffer. Store up to a month at room temperature. By mixing the two buffers according to Table 1 [6], you obtain different buffers with a pH from 2.2 to 8.0 as desired, in steps of 0.2 pH units.
4. 50 mM glycine buffer: Weigh 1.88 g of glycine (MW 75.07 g/mol) and dissolve in 450 mL of water, adjust pH as required using 0.1 M NaOH/HCl and fill up with water to 500 mL. Prepare this for each pH to be assayed between pH 8.0 and 10.0.
5. 0.4 M phosphate buffer at a pH of 11.4 which will be used to stop the enzymatic reaction. Weigh 28.4 g of Na_2HPO_4 (MW 141.96 g/mol), add 450 mL water, adjust pH to 11.4 and fill up to 500 mL with water (*see* **Note 1**)
6. Solutions for testing cation dependence: $MgCl_2$ (e.g., 5 mM, 10 mM), KCl (e.g., 5 mM, 10 mM), and NaCl (e.g., 50 mM, 150 mM) in selected buffer at the pH optimum (e.g., glycine buffer pH 9.0).
7. Readings are performed in 96-well plates using a plate reader Infinite 200 (TECAN, Männedorf, Switzerland)] and catalytic constants can be calculated using software such as the Sigmaplot 12 or Systat Software.

2.2 Test for Substrate Specificity

1. Assay buffers: Buffers with an adequate pH and cation concentration, if needed, according to the optimum conditions for activity as determined during enzymatic characterization (e.g., 50 mM glycine buffer, pH 9.0).
2. Fucosylated substrate solutions: 100 mM 2-fucosyllactose and 3-fucosyllactose (Dextra Laboratories, Reading, UK). 100 mM H-trisaccharide, Lewis A trisaccharide, Fuc-α(1,4)-Gal, and Fuc-α(1,6)-GlcNAc (Carbosynth, Compton, UK). 5 mM H-trisaccharide, Lewis A trisaccharide, and Fuc-α(1,6)-GlcNAc (Carbosynth). Solutions are aliquoted and kept frozen at −20 °C until use. 2 mM monosaccharide standards for HPAEC-PAD are 20× stock solutions in MilliQ water.
3. Eluents for HPAEC-PAD: Eluent A, MilliQ water. Eluent B, 50 mM NaOH, (dilute 250 mL of 200 mM NaOH with

Table 1
McIlvaine buffer preparation (taken from [6])

pH required	0.2 M Na_2HPO_4 [mL]	0.1 M citric acid [mL]
2.2	0.4	19.6
2.4	1.24	18.76
2.6	2.18	17.82
2.8	3.17	16.83
3	4.11	15.89
3.2	4.94	15.06
3.4	5.7	14.3
3.6	6.44	13.56
3.8	7.1	12.9
4	7.71	12.9
4.2	8.28	12.29
4.4	8.82	11.72
4.6	9.35	11.18
4.8	9.86	10.65
5	10.3	10.14
5.2	10.72	9.7
5.4	11.15	9.28
5.6	11.6	8.85
5.8	12.09	8.4
6	12.63	7.91
6.2	13.22	7.37
6.4	13.85	6.78
6.6	14.55	6.15
6.8	15.45	5.45
7	16.47	4.55
7.2	17.39	3.53
7.4	18.17	2.61
7.6	18.73	1.83
7.8	19.15	1.27
8	19.45	0.85

750 mL MilliQ water). Eluent C, 200 mM NaOH (10 mL 50% NaOH and 990 mL MilliQ water). Do not use NaOH pellets. Online degassing of eluents is performed with helium. Working eluents are mixed online.

4. HPAEC-PAD (High Performance Anion Exchange Chromatography with Pulse Amperometric Detection) with an ICS3000 chromatographic system equipped with an Electrochemical Detector ED3000 (Dionex Austria GmbH, Thermo Fisher Scientific) on a CarboPac PA-1 column (*see* **Note 2**).
5. Amicon 3-kDa cutoff spin column (Merck Millipore).

2.3 Activity Using the K-Fucose Kit from Megazyme

1. Buffer with an adequate pH and cation concentration for enzymatic characterization.
2. Solution for each substrate to be tested (*see* Subheading 2.2).
3. Fucose dehydrogenase (FDH), enzyme, and $NADP^+$ are supplied in the kit (Megazyme).
4. Spectrophotometer.

3 Methods

3.1 Enzymatic Characterization with pNP-Substrates

1. Using microtiter 96-well plates, add the following to each well: 1 μL (0.17 μM) of purified recombinant enzyme, 37 μL of each McIlvaine or glycine buffer for a range of different pH values (3.0–10.0), and 2 μL (5 mM) of the colorimetric substrate *p*NP-fucose, to make a total volume of 40 μL [5]. Leave some wells without enzyme to be used as a blank. Incubate at room temperature for 3–5 min (*see* **Note 3**).
2. Stop the enzymatic reaction after 3–5 min by addition of 260 μL of phosphate buffer, pH 11.4, to each well, including the blanks (*see* **Note 4**).
3. The measurement of absorbance at 405 nm (A_{405}) in each well is performed using a plate reader. The A_{405} of the blank wells should be subtracted from the A_{405} of wells containing enzyme. The A_{405} is then plotted against the pH to find the pH optimum of the enzyme (*see* Fig. 1).
4. Once the pH optimum has been established, **steps 1–3** can be repeated by using a fixed buffer (pH optimum) and varying the cation concentrations (buffers from Subheading 2.1, **step 6**) to evaluate the effect of these on the enzyme's activity. In this way, different cations and their concentrations can be screened, e.g., $MgCl_2$ (5 mM, 10 mM), KCl (5 mM, 10 mM) and NaCl (50 mM, 150 mM). For the enzyme shown here as an example, this was assayed in 50 mM glycine buffer, pH 9.0. Different

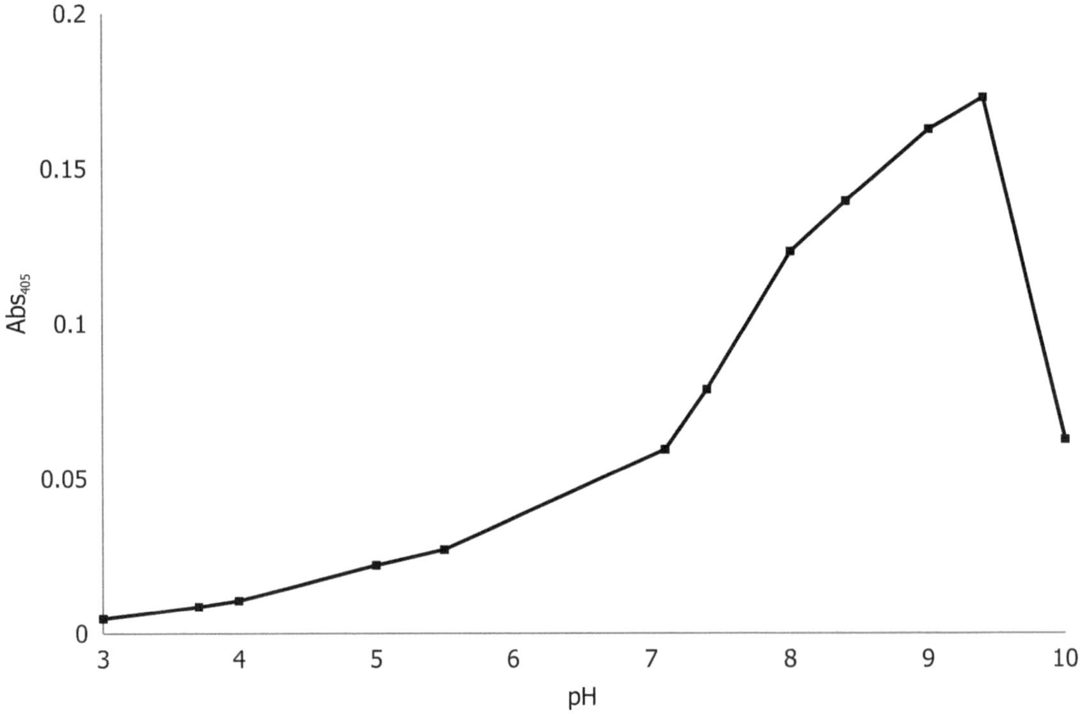

Fig. 1 pH profile of *T. forsythia* fucosidase using *p*NP-fucose as a substrate. Activity was measured as the increase in A_{405} due to the released 4-nitrophenol product. Citrate–phosphate buffer (0.1 M) was used to assay the pH range from 3.0–8.0, 50 mM glycine buffer was used for the pH range from 8.0–10.25. Taken from [1] with permission of Taylor & Francis

temperatures can also be tested by performing several different incubations.

5. For comparative purposes with other studied enzymes or for testing the effect of inhibitors, the catalytic constants (e.g., K_M and V_{max}) are calculated for the *p*NP-substrates in a similar way (**steps 1–3**). In order to do this, the enzymatic reaction or increase in A_{405} should be measured in the presence of the substrate at different concentrations (e.g., from 0.01 to 50 mM) (*see* **Note 5**).
6. In order to translate this ΔA_{405} to μmol of substrate converted over time, which is needed for calculating catalytic constants, a 4-nitrophenol standard curve is made. Measure the A_{405} of known amounts of 4-nitrophenol per well (e.g., 0, 4, 8, 12, 16, 20, and 24 nmol; add 0, 2, 4, 6, 8, 10, 12 μL of the 2 mM 4-nitrophenol solution per well in 300 μL of the buffer normally used to stop the reaction when enzyme is present, in this case, phosphate buffer at pH 11.4). Ensure that the conditions the absorbance is measured in (volume in the well and pH of the solution) are the same as in **steps 2–3**. Quantification of the enzymatic release in each well can be then done directly using

the resulting standard curve A_{405} versus known concentration of 4-nitrophenol. This will give you the reaction rate at each substrate concentration (μmol/min).

7. Plot reaction rate versus substrate concentration. Reaction rate is expressed in activity units (U) defined as μmol substrate converted per minute (μmol/min). Express K_M (concentration of substrate when enzyme achieves half V_{max}) in mM and V_{max} (maximum rate of reaction) in U/mg of enzyme.

3.2 Test for Enzyme Linkage Specificity

1. Incubate substrates with the enzyme overnight at 37 °C in a total volume of 100 μL by mixing 0.34 μM enzyme with 0.5 mM substrate in 50 mM glycine buffer, pH 9.0 (activity optimum).
2. After the incubation, dilute the reaction volume with 400 μL Milli-Q water in order to minimize the effect of the buffer in the next step.
3. Remove the enzyme from the mixture by using an Amicon 3-kDa cut-off spin column (Merck Millipore).
4. Collect the flow-through. Perform HPAEC-PAD by applying 25 μL of this flow-through to a CarboPacPA-1 column equilibrated in 16 mM NaOH using full-loop injection.
5. Elution, washing and equilibration is done at a flow rate of 1 mL/min, using the following program which can be adapted to meet needs: Elute from 0–46 min in 16 mM NaOH (elution of monosaccharides, disaccharides, and trisaccharides), perform column wash from 46–51 min with 150 mM NaOH, re-equilibrate column from 51–65 min with 16 mM NaOH (*see* **Note 6**).
6. Perform **steps 2–5** for each enzymatic reaction. Add enzymatic mixtures with inactivated enzyme to allow for identification of protein interference with the column separation. Add a mixture without enzyme which will be the substrate control.
7. Perform **step 5** with appropriate 100 μM monosaccharide standards (e.g., fucose) to identify elution peaks at specific retention times.
8. Elution profiles of substrates, monosaccharide standards, and components in the reaction mixture are seen in Fig. 2. In this example, the enzyme was active on the substrates from which release of fucose was observed (2-fucosyllactose, H-trisaccharide and Fuc-α(1,6)-GlcNAc).

3.3 Activity Using the K-Fucose Kit from MEGAZYME

1. In a total volume of 250 μL of buffer, e.g., 50 mM glycine buffer at pH 9.0 (buffer shown to be optimal for enzymatic reaction), the substrates 2-fucosyllactose, H-trisaccharide and Fuc-α(1,6)-GlcNAc, are incubated separately at a

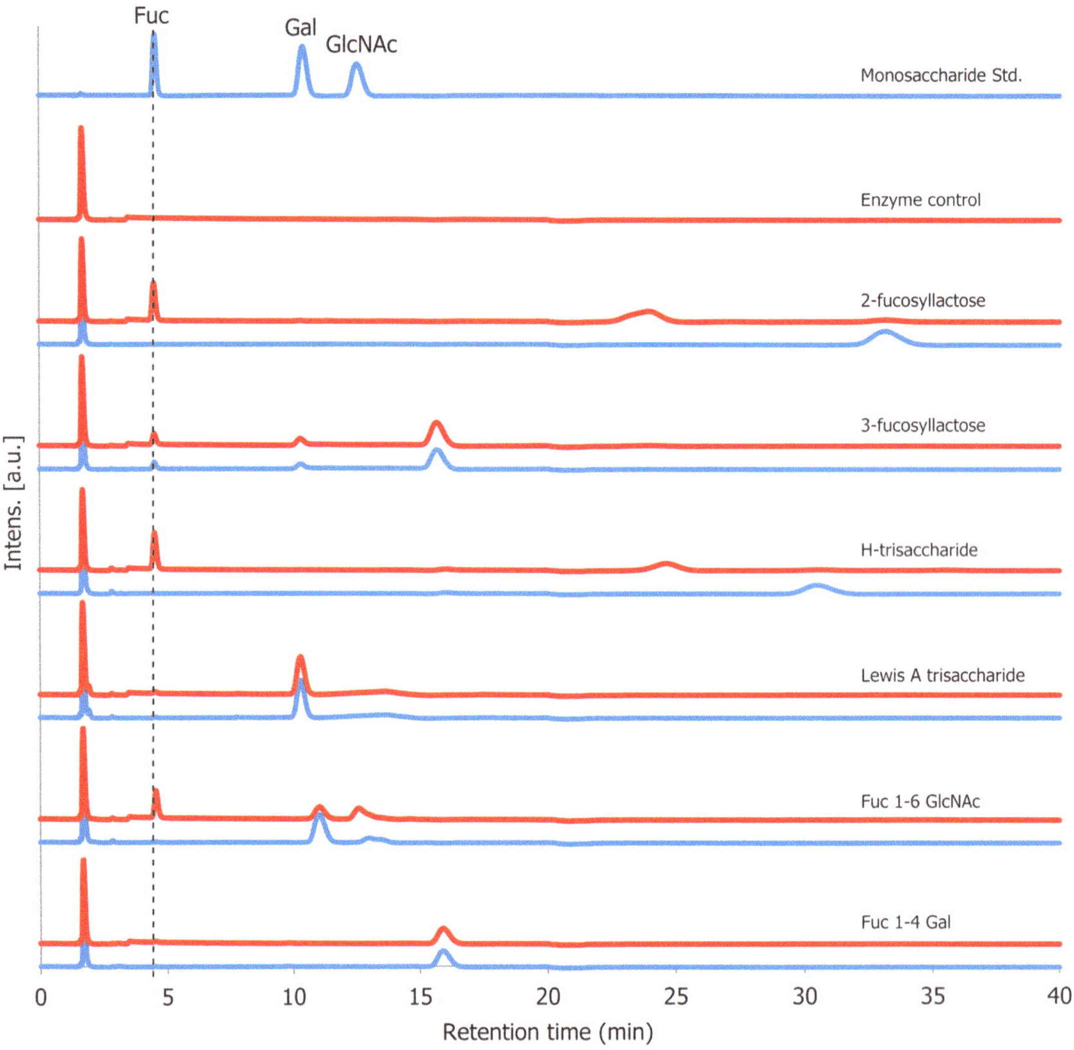

Fig. 2 *Tannerella forsythia* fucosidase activity on standard fucosylated substrates after overnight incubation as determined by HPAEC-PAD. Blue lines represent samples which were incubated in the absence of enzyme (substrate standard) and red lines represent samples incubated in the presence of enzyme. Cleavage of the substrates was determined by the appearance of a fucose peak, as determined by the retention time of the standard monosaccharide. Taken from [1] with permission of Taylor & Francis

concentration of 0.5 mM with 1.83 μL of fucose dehydrogenase (FDH) and 9.15 μL of $NADP^+$ in a cuvette at 37 °C for 10 min. During this incubation time, small amounts of free fucose which could be present in the substrates will be converted by the FDH, reducing $NADP^+$ to NADPH, which will lead to a small increase of A_{340}. This A_{340} can be followed during the incubation period using a spectrophotometer.

2. When the reaction has reached a constant A_{340}, which will be the value for the blank, add recombinant enzyme to the

mixture at a concentration of 0.34 μM. The fucosidase will cleave off fucose residues from the substrates which, in turn, will be converted by the FDH. Follow the formation of NADPH by continuous measurement of the increase in A_{340}.

3. The activity of the enzyme on each substrate should be calculated according to the supplier's specifications from ΔA_{340}/min where the formation of NADPH is linear over time.
4. Repeat the experiment with different enzyme dilutions (1:10; 1:100) to prove the reliability of the method.

4 Notes

1. In this example, the recombinant fucosidase studied had an optimal pH of 9.0 and, therefore, a buffer of pH 11.4 was needed to stop the reaction. However, some *p*NP substrates are not resistant to this high pH, leading to a nonenzymatic cleavage of the bond and release of 4-nitrophenol. If this occurs, alternatively, another buffer can be used as long as the pH is far enough away from the enzymatic optimum and is above 9.0 (as accurate measurement of enzyme activity requires that the 4-nitrophenol product is fully deprotonated), e.g., 0.4 M glycine/NaOH buffer at pH 10.4.
2. Technical **Note 20** (Dionex, Thermo Fisher Scientific, Vienna, Austria).
3. If pipetting is done manually, rather than making triplicates of each condition on a single plate, it is recommended to repeat the study several times, as pipetting time should be kept to a minimum. Depending on the speed of the reaction (some wells should be visibly yellow), more incubation time may be needed or, alternatively, incubations can be performed at 37 °C.
4. This is an important step as the released 4-nitrophenol has a different extinction coefficient depending on the pH of the solution it is dissolved in and accurate measurement of enzyme activity requires that the 4-nitrophenol product is fully deprotonated at an alkaline pH. Therefore, this step serves two purposes, stopping the reaction and putting all wells at the same pH so that the released colorimetric substance can be quantified in all wells equally.
5. The activity on the substrate should be calculated where the plot A_{405} versus time is linear (i.e., short reaction times). In this example, the reaction rates were calculated at pH 9.0 in 50 mM glycine buffer at room temperature.

6. Detection with Electrochemical Detector ED3000. Waveform A: Quadruple Potential Waveform.

Time (s)	Potential (V)	
Time 0.0	Potential 0.1	
Time 0.2	Potential 0.1	
Time 0.4	Potential 0.1	Integration 0.2–0.4 s
Time 0.41	Potential −2.0	
Time 0.42	Potential −2.0	
Time 0.43	Potential 0.6	
Time 0.44	Potential −0.1	
Time 0.5	Potential −0.1	

Acknowledgments

This work was supported by the Austrian Science Fund FWF, projects P24317-B22 (to C.S.) and the Doctoral Programme W1224 Biomolecular Technology of Proteins.

References

1. Megson ZA, Koerdt A, Schuster H et al (2015) Characterization of an α-L-fucosidase from the periodontal pathogen *Tannerella forsythia*. Virulence 6:282–292
2. Guillotin L, Lafite P, Daniellou R (2014) Unraveling the substrate recognition mechanism and specificity of the unusual glycosyl hydrolase family 29 BT2192 from *Bacteroides thetaiotaomicron*. Biochemist 53:1447–1455
3. Berteau O, McCort I, Goasdoue N et al (2002) Characterization of a new α-L-fucosidase isolated from the marine mollusk *Pecten maximus* that catalyzes the hydrolysis of α-L-fucose from algal fucoidan (*Ascophyllum nodosum*). Glycobiology 12:273–282
4. Rodriguez-Diaz J, Monedero V, Yebra MJ (2011) Utilization of natural fucosylated oligosaccharides by three novel α-L-fucosidases from a probiotic *Lactobacillus casei* strain. Appl Environ Microbiol 77:703–705
5. Leonard R, Rendic D, Rabouille C et al (2006) The *Drosophila* fused lobes gene encodes an *N*-acetylglucosaminidase involved in *N*-glycan processing. J Biol Chem 281:4867–4875
6. McIlvaine TA (1921) A buffer solution for colorimetric comparison. J Biol Chem 49:183–186

Chapter 22

In Vitro Synthesis of Bioactive Glycovariants of Enterocin 96, an Antimicrobial Peptide from *Enterococcus faecalis*

Rupa Nagar and Alka Rao

Abstract

Antimicrobial peptides (AMPs) are novel agents for therapeutic application for their inherent broad spectrum of activities against bacteria, fungi, and viruses, as well as anti-inflammatory, anticancerous, and immunomodulatory activities. This chapter presents an enzymatic method to generate glycovariants of one such antimicrobial peptide, namely enterocin 96, using a bacterial protein O- and S-glycosyltransferase, in vitro.

Key words In vitro glycosyltransferase assay, Mass spectrometry, High performance liquid chromatography (HPLC), Chymotrypsin digestion, Exoglycosidase treatment/Deglycosylation assay, Glycodiversification

1 Introduction

Glycoconjugates, namely, glycoproteins, glycopeptides, and glycolipids, play vital roles in several cellular functions and biological processes. Bacterial glycoconjugates express an enormous diversity and possess medicinal, therapeutic, and industrial importance. Among these, the glycosylated proteins are a rather recent discovery in bacteria. The challenge in the area of protein glycosylation in bacteria is due to huge diversity and complexity observed. Yet it is fascinating, primarily for its relevance in pathogenicity of bacteria of medical importance [1]. The in vivo production of bacterial glycoconjugates is complicated as it involves complete pathways. Recent reports on glycosylated bacteriocins, now termed glycocins, present the importance of glycosylation in self-defense of bacteria and their antimicrobial activity against other bacteria. This also offers a novel approach to generate antimicrobials using in vitro systems [2–6]. The key players in the generation of glycoconjugates are glycosyltransferases (GTs). Such bacterial GTs can be useful and amenable tools for various glycoengineering/in vitro applications. In this context, we have recently described a recombinant

Inka Brockhausen (ed.), *Bacterial Polysaccharides: Methods and Protocols*, Methods in Molecular Biology, vol. 1954, https://doi.org/10.1007/978-1-4939-9154-9_22, © Springer Science+Business Media, LLC, part of Springer Nature 2019

glycosyltransferase (EntS) of *Enterococcus faecalis* TX0104 that can diglycosylate Ser, Thr, and Cys residues, wherein two hexose residues are transferred on to a single Ser or Thr or Cys residue in its acceptor peptide substrate EC (custom-synthesized leaderless enterocin 96 peptide), in a sequential iterative manner, in vitro [3]. Enterocin 96 (EC 96) is a known antimicrobial peptide [7]. Upon in vitro glycosylation by EntS, EC peptide exhibits antimicrobial activity against *Listeria monocytogenes* EGD-e, a food borne pathogen. Accordingly, the application of EntS can be exploited towards synthesis of sequence and glycan variants of acceptor substrate peptide (s) of EntS. In this chapter, an elaborate methodology is described to create glycovariants/glycoconjugates of EC by employing the sequential iterative glycosylation capability of EntS. The overall methodology is presented in Fig. 1, wherein once the recombinant protein glycosyltransferase of *Enterococcus faecalis* TX0104 is cloned, overexpressed and purified, the glycosyltransferase activity of EntS is confirmed using in vitro GT assay. The peptide EC 96 serves as an acceptor substrate and UDP-Glc or UDP-Gal as donor substrate in the said assay. Under optimized assay conditions, a diglucosylated or digalactosylated peptide product is formed, whereas under controlled GT assay conditions, it results into a mixture of unglycosylated, monoglycosylated, and diglycosylated peptide variants. Further, RP-HPLC is used to separate these glycoforms. The monoglucosylated or monogalactosylated peptide is finally diglycosylated using enzyme activity of EntS. A second in vitro glycosylation reaction involves monoglucosylated or monogalactosylated peptide as an acceptor substrate and UDP-Gal or UDP-Glc as a donor substrate. The end product is a glycodiversified product which is either diglucogalactosylated or digalactoglucosylated peptide (Fig. 2) with differential anti-listerial activity. Thus In addition, in this chapter, we describe the analysis of EC by mass spectrometry.

2 Materials

Prepare all the solutions using molecular biology, proteomics, or mass spectrometry grade reagents as prerequisite. The distilled water (dH_2O) (with 18 MΩ-cm resistivity at 25 °C) is to be used to prepare all the reagent solutions.

2.1 Bacterial Strains and Plasmids

1. Bacteria: Use *E. coli* TOP10 as a host for cloning and plasmid propagation and *E. coli* Lemo21(DE3) as a host for protein expression (obtained from NEB). Use pET28a as a cloning vector (Novagen).
2. Bacterial growth: Grow *E. coli* strains in Luria–Bertani (LB) medium at 37 °C. Use kanamycin and chloramphenicol

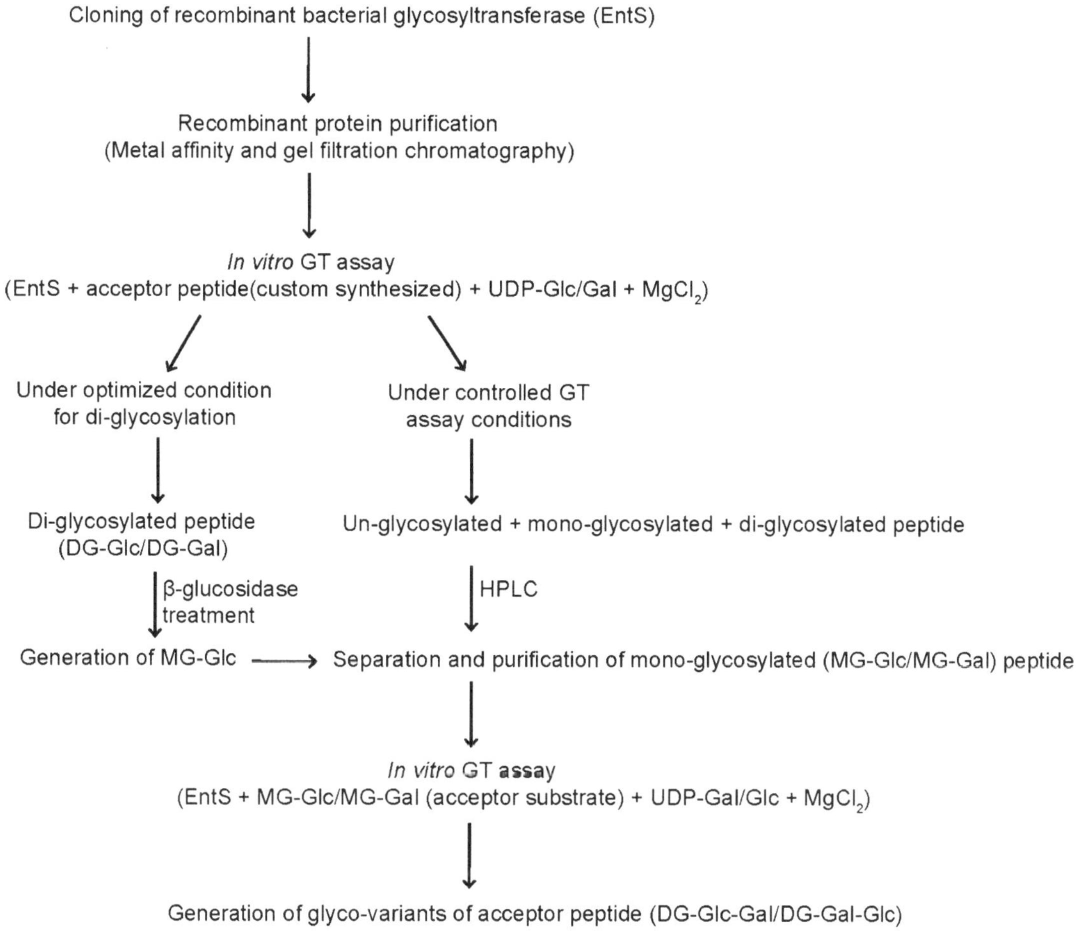

Fig. 1 Schematic representation of the methodology used to create the glycovariants of the acceptor peptide substrate. The schema describes all the steps that are required to generate the different glycoforms of acceptor peptide and create glycodiversified products. In vitro GT assay of the purified recombinant bacterial glycosyltransferase performed under optimized conditions results in completely diglycosylated product, whereas controlled GT assay results in a mixture of three different glycoforms. Further, separation of monoglycosylated product using RP-HPLC and using it as an acceptor substrate in the second GT assay reaction with a different UDP-sugar donor results in generation of glycodiversified peptide products

antibiotics at final concentrations of 50 μg/mL and 35 μg/mL for *E. coli* TOP10 and Lemo21(DE3) strains, respectively.

3. *L. monocytogenes* cells*:* The following reagent was obtained through "The Biodefense and Emerging Infections Research Resources Repository (BEI Resources)" NIAID, NIH as part of the Human Microbiome Project: *Listeria monocytogenes,* Strain EGD-e (NR114). Grow *L. monocytogenes* cells in brain--heart infusion (BHI) medium at 37 °C.
4. Glycerol stocks: Prepare all the glycerol stocks by mixing 500 μL of 50% glycerol to the 500 μL of saturated bacterial cultures (*see* **Notes 1** and **2**). Store the stocks immediately at

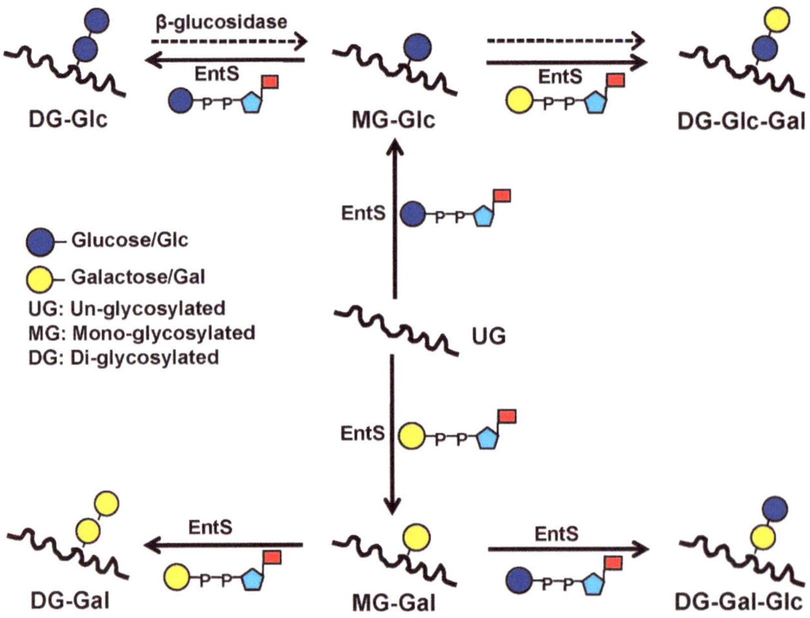

Fig. 2 Pictorial representation of strategy used to generate the glycodiversified peptide products. Under optimized in vitro GT assay conditions with UDP-Glc and UDP-Gal EntS converts the unglycosylated (UG) peptide into completely diglucosylated (DG-Glc) or digalactosylated (DG-Gal) products, respectively. Whereas under controlled GT assay conditions with UDP-Glc and UDP-Gal, EntS is not efficient enough to create the completely glycosylated product and results into a mixture of peptide populations. The mono-glycosylated (MG-Glc or MG-Gal) is used as an acceptor substrate in the second in vitro GT assay reaction which results into a glycodiversified product (DG-Glc-Gal or DG-Gal-Glc), when a different UDP sugar is used as a donor substrate

−80 °C, till further use. Revive the glycerol stocks by scraping the frozen cells and streaking onto agar medium or inoculating in liquid medium with appropriate antibiotics (*see* **Note 3**).

2.2 Culture Media and Supplements

1. Culture media: Prepare all type of media in double distilled water and sterilize by autoclaving at 121 °C (250 °F) at 100 kPa (15 psi) above atmospheric pressure for 15 min.
2. Antibiotics: Prepare the stock solutions of kanamycin by resuspending in water at 50 mg/mL and filter sterilize; prepare chloramphenicol by dissolving in absolute ethanol at 35 mg/mL concentration. Aliquot the stock solutions and store at −20 °C.
3. 25 μM L-rhamnose.
4. Isopropylthiogalactoside (IPTG).
5. Restriction enzymes.

```
>ATGTATTCTGAAAATTTTATTGCTAATGACTGGTTTAATGTAGAGGTATTTAATAAAAATAAGTATACTT
TAACGAACCAAGAGAATAAAGATGTAACAGAATTATGGTTACAAATTTTAAAAGGGCTAAAGTTCCCCAA
CGAATTAAAGGAAACTGTCAGTTACTCTAAAAATTTAAAAGAATTATCTTTAAAAACTCACGCAGAAGTA
TCTGTATGTATTATTGCTAAGAATGAACAGGATTCAATAAGAAAATGTATTAATAGTATCTATGAATTTT
CAGATGAAATTATATTTATTGATACAGGATCAATTGATTCGACAAAAAAAAATAGTAAAAGAAATAGCAAG
CGAAAAAGTAAAAATATTTGATTATACTTGGCAAGATGATTTTTCAGATGCGAGAAATTATTCAATACAA
AAAGCAAGTAAAGAATGGATATTAATTATTGATGCAGATGAATATGTATCTTCAGATGAGCTTATCAAAT
TAAGGCTCTTAATAGATATGTTAGACAGGTTTAAATTTAAAGATTCATTAAGAGTTAGTTGTGCAATATA
TCAATTAGATAATGTTATCACACATGGCCAAAGTCGATTATTTAGAAACAATAATAAAATTAAGTATTAT
GGTCTAATACATGAAGAGTTGAGGAACAACAAAGGATTAGATCCAATTTTTAACGTTGAAAGTGAGATTA
CTTTTTTCCATGACGGTTACAAAGAAATACTTAGGAAAGAGAAGTGTGAAAGAAACATAAGGCTACTAGC
TAAGATGTTAGAAAAAGAGCCAGACAATGTTAGATGGGCATACTTGTATTGTAGAGATTCATTTTCTATA
AATTCCAACATTGATTTTGAAAAAATTCTACTTCCATTTTTAATAAAGAATATGGATGAAAGTATATCAT
GTGAGAATATCCTACTTACAAACTATACTCATTTAATCCTATTTCTTATTACTAAGAAATATATAATTGA
TGGGAAAAGCTCACTTGCAAGTAAATGTATAGAGGTGTTAGAAAAAATGCTACCTAACTCTTCTGATGTT
ACTTTTTACAAATTTTTAAATAAACAGCATAGTTTGTATGAACAACAATTTGAATTTTTAAAAGAAGTAA
TTCAATTTAGAAAAAATAATGAATATGATCAATATAGCCAAATAGGGTGTAATTTATTACACTATGATTT
ATTAATTTCAGGATTACTTTTTGATGTTAAGTCTTATGATTATTCATATCAATACTTTTTAAAATTAGAT
TTAGCTAACTATTTTTCTGAATTAGAGATTCCTGATGAATACAAAATGTTAATAAATAAGTATCGGGAGA
ATGAATCATGA
```

Fig. 3 Nucleotide sequence of recombinant bacterial protein glycosyltransferase (ORF0417, NZ_GG668924.1, GI: 488313438) of *Enterococcus faecalis* TX0104

2.3 Custom Synthesis of Codon-Optimized Bacterial Glycosyltransferase Gene

Use the custom-synthesized, codon-optimized bacterial glycosyltransferase gene (ORF 0417, Protein ID: WP_002382823.1 of *Enterococcus faecalis* TX0104, obtained from GenScript, USA, cloned in pUC57 vector between NcoI and XhoI restriction sites). The nucleotide sequence of the gene is given in Fig. 3.

2.4 Chemically Synthesized Peptides

Use the custom-synthesized enterocin 96 core peptide termed EC (Lifetein, USA) having sequence: "MSKRDCNLMKACCAGQAVTYAIHSLLNRLGGDSSDPAGCNDIVRKYCK" (*see* **Note 4**). Aliquot the lyophilized powder of peptide and store at conditions described by manufacturer. Resuspend one aliquot or minimum amount required for the experiment in water. Store the remaining peptide solution at −20 °C.

2.5 Reagent Stock Solutions

Prepare the stock solutions of buffers and reagents as follows.

1. 1 M Tris–HCl: Resuspend 12.11 g of Tris base in 90 mL dH_2O and adjust the pH to 8.0 or 7.5 using conc. HCl. Make up the final volume up to 100 mL with dH_2O and autoclave.
2. 1 M HEPES: Dissolve 23.08 g of HEPES, Free acid in dH_2O and adjust the pH to 7.5 using 5 M NaOH. Make up the final volume up to 100 mL with dH_2O and autoclave.
3. 1 M Sodium acetate: 6.804 g of sodium acetate was dissolved in 40 mL of dH_2O and pH adjusted to 5.0 or 5.5 with acetate

buffer. The final volume was made up to 50 mL with dH_2O and autoclaved.

4. 5 M NaCl: Dissolve 292.2 g of NaCl in dH_2O by making the final volume up to 1 L and autoclave.
5. 5 M NaOH: Dissolve 20 g pellets of NaOH in 100 mL of dH_2O.
6. 1 M dithiothreitol (DTT): Dissolve 1.54 g of DTT in 10 mL of ddH_2O, aliquot and store at −20 °C.
7. 5 M imidazole: Dissolve 17.02 g of imidazole in ddH_2O and make up the final volume up to 50 mL and store at 4 °C.
8. 1 M Tris (2-carboxyethyl)phosphine (TCEP): Dissolve 2.86 g of TCEP in 10 mL of ddH_2O, aliquot and store at −20 °C.
9. 0.1 M $MgCl_2$: Dissolve 1 g of $MgCl_2$ in 50 mL of dH_2O and autoclave.
10. 0.1 M $CaCl_2$: Dissolve 0.29 g of $CaCl_2$ in 20 mL of dH_2O and autoclave.

2.6 Recombinant Bacterial Glycosyltransferase Purification

1. Ni-Sepharose (GE Healthcare life sciences).
2. Superdex-200 10/300 GL (GE Healthcare life sciences).
3. Lysis buffer: 20 mM HEPES, pH 7.5, 500 mM NaCl, 2 mM DTT, and 10% glycerol.
4. Wash buffer: 20 mM HEPES, pH 7.5, 1 M NaCl, 2 mM DTT, 30 mM imidazole, and 10% glycerol.
5. Elution buffer: 20 mM HEPES, pH 7.5, 200 mM imidazole, 200 mM NaCl, and 2 mM DTT.
6. Wash and equilibration buffer for gel filtration: 20 mM HEPES, pH 7.5, 200 mM NaCl, and 2 mM DTT.
7. HRP conjugated anti-6× His tag antibody.

2.7 In Vitro Glycosyltransferase Assay

1. Enzyme: 0.2 μM Purified recombinant bacterial glycosyltransferase.
2. Peptide acceptor: 5 μM custom-synthesized peptide as acceptor.
3. Donor substrate: 100 μM NDP-Sugar (UDP-Glc/UDP-Gal).
4. 1 mM $MgCl_2$.
5. GT assay buffer: 20 mM HEPES, pH 7.5, 50 mM NaCl, 2 mM DTT/1 mM TCEP.
6. 5% trifluoroacetic acid (TFA).
7. $ZipTipC_{18}$.

2.8 Chymotrypsin Digestion

1. Chymotrypsin.
2. Reconstitution buffer: 1 mM HCl, 2 mM $CaCl_2$. Resuspend the lyophilized Chymotrypsin powder in reconstitution buffer to prepare a stock solution of 0.5 μg/μL. Prepare aliquots and store at −20 °C.
3. Digestion buffer: 100 mM Tris–HCl, pH 7.5, 10 mM $CaCl_2$, and 5 mM TCEP.

2.9 β-Glucosidase Treatment/ Deglycosylation Assay

1. β-glucosidase.
2. Reconstitution buffer: 0.1 M Sodium acetate, pH 5.0.

   Resuspend the lyophilized enzyme powder in reconstitution buffer to prepare a stock solution of 5 mg/mL. Prepare aliquots and store at −20 °C.
3. Digestion buffer: 0.1 M sodium acetate buffer, pH 5.0 and 1× (0.1 mg/mL) purified bovine serum albumin (BSA).

2.10 ZipTipC_{18} Cleanup

1. ZipTipC_{18} (Merck, Millipore).
2. Wetting solution: 50% acetonitrile (ACN).
3. Equilibration solution or washing solution: 0.1% TFA.
4. Elution solution: 50% ACN in 0.1% TFA.

2.11 Sep-Pak tC18 Cleanup

1. Sep-Pak tC18, 1 cc cartridge (Waters).
2. Wetting solution 1: 100% ACN.
3. Wetting solution 2: 50% ACN in 0.5% acetic acid.
4. Equilibration and washing solution 1: 0.1% TFA.
5. Washing solution 2: 0.5% acetic acid.
6. Elution solution: 50% ACN in 0.5% acetic acid.

2.12 HPLC to Separate Glycoforms of Peptide

1. HPLC Column: Analytical: Jupiter C18 column (250 × 4.6 mm, 5 μm, 300 Å, Phenomenex) and Preparative: Jupiter Proteo C12 column (250 × 10 mm, 10 μm, 90 Å, Phenomenex).
2. Solvent A: 0.1% TFA in water (Type 1).
3. Solvent B: 0.1% TFA in acetonitrile.
4. Washing solvent: 100% methanol.

2.13 Mass Spectrometry

1. Calibration kit (AB Sciex).
2. Matrix: α-cyano-4-hydroxycinnamic acid (CHCA) (Sigma).

   Resuspend the lyophilized matrix in 50% ACN and 0.1% TFA to prepare a stock solution of 10 mg/mL. Prepare aliquots in amber tubes and store at −20 °C.
3. LC column: Zorbax Eclipse Plus C8 column (3.0 × 150 mm, 5 μm, 95 Å, Agilent Technologies).

4. Solvent A: 0.1% formic acid in water.
5. Solvent B: 0.1% formic acid in acetonitrile.

2.14 General Instruments Required

1. PCR: Mastercycler (Eppendorf).
2. DNA sequencing: Automated DNA sequencer (ABI).
3. Gel filtration chromatography: AKTA purifier (GE Healthcare).
4. Analytical HPLC: Prominence UFLC (quaternary LC) system (Shimadzu).
5. Preparative HPLC: Agilent 1260 infinity binary LC system (Agilent Technologies).
6. Matrix-assisted laser desorption ionization time-of-flight mass spectrometry (MALDI-TOF-MS): AB Sciex TOF/TOF 5800.
7. Liquid chromatography electrospray ionization mass spectrometry (LC-ESI-MS): 6550 iFunnel Q-TOF LC-MS.

3 Methods

3.1 SDS-PAGE

Use SDS-PAGE to separate and analyze the quality of protein fractions.

1. Prepare SDS-PAGE gel with 12.5% or 15% resolving and 5% stacking gels.
2. Mix the protein samples to be analyzed with 1× loading dye, heat at 95 °C for 10 min and subject to a short spin to pellet down the debris.
3. Load the prepared protein samples into the wells along with protein molecular weight marker.
4. Run the gel in 1× running buffer at constant voltage until the dye front reaches the bottom.
5. Stain the gel with Coomassie Brilliant Blue and compare the size of observed protein bands with molecular weight standards.

3.2 Protein Estimation Assay Using Bradford Reagent

Estimate the protein concentration in the purified samples using Bradford reagent in a 96-well plate.

1. Prepare a BSA standard curve by estimating known concentrations of BSA with Bradford reagent at A_{595}.
2. Mix the protein sample and Bradford reagent in the ratio of 1:30 (5 μL protein sample and 150 μL Bradford), incubate for 5 min at room temperature, and record the absorbance at 595 nm.

3. Put an assay blank, having buffer in place of protein sample.
4. Subtract the absorbance of blank from the protein sample and estimate the protein concentration using the BSA standard curve.

3.3 Subcloning of Codon-Optimized and Custom-Synthesized Bacterial Glycosyltransferase (EntS) in pET28a Vector

Clone the bacterial glycosyltransferase gene or subclone the codon-optimized custom-synthesized gene into a cloning vector to create a recombinant glycosyltransferase gene construct that overexpresses with a purification tag (in this case, EntS gene from EntSpUC57 subcloned into pET28a cloning vector between NcoI and XhoI restriction sites to create EntSpET28a with a C-terminal 6×-His tag). Figure 3 provides the nucleotide sequence of the protein glycosyltransferase of *Enterococcus faecalis* TX0104.

1. Digest the insert EntSpUC57 and vector pET28a with NcoI and XhoI and purify the digested fragments using QIAquick gel extraction kit.
2. Ligate the purified insert and vector using T4 DNA ligase.
3. Transform the ligation mixture into *E. coli* TOP10 cells using heat shock method (*see* **Note 5**).
4. Select the transformants on LB-kanamycin agar plate, incubated at 37 °C for 12 h.
5. Confirm the positive transformants carrying EntSpET28a construct gene-specific PCR.
6. Isolate the recombinant construct EntSpET28a using QIAprep Spin Miniprep kit and verify the transformant by restriction digestion (*see* **Note 6**).
7. Confirm the sequence of the gene construct by DNA sequencing.

3.4 Overexpression and Purification of Recombinant Bacterial Glycosyltransferase Protein (EntS)

The recombinant bacterial glycosyltransferase gene can be overexpressed by transforming a construct into any *E. coli* expression cells and protein can be purified using a suitable combination of chromatographic techniques, namely metal affinity, gel filtration and ion exchange chromatography. Here, the purification method is described for EntS using metal affinity and gel filtration chromatography. In routine, from 3 L culture, approximately 10–12 mg of recombinant bacterial glycosyltransferase protein (EntS) is purified.

1. Transform the EntSpET28a construct into *E. coli* Lemo21 (DE3) cells for overexpression of EntS.
2. Confirm the protein expression using western blotting upon labeling with HRP conjugated anti-6× His tag antibody (*see* **Note 7**).
3. Inoculate primary culture in 30 mL of LB broth containing 35 μg/mL chloramphenicol and 50 μg/mL kanamycin, with a

single colony of EntS transformant and grow it overnight at 37 °C.

4. Use 1% of this saturated primary culture to inoculate 3 L LB broth containing 35 μg/mL chloramphenicol, 50 μg/mL kanamycin and 25 μM L-rhamnose. Incubate the culture at 37 °C until OD_{600} reaches 0.6.
5. Induction of expression using 0.4 mM IPTG and incubate at 25 °C for 12 h and harvest the cells by centrifugation at 8000 × *g* for 20 min at 4 °C (*see* **Note 8**).
6. Resuspend the cell pellet in 150 mL of lysis buffer (*see* **Note 9**).
7. Lyse the cells using sonication with 10 s on and 10 s off pulse and 30% amplitude. Remove the cell debris by centrifugation at 12,000 × *g* for 30 min at 4 °C.
8. Perform metal ion affinity chromatography using Ni-Sepharose beads using the supernatant (*see* **Note 10**).
9. Wash and equilibrate the Ni-Sepharose resin with lysis buffer, 2 mL beads (binding capacity: 40 mg/mL) and mix with supernatant and leave for binding at 4 °C for 30 min with gentle mixing.
10. Wash the beads with washing buffers with gradual reduction in the salt concentration (*see* **Note 11**).
11. Perform elution using half column volume of elution buffer.
12. Run all elution fractions on 12.5% SDS-PAGE to check the quality of purified fractions.
13. Pool and concentrate all the fractions containing EntS using Amicon ultra membrane filters (30 kDa MWCO, Millipore).
14. Subject the concentrated fraction to gel filtration chromatography using FPLC system (AKTA, GE Healthcare life sciences) equipped with Superdex-200 10/300 GL (GE Healthcare life sciences) (*see* **Note 12**).
15. Wash and equilibrate the column with gel filtration buffer.
16. Perform elution at a flow rate of 0.5 mL/min and monitor the absorbance at 280 nm.
17. Collect the desired fractions, pool and concentrate using Amicon ultra membrane filters (30 kDa MWCO, Millipore).
18. Estimate the concentration of purified protein fractions using Bradford reagent.
19. Store the purified protein in buffer containing 20 mM HEPES, pH 7.5, 50 mM NaCl and 2 mM DTT at −80 °C until further use (*see* **Note 13**).

3.5 In Vitro Glycosyltransferase (GT) Assay of Recombinant Bacterial Glycosyltransferase Protein (EntS)

1. Determine the in vitro GT activity of purified recombinant bacterial glycosyltransferase protein (EntS) by incubating 0.2 μM of purified EntS with synthetic acceptor peptide substrates (5 μM) and UDP-glucose (100 μM) in GT assay buffer at 25 °C for 12 h (*see* **Note 14**).
2. Use the following assay control reactions: (a) buffer control, containing only buffer, (b) negative control containing all components except purified bacterial glycosyltransferase (EntS), and (c) enzyme control containing only purified bacterial glycosyltransferase (EntS) in assay buffer (set with every new lot of purified protein to rule out any artifact).
3. Quench all the assay reactions with 5% TFA to pH 2.0, cleanup using ZipTip C_{18} and vacuum-dry.
4. Reconstitute the samples either in 1:1 (v/v) ratio with CHCA matrix solution for MALDI-TOF MS analysis or in 0.1% formic acid for LC-ESI-Q-TOF analysis.

3.6 β-Glucosidase Treatment/Deglycosylation Assay

1. Incubate the diglucosylated enterocin core (EC) peptide (5 μM) with β-glucosidase keeping the final concentration of 2 μg/μL in a reaction buffer containing 100 mM sodium acetate buffer, pH 5.0 and 1× (0.1 mg/mL) purified BSA at 37 °C for 18 h (*see* **Note 15**).
2. Use a control assay reaction by incubating the diglucosylated EC peptide with all reaction components in the absence of β-glucosidase enzyme.
3. Quench the reactions with 5% TFA to pH 2.0 and subject to LC-ESI-Q-TOF MS analysis without desalting.

3.7 In Vitro Synthesis of Glycovariants/Glycoconjugates (Glycodiversification) Using Controlled GT Assay

1. Use a controlled GT assay of EC peptide and separate the monoglucosylated/monogalactosylated EC peptide species on HPLC.
2. Use the purified monoglucosylated/monogalactosylated EC peptide species as acceptor.
3. Substrate for the second GT assay reaction (by employing the glycan extending property of recombinant bacterial glycosyltransferase (EntS) to get a diglycosylated (diglucosylate/digalactosylated/glucogalactosylated/galactoglucosylated) EC peptide species.
4. Incubate monoglucosylated (MG-Glc) or monogalactosylated (MG-Gal) (5 μM) with EntS (0.2 μM) in the presence of UDP-Gal or UDP-Glc (100 μM), respectively, in GT assay buffer, at 25 °C for 12 h.
5. Quench the assay reactions with 5% TFA to pH 2.0.

6. Desalt the samples using ZipTipC_{18} and vacuum-dry.
7. Subject the samples to LC-ESI-QTOF-MS analysis to confirm the mass of glycodiversified species.

3.8 In Vitro Synthesis of Glycovariants/Glycoconjugates (Glycodiversification) Using Controlled GT Assay in Combination with β-Glucosidase

1. Generate the monoglucosylated EC (MG-Glc) peptide by incubating the diglucosylated EC (DG-Glc) peptide with β-glucosidase as mentioned above.
2. Use β-glucosidase digested MG-Glc EC peptide as acceptor substrate for the second GT assay reaction with UDP-Gal to generate a diglycosylated /glycodiversified EC peptide.
3. Incubate MG-Glc (5 μM) with EntS (0.2 μM) in the presence of UDP-Gal (100 μM) in GT assay buffer, at 25 °C for 12 h.
4. Quench the assay reactions with 5% TFA to pH 2.0.
5. Desalt the samples using ZipTipC_{18} and vacuum-dry.
6. Subject the samples to LC-ESI-QTOF-MS analysis to confirm the mass of glycodiversified species.

3.9 RP-HPLC Method to Separate Glycoforms/Glycovariants of EC Peptide

Optimize the controlled GT assay conditions for partial diglycosylation and to get a mixture of unglycosylated (UG) EC, monoglycosylated (MG) EC and diglycosylated (DG) EC peptides. The optimized MS profile of EC peptide species confirms that diglucosylated EC elutes first, followed by monoglucosylated EC and then unglucosylated EC.

1. Incubate EC peptide (100 μM) and EntS (0.25 μM) in [E]:[S] = 1:400 ratio along with 100 μM UDP-Glc/UDP-Gal in GT assay buffer for 1 h at 25 °C.
2. Analyze the reactions on analytical HPLC system (Shimadzu) equipped with a reverse phase HPLC column.
3. Optimize a RP-HPLC method to separate unglycosylated EC, monoglycosylated EC, and diglycosylated EC peptide populations from the reaction mixture.
4. Equilibrate the HPLC column with 5% solvent B (*see* **Note 16**).
5. Inject 20 μL of the reaction mixture on the column and fractionate the sample by employing a gradient of solvent B (5–100%) over 26 min with a flow rate of 1 mL/min and monitor at 220 nm (*see* **Notes 17** and **18**).
6. Confirm the identity of separated species by observing the mass on MALDI-TOF-MS (*see* **Note 19**).
7. For large scale production of peptide species, put the assay in 10× volume (1 mg of peptide substrate) under controlled GT assay conditions, perform reaction cleanup using tC18 cartridges. Now, translate the HPLC purification method on

semipreparative HPLC column and purify the peptide species in higher amount (*see* **Note 20**).

8. Lyophilize the purified HPLC fractions containing single population and store at −80 °C (*see* **Note 21**).

3.10 Glycosite Annotation of Glycovariants Using Chymotrypsin Digestion

1. Glycosylate the acceptor peptide (50 μM) using EntS under optimized GT assay conditions and desalt then using ZipTipC_{18}.
2. Digest the glycosylated peptide with chymotrypsin keeping enzyme and substrate ([E]: [S]) ratio 1: 25 and incubate in chymotrypsin digestion buffer at 30 °C for 5 h.
3. Use the following controls: (a) Unglycosylated peptide with all the reaction components including chymotrypsin, (b) Chymotrypsin alone in digestion buffer to test any degradation in the enzyme.
4. Quench the reaction digests with 5% TFA to pH 2.0, desalt using ZipTipC_{18} and vacuum-dry.
5. Subject all the digests to MALDI-TOF MS and tandem MSMS analysis.
6. Identify the site of glycosylation by de novo sequencing of digested fragments.

3.11 ZipTip C18 Cartridge Cleanup

Use ZipTipC18 or Sep-Pak tC18 (for large scale peptide) to clean up the assay reactions before subjecting them to mass spectrometry.

1. Wetting: Aspirate and discard the wetting solution 5–6 times.
2. Equilibration: Aspirate and discard the equilibration solution 5–6 times.
3. Sample binding: After quenching the reaction with 5% TFA (pH ≈ 2.0), aspirate and dispense the sample 20–25 times.
4. Washing: Aspirate and discard the washing solution 5–6 (*see* **Note 22**).
5. Elution: Elute the bound sample by aspirating and dispensing into elution solution for 20–25 times.
6. Repeat the **steps** from **1–5** for **4–5** times for maximum recovery (*see* **Note 23**).
7. Vacuum-dry the eluates by vacuum centrifugation.

3.12 Sep-Pak tC18 Cartridge Cleanup

The tC18 cartridge (1 cc) cleanup is for 1 mg peptide use.

1. Wetting: Wash the column with 3 mL of wetting solution 1 and 3 mL of wetting solution 2 sequentially.
2. Equilibration: Wash the column with 3 mL of equilibration solution.

3. Sample binding: Load the sample on to the column and collect the flow through.
4. Washing: Wash the column with washing solution 1 and washing solution 2, with 3 mL of each.
5. Elution: Elute the bound peptides in 1 mL elution solution.
6. Vacuum-dry the eluates by vacuum centrifugation.

3.13 Mass Spectrometry (MS) and Tandem Mass Spectrometry (MSMS) Analysis Using MALDI-TOF

Carry out matrix-assisted laser desorption ionization time--of-flight mass spectrometry (MALDI-TOF) at any MALDI-TOF system (Here, the method is described for AB Sciex TOF/TOF 5800 system equipped with 1000 Hz frequency, Nd:YAG 349 nm laser).

1. Before acquiring the data, calibrate the analyzer externally with a mixture of the peptides (bradykinin, angiotensin, Glu-1-Fibrinopeptide, ACTH fragment 7–38, ACTH fragment 1–17 and ACTH fragment 18–39) as per manufacturer's protocol.
2. For MS and MSMS analysis of samples, mix 1 μL aliquot of salt-free analyte in 1 μL of CHCA matrix solution in 50% ACN–50% water with 0.1% TFA).
3. Spot 1 μL of the total mixture onto a 384-well MALDI target plate (*see* **Note 24**).
4. Dry the spots under ambient conditions (at room temperature) prior to analysis.
5. Acquire MS and MSMS data in reflector and positive ion mode keeping the laser intensity in the range of 3000–3400 V.
6. Collect the MS data in mass range 700–4000 Da with total number of shots 2000 (2000 spectra/spot), bin size 0.5 ns and pulse rate 400 Hz for each spot.
7. Collect the MSMS data in mass range 10–4000 Da, keeping the laser intensity in the range 4000–4200 V and adduct tolerance 0.03.
8. Exclude the precursor ions showing S/N below 20 and resolution below 200 in MS for MSMS.
9. Post-acquisition, import the collected raw data (t2d file format) in the Data Explorer software, version 4.9 of Applied Biosystems and convert it into text file format.
10. Import and plot the text files having the peaks list in Origin software (OriginPro 2015 b.9.2.214).

3.14 Mass Spectrometry (MS) Analysis Using LC-ESI-Q-TOF

Perform liquid chromatography electrospray ionization mass spectrometry (LC-ESI-Q-TOF MS) at any LC-MS system. Here, the method is described for 6550 iFunnel Q-TOF LC-MS system (Agilent Technologies) equipped with Agilent Dual jet-stream ESI source.

1. For MS analysis, reconstitute the dried sample analyte in 0.1% formic acid.
2. Equilibrate the reverse phase column with 40% solvent B.
3. Inject 5 μL volume of reconstituted sample on UHPLC system and employ a gradient of solvent B (40%–100%) over 12 min with 0.4 mL/min flow rate to fractionate the sample and subject to ESI-Q-TOF MS.
4. Collect the data using Dual AJS ESI positive ion mode with the capillary voltage set to 4.5 kV and nitrogen as sheath gas (11 L/min).
5. Heat the ionization source and sheath gas to 200 °C and 300 °C, respectively.
6. Keep the nozzle and fragmentor voltage at 1000 V and 280 V, respectively.
7. Acquire the MS data (in .d file format) using MassHunter workstation software, version B.05.00 by Agilent Technologies.
8. Post-acquisition, import the acquired raw data files to MassHunter qualitative analysis software, version B.05.00 for further processing.
9. Observe the expected multiple charged ions of the peptides in the acquired LC chromatogram. Integrate and deconvolute the total ion chromatogram (TIC) corresponding to the expected multiple charged ions using maximum entropy or resolved isotope method to obtain MS spectrum.
10. Export the integrated multiple charged ions and deconvoluted mass data to csv file format.
11. Import and plot the csv files having the peaks list in Origin (OriginPro 2015 b.9.2.214).

3.15 Bioactivity Assay of In Vitro-Synthesized Glycoconjugates/Glycovariants of Peptide

1. Prepare DG-EC, MG-EC and UG-EC peptide species under optimized GT assay conditions as described above.
2. Separate and purify the individual peptide species using standardized RP-HPLC based method as described above.
3. Prepared the peptide species for bioactivity assay by resuspending the lyophilized species in 50 mM Tris–HCl, pH 7.5.
4. Prepare the glycovariants using optimized method of glycodiversification and use them for bioactivity assay (*see* **Note 25**).
5. Grow *L. monocytogenes* EGD-e culture in BHI media at 37 °C for 7–8 h and prepare the agar plates by mixing 25 μL of saturated culture ($A_{600} \approx 1.5$) and 100 mL of molten BHI agar (cooled to 40–45 °C) (*see* **Note 26**).

6. Pour the mixture into sterile petri plates and allow the seeded plates to solidify for 30–45 min.
7. Prepare the agar wells using a sterile cork borer.
8. Pour 50 μM of all peptide species to the wells and incubate the plates at 37 °C for 12 h.
9. Determine the antimicrobial activity by the presence and absence of zone of growth inhibition.

4 Notes

1. Before preparing bacterial culture stocks, clean the biosafety cabinet with 70% ethanol, switch on UV light and leave for at least 15 min for decontamination. Using autoclaved filtered pipette tips would further reduce the chances of contamination.
2. Add glycerol to the saturated bacterial culture slowly and drop-wise. Also, prepare glycerol stocks of at least 2–3 different bacterial colonies, to avoid the loss of culture stock in case of contamination of a stock vial.
3. Keep reviving the culture stocks periodically to check and avoid contamination.
4. Avoid suspending large amount of peptide at a time as peptide tend to degrade in solution condition upon storing for long duration.
5. Use high efficiency ($\approx 10^7$) *E. coli* TOP10 chemical competent cells using $CaCl_2$ method and be extra gentle while preparing the cells. Also, keep checking the contamination of cells by growing the cells on LB plates with and without antibiotics.
6. Use vector (pET28a alone, without a gene insert) as a negative control while performing the restriction digestion of selected transformants. Try to use at least 2–3 restriction enzymes to get a unique pattern of the transformant.
7. Optimize protein expression by inducing the expression with different concentration of IPTG (0.1 mM–0.5 mM) and incubating at different temperature (20 °C, 25 °C, 30 °C, 37 °C) for different time-period (4–18 h). Use uninduced and untransformed cells as negative control.
8. After harvesting, wash the cells with 1× PBS or 20 mM Tris–HCl, pH 8.0 to remove the remaining media components from the cells. To wash centrifuge the cells at 8000 × *g* for 10 min at 4 °C.
9. For good cell lysis, keep the lysate as diluted as possible.

10. Use freshly prepared buffers (lysis, wash and elution buffer) for all the chromatographic steps.
11. Always use fresh/regenerated/washed beads for binding to avoid contamination. After binding, wash the beads extensively with wash buffers to remove maximum nonspecifically bound proteins. To maximize the recovery of overexpressed protein, collect the flow-through of first step, rebind it to the fresh/washed beads and perform all the steps washing and elution.
12. Calibrate the gel filtration column with gel filtration calibration kit to estimate the size of the eluted proteins.
13. Aliquot the purified protein and store at –80 °C. This is to avoid the multiple rounds of freeze thaw that leads to degradation of purified protein.
14. Use freshly prepared UDP-Glc/Gal solution as UDP-sugars tend to degrade upon storing in solution condition.
15. Avoid multiple freeze thaw rounds of β-glucosidase stock solution. BSA helps in stabilization of peptide and protect it from degradation while incubation at 37 °C. Therefore, optimize the BSA concentrations (1×, 2×, 5×, 10×) required for the stability of peptide during glycosidase digestion.
16. Before injecting the peptide sample on HPLC column, clean and equilibrate it with solvents. Wait until the base line is stable. Also, give a blank injection (buffer without peptide sample) before and after injecting the sample onto the column to remove all the nonspecifically bound material. The column cleaning is mandatory after every 3–4 rounds of injections to avoid the column blockage.
17. Avoid collecting a single peak into a single vial. Rather, collect the peak in multiple fractions (each fraction of 500 μL–1 mL) to avoid mixing of nonspecific eluates, especially at the beginning and ending of the peak.
18. Avoid overloading of sample on to the column. Optimize the sample concentration that eluates in single run, else the overloaded amount will keep on eluting in subsequent rounds of elution.
19. The HPLC fractions are diluted, therefore may not give a significant signal in mass spectrometry. Hence, pool the fractions of one peak, concentrate using vacuum centrifugation and then subject to MS analysis.
20. Translating the analytical HPLC method to semipreparative system requires further optimization of HPLC method for flow rate, loading concentration and gradient. To avoid blockage of the column, do not load the whole assay reaction on the column. Therefore, sample cleanup using tC18 cartridges is suggested.

21. Aliquot the fractions before lyophilization would help in avoiding freeze thaw of purified peptide sample.
22. Wash the ZipTip thoroughly and remove all the remaining wash buffer from the tip before the elution step.
23. The binding capacity of a ZipTip can be improved by cleaning it well with 50% ACN with 0.1% TFA after use. Therefore, in practical, a single tip can be used to recover approximately 15–20 μg of peptide.
24. Spot multiple dilutions (1:1, 1:5, 1:10, and 1: 20) of peptides to acquire good intensity spectra.
25. Properly sterile biosafety cabinet before performing the bioactivity assay. Use all autoclaved tips and microcentrifuge tubes to avoid contamination.
26. Do not use high concentration of saturated indicator culture (*L. monocytogenes* EGD-e), as it will lead to overgrowth of cells.

References

1. Schäffer C, Messner P (2017) Emerging facets of prokaryotic glycosylation. FEMS Microbiol Rev 41:49–91
2. Maky MA, Ishibashi N, Zendo T et al (2015) Enterocin F4-9, a novel O-linked glycosylated bacteriocin. Appl Environ Microbiol 81:4819–4826
3. Nagar R, Rao A (2017) An iterative glycosyltransferase EntS catalyzes transfer and extension of O- and S-linked monosaccharide in enterocin 96. Glycobiology 27:766–776
4. Wang H, Oman TJ, Zhang R et al (2014) The glycosyltransferase involved in thurandacin biosynthesis catalyzes both O- and S-glycosylation. J Am Chem Soc 136:84–87
5. Stepper J, Shastri S, Loo TS et al (2011) Cysteine S-glycosylation, a new post-translational modification found in glycopeptide bacteriocins. FEBS Lett 585:645–650
6. Oman TJ, Boettcher JM, Wang H et al (2011) Sublancin is not a lantibiotic but an S-linked glycopeptide. Nat Chem Biol 7:78–80
7. Izquierdo EC, Wagner C et al (2009) Enterocin 96, a novel class II bacteriocin produced by enterococcus faecalis WHE 96, isolated from Munster cheese. Appl Environ Microbiol 75:4273–4276

Chapter 23

High-Throughput Screening for Inhibitors of Wall Teichoic Acid Biosynthesis in *Staphylococcus aureus*

Omar M. El-Halfawy and Eric D. Brown

Abstract

The world is heading toward a dangerous post-antibiotic era where antibiotics fail to treat infections. *Staphylococcus aureus* is the leading cause of healthcare-associated infections worldwide, and an ever-increasing percentage of them are methicillin-resistant (MRSA). New strategies are urgently needed to combat this pathogen. Wall teichoic acids (WTA) in *S. aureus* are polyribitol phosphate polymers that play important roles in virulence and resistance to β-lactam antibiotics. Here, we describe a high-throughput whole-cell screening platform for inhibitors targeting WTA biosynthesis. This platform takes advantage of the unique dispensability patterns of genes encoding WTA biosynthesis. We further describe follow-up dose-response assays to identify WTA inhibitors among the primary bioactives. WTA inhibitors offer an exciting opportunity for the development of novel antibacterial leads of unique mechanism in the fight against drug-resistant staphylococcal infections.

Key words Small-molecule screening, Small-molecule library, Wall teichoic acid, Antagonism screen, Antibiotic, Antibacterial, *Staphylococcus aureus*

1 Introduction

The world is heading towards a dangerous post-antibiotic era where existing antibiotics fail to treat infections [1]. Ineffective antibiotic killing exposes bacteria to sublethal concentrations of antibiotics, which in turn elicits adaptive stress responses leading to increased antibiotic resistance and tolerance [2–5]. An especially troublesome pathogen in both the clinic and community is the gram-positive bacterium *Staphylococcus aureus*. An ever-increasing percentage of *S. aureus* strains are methicillin-resistant (MRSA), the leading cause of healthcare-associated infections worldwide [6]. It is a leading cause of bacteremia, infective endocarditis, osteoarticular, skin and soft tissue, respiratory tract, and device-related infections among others [7]. Further, *S. aureus* is the most common pathogen in children with cystic fibrosis (CF) and the second most common in the adult CF population [8]. The burden of

Inka Brockhausen (ed.), *Bacterial Polysaccharides: Methods and Protocols*, Methods in Molecular Biology, vol. 1954,
https://doi.org/10.1007/978-1-4939-9154-9_23,

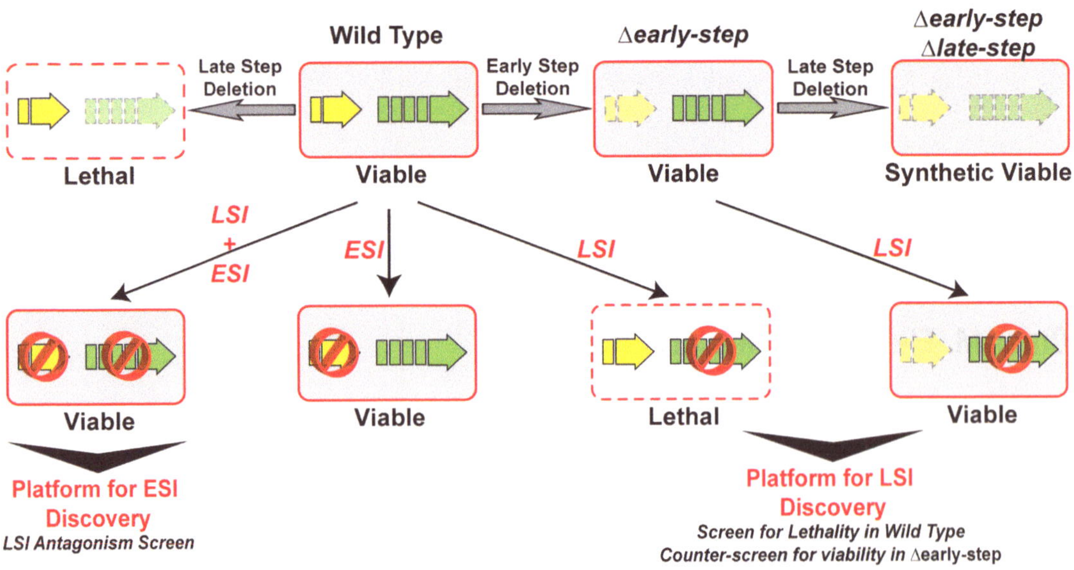

Fig. 1 A screening platform to identify WTA inhibitors based on paradoxical dispensability patterns of genes encoding WTA biosynthesis. *ESI* early-step inhibitor, *LSI* Late-step inhibitor

staphylococcal infections is still on the rise despite concerted efforts in recent years to combat this pathogen and hence new strategies are needed.

Wall teichoic acids (WTA) are phosphate-rich polymers that make up a large proportion of the mass of the cell wall of gram-positive bacteria. In *S. aureus*, WTA is composed of a polymer (n~40) of 1–5 linked ribitol-phosphate residues that are covalently attached via a disaccharide linkage unit to peptidoglycan (PG). WTA polymers underpin host colonization, coordination of peptidoglycan synthesis, and resistance to β-lactam antibiotics [9]. Thus, WTA biosynthesis constitutes an exciting target for antibacterial chemical matter of unique mechanism.

The Brown research group has shown that WTA biosynthetic proteins are encoded by genes that exhibit paradoxical gene dispensability patterns [10–12] (reviewed in [9], Fig. 1). WTA synthesis is initiated by the action of two early-step gene products; TarO, which transfers an N-acetyl-glucosamine-1-phosphate to an undecaprenyl phosphate carrier lipid, followed by the transfer of N-acetylmannosamine by TarA [9]. The late-step gene products, TarBFL then extend the polymer with ribitol-phosphate repeats, followed by TarM/S, which glycosylate the WTA polymer, followed by TarGH, the ABC transporter which exports it to the external surface leading to its final attachment onto PG [9]. Deletions in early genes in the pathway (*tarO* and *tarA*) are viable but avirulent, due to dysregulated cell division and defects in host adhesion [10, 13, 14]. Inactivation of TarO or TarA compromises in vivo viability, rendering early steps as potential targets for compounds to attenuate virulence by compromising viability in the host. Further, the Brown group has shown the utility of early-step inhibitors in resensitizing MRSA to β-lactam antibiotics by discovering the first TarO inhibitor, ticlopidine, otherwise renowned as an antiplatelet drug, exposing WTA polymers as key resistance determinants in MRSA [15].

In contrast, late-step genes cannot be deleted, unless flux into the pathway is prevented, for example, by deleting either *tarO* or *tarA* (Fig. 1). The conditional essentiality of the late acting genes is thought to result from an accumulation of toxic intermediates and/or depletion of the bactoprenol-phosphate carrier lipid, which is also used for the synthesis of PG. The essentiality of the late-steps positions them as traditional antibiotic targets. The WTA transporter protein, TarG, has been validated as a highly druggable target, with the discovery of targocil by the Walker laboratory [16] and other inhibitors discovered by researchers at Merck [17]. TarG inhibitors appear to be abundant in chemical collections but these compounds have relatively high frequencies of resistance ($>10^{-6}$). Interestingly, suppressors commonly lead to the inactivation of early steps in WTA synthesis and are thought to be compromised for virulence.

Indeed, the unique dispensability patterns of WTA genes provides the basis of a sophisticated cell-based high throughput screening platform for discovery of compounds active against *S. aureus* cells that target early or late steps in the WTA pathway (Fig. 1): (1) Early-step inhibitors (ESIs) can be identified in a screen for antagonism of the lethal action of late-step inhibitors abolishing flux into the WTA pathway. (2) Late-step inhibitors (LSIs) can be identified by screening for lethality of wild-type *S. aureus* followed by counter-screening in a mutant with a deletion of an early-step encoding gene in which lethality should be lost. The latter screen for the discovery of LSIs can follow the standard high throughput screening protocol described in [18]. Herein, we describe the antagonism screen, pioneered by the Brown group [19, 20], for the discovery of early-step WTA inhibitors which can serve as adjuvants to resensitize MRSA to β-lactam antibiotics. Overall, WTA inhibitors offer an exciting opportunity for the development of novel antibacterial leads of new chemical class and mechanism in the fight against drug-resistant staphylococcal infections.

2 Materials

2.1 Preparation of Bacterial Culture

1. Appropriate frozen stocks of bacterial strains such as the methicillin-sensitive *S. aureus* (MSSA) Newman or the methicillin-resistant *S. aureus* (MRSA) USA300.
2. Sterile wooden sticks or loops for colony picking.

3. Sterile culture test tubes.
4. Cation-adjusted Mueller-Hinton Broth (CAMHB) media: Dissolve CAMHB powder in deionized water according to the manufacturer's directions. Distribute the media into 100–500 mL aliquots in screw-cap glass bottles or capped glass flasks, then sterilize by autoclaving. Let the media cool down before use.
5. CAMHB Agar plates—Dissolve agar in CAMHB medium (1.5% w/v) and sterilize by autoclaving. Cool the media to ~50 °C. Pour ~20 mL of sterile medium per Petri dish (100 mm diameter) and cool it till the agar solidifies. Store the plates at 4 °C.
6. Temperature controlled incubator.
7. Temperature controlled shaking incubator.
8. 70% ethanol.
9. Sterile screw top tubes and flasks.

2.2 Determination of Appropriate Incubation Time

1. Working bacterial culture as described in Subheading 3.1.
2. A multichannel pipettor or a liquid handler.
3. Neat dimethylsulfoxide (DMSO).
4. Solutions of an early step inhibitor (e.g., ticlopidine, Sigma-Aldrich) and a late-step inhibitor (e.g., targocil, synthesized in house according to [21]) in neat DMSO stored in aliquots at −20 °C.
5. Sterile, lidded, 384-well, flat-bottom, clear plates.
6. A plate reader that measures absorbance at 600 nm.
7. Temperature-controlled incubator.

2.3 Primary Screen

1. Working bacterial culture as described in Subheading 3.1.
2. Compound library in 96-well polypropylene plates stored at −20 °C.
3. Solution of a late-step inhibitor (e.g., targocil) in neat DMSO stored in aliquots at −20 °C.
4. Appropriate antimicrobial compounds for low growth controls; for the example here, erythromycin.
5. Neat DMSO.
6. Sterile, lidded, 384-well, flat-bottom, clear plates.
7. Reservoirs to hold bacterial cultures.
8. 20 μL and 200 μL 96-racked sterile tips for the liquid handler or multichannel pipettors.
9. A plate reader that measures absorbance at 600 nm.
10. Temperature controlled shaking incubator.

2.4 Follow-Up Tests of Primary Actives

1. Working bacterial culture as described in Subheading 3.1. (an MRSA strain is preferred for the checkerboard assays with β-lactam antibiotics in Subheading 3.4.2 to increase the window to detect synergy).
2. Compounds identified as actives from Subheading 3.3.2 stored at −20 °C.
3. Solution of a late-step inhibitor (e.g., targocil) in neat DMSO stored in aliquots at −20 °C.
4. A β-lactam antibiotic (cefuroxime in this example, Sigma-Aldrich).
5. Neat DMSO.
6. Sterile, lidded, 96-well, flat-bottom, clear plates.
7. Low-volume polypropylene plates to reformat active compounds.
8. Reservoirs to hold bacterial cultures.
9. 20 μL and 200 μL 96-racked sterile tips for liquid handler or multichannel pipettors.
10. A plate reader that measures absorbance at 600 nm.
11. Temperature controlled shaking incubator.

3 Methods

3.1 Preparation of Bacterial Culture

1. Under sterile conditions, streak out the frozen bacterial stock on a CAMHB Agar plate with appropriate antibiotics using sterile wood applicators (*see* **Note 1**).
2. Incubate the plate at 37 °C in a stationary incubator for 16–18 h and store it at 4 °C for further use (*see* **Note 2**). This plate can be used for 2 weeks to pick colonies.
3. Using a sterile wood applicator, inoculate 5 mL of CAMHB in a sterile test tube with a single colony from the CAMHB Agar plate.
4. Incubate the 5 mL CAMHB culture overnight (16–18 h) at 37 °C in a shaking incubator at 250 rpm.
5. Dilute the overnight liquid culture 1:100 in fresh CAMHB in a sterile culture tube to make a *subculture*. Subculture is generally 5 mL of CAMHB with 50 μL of the overnight culture.
6. Grow the subculture at 37 °C in a shaking incubator at 250 rpm until it reaches mid-log growth phase (Absorbance at 600 nm, $A_{600} = 0.35–0.5$).
7. For screen optimization and the primary screen, dilute the mid-log subculture in fresh sterile CAMHB to $A_{600} = 0.001$ to prepare the *working bacterial culture*.

3.2 Determination of Appropriate Incubation Time

1. Prepare a working bacterial culture as in Subheading 3.1.
2. Using either a multichannel pipettor or a liquid handler, aliquot 48 μL of the culture to two 384-well plates.
3. Add the late-step inhibitor (e.g., targocil) at a final concentration equivalent to 8×MIC to all wells of each plate (MIC of the late-step inhibitor should be determined at the same conditions prior to the screen).
4. Add neat DMSO to one-half of the wells of each plate (negative controls yield 0% growth). The final concentration of DMSO is the same as that to be used in the primary screen (typically, 0.2–4%).
5. Add an early step inhibitor (e.g., ticlopidine) or any other known late-step antagonist at a final concentration of 20 μM to the other half of each plate (positive controls yield 100% growth). Ensure the same final concentration of DMSO is used as in the previous step and that the final volume in each well is 50 μL.
6. Measure the absorbance at 600 nm for each plate.
7. Incubate both plates at 37 °C and 250 rpm sealed in plastic bags. Take absorbance readings every 2 h or as manageable, until the growth of the microbe reaches the stationary phase, as indicated by little comparable change in growth over several hours.
8. Plot growth versus time. Set the most appropriate incubation time for use in the primary screen by choosing the time at which microbial growth in the positive control is still in log phase while no suppressor mutants emerge against the late-step inhibitors in the negative control (*see* **Note 3**). At the selected incubation time, the calculated statistical parameter Z' should be >0.5 (*see* **Note 4**).

3.3 Primary Screen

3.3.1 Labware Preparation

1. Prepare a working bacterial culture as in Subheading 3.1.
2. Thaw small molecule compound libraries at room temperature while still sealed in storage plates. Unseal the plates right before use and reseal immediately after use to minimize uptake of atmospheric water contamination (*see* **Note 5**).
3. Label all assay plates adequately.
4. Sterilize a reservoir with 70% ethanol and exposure to UV light for 10 min.
5. Transfer the required volume of working bacterial culture to the reservoir.

3.3.2 Liquid Handling

An automated liquid handler or a multichannel pipettor can be used to set up the assay plates in the following manner.

1. Using the 96-racked 20 μL sterile tips, transfer appropriate volume of targocil (1 μL of 0.8 mg/mL stock in 100% DMSO for a final screening concentration of 16 μg/mL in this case) (*see* **Note 6**) followed by the systematic addition of small molecules from storage compound plates (1 μL of 1 mM stock in 100% DMSO for a final screening concentration of 20 μM in this case) to two 384-well assay plates, replicates 1 (R1), and 2 (R2).
2. In lieu of targocil and small molecules, the two outermost columns (columns 1 and 24) were used for high and low growth controls. Dispense erythromycin at final concentration of 16 μg/mL (2 μL from 0.4 mg/mL stock solution) as a low growth control and neat DMSO (2 μL for a final concentration of 4% in this case) as a high growth control.
3. Repeat **steps 1** and **2** to prepare assay plates for all the compound plates. Change tips after each compound plate.
4. Using 200 μL sterile tips, dispense the required volume of microbial culture (48 μL per well for our study purposes) from the reservoir into all wells of the 384-well assay plates. Wash/change tips between handling different assay plates.
5. After the completion of liquid transfer for all the assay plates, measure the absorbance of the assay plates at 600 nm for $t = 0$ measurements.
6. Incubate the assay plates at 37 °C with shaking at 250 rpm for the appropriate incubation time as determined in Subheading 3.3 (12 h in this case), and measure the absorbance at 600 nm at the end of incubation for $t = 12$ measurements.
7. Calculate the growth of bacteria (G_i) in each well as:

$$G_i = A_{600\ (t=12)} - A_{600\ (t=0)},$$

   where i represents a well in the assay plate.
8. Normalize the data to identify actives, as described in Chand et al [21]. Briefly, normalize data as a percentage of rescued growth for each test well using the following: (Sample A_{600}–IQ mean)/(mean high control—IQ mean) × 100; where IQ mean is the mean of the middle 50% of the ordered raw data on the corresponding plate. A value of 0% represents no growth on the plate and 100% represents full rescue of targocil activity. A hit cut-off is determined by calculating the mean and standard deviation of the full dataset. A statistical cutoff equal to the mean + 4 standard deviations was established (*see* **Note 7**).

3.4 Follow-Up Tests of Primary Actives

3.4.1 Dose–Response Determination

1. Prepare the working bacterial culture as in Subheading 3.1.
2. Reformat active compounds from Subheading 3.3.2 into polypropylene plates with one compound per well, aliquot 15 μL of 8 compounds into wells A1–H1, and 15 μL neat DMSO to all wells in columns 2–11. Using a multichannel pipettor or liquid handler, transfer 15 μL of the compounds in column 1 into the DMSO into column 2. Continue the two-fold serial dilution across the plate to column 11 (*see* **Note 8**). Aliquot 15 μL neat DMSO to wells A12–H12.
3. Using the 96-racked 20 μL sterile tips, transfer appropriate volume of targocil (2 μL of 0.8 mg/mL stock in 100% DMSO for a final screening concentration of 16 μg/mL in this case) to all wells of two 96-well assay plates, except wells E12-H12 to which 2 μL of neat DMSO were added.
4. Transfer 2 μL from each well in the primary actives plates prepared in **step 2** into the corresponding wells in the two 96-well assay plates that received targocil and another two assay plates that did not.
5. Using 200 μL sterile tips, dispense 96 μL of microbial culture from the reservoir into all wells of the 96-well assay plates. Wash/change tips between handling different assay plates.
6. Incubate the plates for 16 h at 37 °C with shaking at 250 rpm.
7. Plot growth versus concentration. Compounds showing dose-dependent targocil antagonism but no bioactivity on their own should be further considered (*see* **Note 9**).

3.4.2 Checkerboard Assays

1. Prepare dilution series of the prioritized actives as described in Subheading 3.4.1, **step 2**.
2. Prepare similar dilution series of a β-lactam antibiotic such as cefuroxime in another polypropylene plate with the exception that the highest concentration be placed in well A1 and the lowest in well G1 with well H1 containing neat diluent (sterile water in this case); that is, this dilution series is in the second dimension of the plate compared to the dilution series of the prioritized actives.
3. Transfer 2 μL from each well in the dilution series of the prioritized actives to all 8 wells in the corresponding column in the assay plate. Each bioactive will be tested in a different assay plate.
4. Transfer 2 μL from each well in the dilution series of the β-lactam antibiotic to all 12 wells in the corresponding row in the assay plate.
5. Dispense 96 μL of microbial culture (prepared as in Subheading 3.1 or as described by CLSI [22]) from the reservoir into all wells of the 96-well assay plates. Wash/change tips between handling different assay plates.

6. Incubate the plates for 16 h at 37 °C with shaking at 250 rpm.
7. Calculate fractional inhibitory concentration indices (FICI) (*see* **Note 10**). Bioactives showing synergism with β-lactams are potential early-step inhibitors (*see* **Note 11**) and should be subjected to further follow-up assays to confirm WTA inhibition (*see* **Note 12**).

4 Notes

1. Spray working counters with a generous amount of 70% ethanol to ensure sterile conditions. Working in a flow hood or close to a Bunsen burner also helps in preventing contamination.
2. The optimum growth temperature for the bacterial strain used in this study is 37 °C. Depending on the bacterial strain being investigated, the incubation temperature and time may need to be optimized.
3. Generally, when adding microbes to assay plates with test compounds, it is desirable to use a culture concentration in which the organism is in lag phase or early log phase. At the time of final detection, after an incubation period, the microbes should be in log phase; if the stationary phase is reached, the assay will be much less sensitive to subtle effects of small molecules tested. On the other hand, late-step inhibitors such as targocil show a high frequency of resistance [16, 17]. Suppressor mutants frequently emerge and their rates of detection increase by time. Hence, the incubation time should not be extended enough to observe suppressor mutants.
4. Z' is a simple, dimensionless statistical measure of the quality of a small-molecule screening campaign. It is defined as:

$$Z' = 1 - \left(\frac{3\sigma_{c+} + 3\sigma_{c-}}{|\mu_{c+} - \mu_{c-}|} \right)$$

   where σ_{c+}, σ_{c-}, μ_{c+}, and μ_{c-} are the standard deviations (σ) and means (μ) of the positive (c+) and negative (c−) controls [23]. A Z' value of 0.5 or greater represents a set of data with a good screening window and clear separation between the positive and the negative controls.
5. Compound libraries are generally dissolved in DMSO. As DMSO is highly hygroscopic in nature, minimize exposure to air to avoid uptake of atmospheric moisture.
6. DMSO can affect the growth of bacteria. Therefore, care should be taken to add as low as possible of DMSO. For *S. aureus*, DMSO concentrations of 0.2–4% (v/v) were found to have negligible effects on bacterial growth.

7. The hit rate can be modified by adjusting the stringency of the standard deviations in case of getting too many/few hits.
8. Compounds should be tested at concentrations between 0.4 μM and 50 μM.
9. One of the possible targets for compounds that display intrinsic growth inhibitory activity while showing dose-dependent targocil antagonism is undecaprenyl diphosphate synthase (UppS) [19, 20]. Late stage WTA biosynthesis inhibition leads to a lethal accumulation of undecaprenol-linked WTA biosynthetic intermediates [19]. UppS inhibition reduces the pool of undecaprenyl-phosphate, hence less of this material is available for WTA synthesis which would reduce the effectiveness of late-stage inhibitors [19]. Peptidoglycan, same as WTA, is synthesized on the lipid carrier undecaprenyl phosphate, hence UppS inhibition is growth inhibitory.
10. $FICI = A/MIC_A + B/MIC_B$, where A and B are the concentrations of two antibiotics required in combination to inhibit bacterial growth and MIC_A and MIC_B are the MIC values for drugs A and B alone. FICI data were interpreted as 'synergy' ($FICI \leq 0.5$), 'antagonism' ($FICI > 4.0$), and 'no interaction or indifference' (FICI 1–4.0). At least three replicates should be performed for each combination and the means used for calculation.
11. WTA polymers are involved in resistance to β-lactam antibiotics in MRSA [9]. Known early-step inhibitors such as ticlopidine (a TarO inhibitor) synergize with β-lactam antibiotics [15]. Hence, bioactives should synergize with β-lactams to be considered potential early-step inhibitors.
12. Synergy with β-lactams should be suppressed in a deletion background targeted at early steps (i.e., Δ*tarO* or Δ*tarA*). The bioactives may be subjected to further phenotypic assays to confirm WTA inhibition, such as interfering with phage infectivity (WTA are receptors for phage binding), increased sensitivity to lysozyme degradation (WTA protects against lysozyme) and depletion of phosphate content contributed by WTA in isolated cell walls.

Acknowledgments

The authors would like to thank Dr. Maya Farha for her critical reading of the manuscript. Wall teichoic acid related research in the authors' laboratory was supported by grants from the Canadian Institutes of Health Research and from the Canadian glycomics network (GlycoNet) and a Tier I Canada Research Chair award to E.D.B. O.M.E. was supported by a Michael G. DeGroote Fellowship Award in Basic Biomedical Science.

References

1. WHO (2014) Antimicrobial resistance: global report on surveillance. World Health Organization reports. World Health Organization
2. Li XZ, Plesiat P, Nikaido H (2015) The challenge of efflux-mediated antibiotic resistance in gram-negative bacteria. Clin Microbiol Rev 28 (2):337–418. https://doi.org/10.1128/CMR.00117-14
3. Starosta AL, Lassak J, Jung K, Wilson DN (2014) The bacterial translation stress response. FEMS Microbiol Rev 38 (6):1172–1201. https://doi.org/10.1111/1574-6976.12083
4. Marrakchi M, Liu X, Andreescu S (2014) Oxidative stress and antibiotic resistance in bacterial pathogens: state of the art, methodologies, and future trends. Adv Exp Med Biol 806:483–498. https://doi.org/10.1007/978-3-319-06068-2_23
5. de la Fuente-Nunez C, Reffuveille F, Fernandez L, Hancock RE (2013) Bacterial biofilm development as a multicellular adaptation: antibiotic resistance and new therapeutic strategies. Curr Opin Microbiol 16 (5):580–589. https://doi.org/10.1016/j.mib.2013.06.013
6. WHO (2017) Prioritization of pathogens to guide discovery, research and development of new antibiotics for drug-resistant bacterial infections including tuberculosis. World Health Organization, Geneva
7. Tong SY, Davis JS, Eichenberger E, Holland TL, Fowler VG Jr (2015) *Staphylococcus aureus* infections: epidemiology, pathophysiology, clinical manifestations, and management. Clin Microbiol Rev 28(3):603–661. https://doi.org/10.1128/CMR.00134-14
8. The Canadian Cystic Fibrosis Registry 2013 Annual Report (2015). Cystic Fibrosis Canada
9. Sewell EW, Brown ED (2014) Taking aim at wall teichoic acid synthesis: new biology and new leads for antibiotics. J Antibiot 67 (1):43–51. https://doi.org/10.1038/ja.2013.100
10. D'Elia MA, Henderson JA, Beveridge TJ, Heinrichs DE, Brown ED (2009) The N-acetylmannosamine transferase catalyzes the first committed step of teichoic acid assembly in *Bacillus subtilis* and *Staphylococcus aureus*. J Bacteriol 191(12):4030–4034. https://doi.org/10.1128/JB.00611-08
11. D'Elia MA, Millar KE, Beveridge TJ, Brown ED (2006) Wall teichoic acid polymers are dispensable for cell viability in *Bacillus subtilis*. J Bacteriol 188(23):8313–8316. https://doi.org/10.1128/JB.01336-06
12. D'Elia MA, Pereira MP, Chung YS, Zhao W, Chau A, Kenney TJ, Sulavik MC, Black TA, Brown ED (2006) Lesions in teichoic acid biosynthesis in *Staphylococcus aureus* lead to a lethal gain of function in the otherwise dispensable pathway. J Bacteriol 188(12):4183–4189. https://doi.org/10.1128/JB.00197-06
13. Campbell J, Singh AK, Santa Maria JP Jr, Kim Y, Brown S, Swoboda JG, Mylonakis E, Wilkinson BJ, Walker S (2011) Synthetic lethal compound combinations reveal a fundamental connection between wall teichoic acid and peptidoglycan biosyntheses in *Staphylococcus aureus*. ACS Chem Biol 6(1):106–116. https://doi.org/10.1021/cb100269f
14. Weidenmaier C, Kokai-Kun JF, Kristian SA, Chanturiya T, Kalbacher H, Gross M, Nicholson G, Neumeister B, Mond JJ, Peschel A (2004) Role of teichoic acids in *Staphylococcus aureus* nasal colonization, a major risk factor in nosocomial infections. Nat Med 10 (3):243–245. https://doi.org/10.1038/nm991
15. Farha MA, Leung A, Sewell EW, D'Elia MA, Allison SE, Ejim L, Pereira PM, Pinho MG, Wright GD, Brown ED (2013) Inhibition of WTA synthesis blocks the cooperative action of PBPs and sensitizes MRSA to beta-lactams. ACS Chem Biol 8(1):226–233. https://doi.org/10.1021/cb300413m
16. Campbell J, Singh AK, Swoboda JG, Gilmore MS, Wilkinson BJ, Walker S (2012) An antibiotic that inhibits a late step in wall teichoic acid biosynthesis induces the cell wall stress stimulon in *Staphylococcus aureus*. Antimicrob Agents Chemother 56(4):1810–1820. https://doi.org/10.1128/AAC.05938-11
17. Wang H, Gill CJ, Lee SH, Mann P, Zuck P, Meredith TC, Murgolo N, She X, Kales S, Liang L, Liu J, Wu J, Santa Maria J, Su J, Pan J, Hailey J, McGuinness D, Tan CM, Flattery A, Walker S, Black T, Roemer T (2013) Discovery of wall teichoic acid inhibitors as potential anti-MRSA beta-lactam combination agents. Chem Biol 20(2):272–284. https://doi.org/10.1016/j.chembiol.2012.11.013
18. Zlitni S, Blanchard JE, Brown ED (2009) High-throughput screening of model bacteria. Methods Mol Biol 486:13–27. https://doi.org/10.1007/978-1-60327-545-3_2
19. Farha MA, Czarny TL, Myers CL, Worrall LJ, French S, Conrady DG, Wang Y, Oldfield E, Strynadka NC, Brown ED (2015) Antagonism

screen for inhibitors of bacterial cell wall biogenesis uncovers an inhibitor of undecaprenyl diphosphate synthase. Proc Natl Acad Sci U S A 112(35):11048–11053. https://doi.org/10.1073/pnas.1511751112

20. Czarny TL, Brown ED (2016) A small-molecule screening platform for the discovery of inhibitors of undecaprenyl diphosphate synthase. ACS Infect Dis 2(7):489–499. https://doi.org/10.1021/acsinfecdis.6b00044
21. Lee K, Campbell J, Swoboda JG, Cuny GD, Walker S (2010) Development of improved inhibitors of wall teichoic acid biosynthesis with potent activity against *Staphylococcus aureus*. Bioorg Med Chem Lett 20(5):1767–1770. https://doi.org/10.1016/j.bmcl.2010.01.036
22. CLSI (2012) Methods for dilution antimicrobial susceptibility tests for bacteria that grow aerobically; approved standard—Ninth Edition, CLSI document M07-A9. Clinical and Laboratory Standards Institute, Wayne, PA
23. Zhang JH, Chung TD, Oldenburg KR (1999) A simple statistical parameter for use in evaluation and validation of high throughput screening assays. J Biomol Screen 4(2):67–73. https://doi.org/10.1177/108705719900400206

Index

Inka Brockhausen (ed.), *Bacterial Polysaccharides: Methods and Protocols*, Methods in Molecular Biology, vol. 1954, https://doi.org/10.1007/978-1-4939-9154-9, © Springer Science+Business Media, LLC, part of Springer Nature 2019

V

W

Y

Z

MIX
Papier aus verantwortungsvollen Quellen
Paper from responsible sources
FSC® C105338
FSC
www.fsc.org

If you have any concerns about our products,
you can contact us on
ProductSafety@springernature.com

In case Publisher is established outside the EU,
the EU authorized representative is:
Springer Nature Customer Service Center GmbH
Europaplatz 3, 69115 Heidelberg, Germany

Printed by Libri Plureos GmbH
in Hamburg, Germany